TRENDS IN CELLULAR SIGNALING

TRENDS IN CELLULAR SIGNALING

DAVE E. CAPLIN
EDITOR

Nova Biomedical
New York

For permission to use material from this book please contact us:
Telephone 631-231-7269; Fax 631-231-8175
Web Site: http://www.novapublishers.com

LIBRARY OF CONGRESS CATALOGING-IN-PUBLICATION DATA

Trends in cellular signaling / Dave E. Caplin (editor).
p. cm.
Includes index.
ISBN 1-59454-633-9
1. Cellular signal transduction. I. Caplin, Dave E.
QP517.C45T74 2006
571.7'4--dc22 2005032470

Published by Nova Science Publishers, Inc. ✢ New York

CONTENTS

PREFACE

Cell signaling is the communication inside a cell or with its environment. It plays an important role in cellular growth, survival and apoptosis. For instance, the growth of a cell requires production of certain proteins. The genes that encode these proteins are located in DNA inside the nucleus while the initial signal for cellular growth may come from outside the cell. To induce the expression of these genes, the external signal must pass through specific pathways to reach these genes in DNA. The signaling pathway involves several proteins. If any one of them becomes abnormal or non-functional, the cellular process would be altered, which may cause disease. Therefore, knowing the detailed pathway is very important in understanding the pathogenesis of diseases.

Chapter 1 focuses on the activation of transduction pathways. Human neutrophils are phagocytic cells involved in host defence mechanisms against bacterial infections. It has been demonstrated that small formyl-peptide derivatives, obtained as bacterial metabolites or derived from disrupted mitochondria, can be potent chemoattractants for phagocytes. for-Met-Leu-Phe (fMLP), together with its synthetic methyl ester derivative for-Met-Leu-Phe-OMe (fMLP-OMe), is used as a model chemoattractant due to its highly effective ability to activate all physiological functions of neutrophils through binding with specific G-protein coupled receptors (FPR) located on the neutrophil membrane. The interaction of fMLP with its receptor expressed on neutrophils triggers multiple second messengers through the activation of phospholipase C (PLC), PLD and PLA_2 and rapidly stimulates phosphatidylinositol-3-kinase (PI3-K), as well as activating tyrosine phosphorylation. An increase in intracellular levels of cyclic AMP (cAMP) and the involvement of kinases, such as protein kinase C (PKC), mitogen-activated protein kinases (MAPKs) [Jun N-terminal kinases (JNK), p38 and extracellular response kinases 1 and 2 (ERK1/2)], has also been demonstrated.

The activation of these transduction pathways is known to be responsible for various biochemical responses which contribute to the physiological defence against bacterial infections and cell disruption. In fact, it has long been known that the transduction pathway underlying the chemotactic response is different from those responsible for cytotoxic functions, and several previous experiments carried out utilizing pharmacological manipulation of the signal transduction pathway have highlighted the fact that distinct mechanisms are involved in each of these neutrophil responses. This can be rationalized on the basis of the existence of at least two different functional receptor subtypes or isoforms; low doses of a full agonist (or a "pure" chemoattractant) are required to interact with a high-affinity receptor subtype (FPR) that activates the transduction pathway responsible for the chemotactic response, while the increase of the full agonist concentration - typical of infections sites - allows binding with the its low-affinity subtype (FPRLike-1), able to activate the transduction pathways responsible for superoxide anion production and lysozyme release.

The use of selective analogues (ligands able to discriminate between different biological responses) allowed the authors to confirm the idea that fine tuning of neutrophil activation occurs through differences in activation of a spectrum of signalling pathways. For each stimulus capable of a unique set of cellular responses, a distinctive imprint of signal protein activation may exist. Through more complete understanding of intracellular signalling, new drugs could be developed for the selective inflammatory blockade.

The review presented in chapter 2 discusses several mechanisms that contribute to the outcome of the responsiveness of CREB target genes to increased cAMP levels. The general accepted paradigm for cAMP-induced transcription of genes presages that cAMP activates protein kinase A, which in turn phosphorylates the cAMP responsive element-binding protein CREB at the serine-133 residue. Phosphorylated CREB, bound to cAMP responsive element(s) in the promoter of such genes, will subsequently recruit the co-activator CBP or its paralog p300. These co-activators enhance CREB-mediated transcription through linking CREB to the basal RNA polymerase II transcription complex and through their intrinsic acetyltransferase activity, which enables the formation of an open chromatin structure. However, CREB-mediated transcription of cAMP-responsive genes remains enigmatic in multiple ways. Despite the ubiquitous expression of protein kinase A, CREB and CBP, many genes display a tissue-specific responsiveness to cAMP. In fact, genes with cAMP responsive element(s) in their promoters can be up-regulated, refractory, or even repressed by increased cAMP levels.

In chapter 3, the authors discuss several approaches used to understand the molecular mechanisms induced by Ret mutants that dictate the clinical phenotype of these two closely related but distinct neoplastic diseases. The multiple endocrine neoplasia (MEN) type 2A and type 2B syndromes which are transmitted as autosomal dominant disorders. Even sharing several overlapping features, as the development of medullary thyroid carcinomas and pheochromocytomas, MEN2A and MEN2B display several clinical differences both as concerns severity and tissues affected thus representing two different diseases. Distinct mutations in the Ret proto-oncogene, a transmembrane receptor-tyrosine kinase, have been identified as the cause of either MEN2A or MEN2B syndromes. These mutations result in constitutive activation of Ret by two distinct mechanisms that involve the generation of intermolecular disulfide bridges in the case of MEN2A, whereas a mutation in the kinase catalytic domain is the most frequent change in MEN2B-associated mutants.

Therefore MEN2A and MEN2B are paradigmatic of inherited syndromes caused by the same gene, in which different mutations give rise to distinct diseases. Indeed, the intracellular domain of Ret contains consensus binding sites for the Src homology-2 (SH2) and the phosphotyrosine-binding (PTB) domains for multiple signalling partners leading to downstream activation of nuclear effectors. However, how specific RET mutations predispose to distinct disease phenotypes is still unclear. Studies *in vitro* indicate that because of the different mechanism of activation, MEN2A- and MEN2B-associated Ret mutants differ in their substrate specificity. Moreover, using cell culture experiments allowed to further insight in the intracellular signalling modalities used by these RET mutants. Although MEN2A- and MEN2B-associated RET mutants involve similar intracellular signalling pathways they differ in the extent of induction of enzymatic cascades and in the gene expression pattern. Indeed, a recent report on expression profiling on MEN2-associated medullary thyroid carcinomas provide molecular evidence on the clinical differences between the two syndromes.

Chapter 4 discusses hypoxia-inducible factor-1 (HIF-1) which is a transcription factor responsible for hypoxia and up-regulates more than 100 target genes to adapt to oxygen deprivation. However, even under non-hypoxic conditions HIF-1 is activated by a number of stimuli and accelerates gene expression to affect cellular function. A variety of signal transduction pathways are involved in hypoxia- or extracellular stimuli-induced HIF-1 activation in a stimulus and cell type-specific manner. Furthermore, HIF-1 is implicated in the pathogenesis of some diseases.

In chapter 5, the authors summarize the molecular mechanisms involved in the transduction of the Wnt signals, and their widespread implication in development and malignancy, with special emphasis on stem cell biology and cancer progession. The secreted proteins of the multimember Wnt family activate Frizzled, G-protein-coupled serpentine receptors (GPCR), and regulate a multitude of cellular functions implicated in normal development, stem cell fate determination, proliferation of embryonic stem cells and progenitor cells in adult life, postnatal woundhealing and tissue regeneration, as well as neoplastic transformation. Wnt family signalling factors evolved and diversified early in multicellular animals about 650 million years ago. In molecular terms, multiple signalling pathways are transduced downstream of the Fz receptors, resulting in regulation of protein kinases and transcription of specific Wnt target genes. These canonical β-catenin -dependent and noncanonical Wnt pathways involve numerous intracellular signalling cascades, molecular scaffolds at cell-cell contacts, as well as secreted and membrane-bound proteins which reinforce or impede Wnt signals. The intricacy of the Wnt functions is finely tuned through the appropriate expression of Wnt ligands (19 genes identified in man), Wnt receptors (9 human Frizzled genes), Wnt cofactors, antagonists, and other proteins that participate in signal transduction. Multiple crosstalks between Wnt and Hedgehog, Notch, bone morphogenic proteins (BMP) and transforming growth factor-β (TGFβ) signalling pathways, as well as receptor and nonreceptor kinases are integrated with critical networks involved in cell growth and differentiation, survival, migration, angiogenesis, and invasion. Dysregulation of the activities of the Wnt signalling elements by activating /inactivating mutations and epigenetic changes is linked to several pathologies. Several molecular alterations targeting Wnt ligands and internal elements of the Wnt and APC/ β-catenin pathways are frequently detected in sporadic cancers of the digestive and urogenital systems.

In: Trends in Cellular Signaling
Editor: Dave E. Caplin, pp. 1-40

ISBN 1-59454-633-9

Chapter I

FMLP-OME ANALOGUES TRIGGER SPECIFIC SIGNALLING PATHWAYS IN THE PHYSIOLOGICAL FUNCTIONS OF HUMAN NEUTROPHILS

***Susanna Spisani*[1*] *and Rita Selvatici*[2]**
[1]Dipartimento di Biochimica e Biologia Molecolare, Università degli Studi di Ferrara, 44100 Ferrara, Italy,
[2]Dipartimento di Medicina Sperimentale e Diagnostica, Sezione Genetica Medica, Università degli Studi di Ferrara, 44100 Ferrara, Italy.

ABSTRACT

Human neutrophils are phagocytic cells involved in host defence mechanisms against bacterial infections. It has been demonstrated that small formyl-peptide derivatives, obtained as bacterial metabolites or derived from disrupted mitochondria, can be potent chemoattractants for phagocytes. for-Met-Leu-Phe (fMLP), together with its synthetic methyl ester derivative for-Met-Leu-Phe-OMe (fMLP-OMe), is used as a model chemoattractant due to its highly effective ability to activate all physiological functions of neutrophils through binding with specific G-protein coupled receptors (FPR) located on the neutrophil membrane.

The interaction of fMLP with its receptor expressed on neutrophils triggers multiple second messengers through the activation of phospholipase C (PLC), PLD and PLA_2 and rapidly stimulates phosphatidylinositol-3-kinase (PI3-K), as well as activating tyrosine phosphorylation. An increase in intracellular levels of cyclic AMP (cAMP) and the involvement of kinases, such as protein kinase C (PKC), mitogen-activated protein

* Correspondence concerning this article should be addressed to Susanna Spisani, Dipartimento di Biochimica e Biologia Molecolare, Via L. Borsari 46, Università degli Studi di Ferrara, 44100 Ferrara, Italy, tel. +39-0532291424, fax +39-0532202723; e-mail: sps@unife.it.

kinases (MAPKs) [Jun N-terminal kinases (JNK), p38 and extracellular response kinases 1 and 2 (ERK1/2)], has also been demonstrated.

The activation of these transduction pathways is known to be responsible for various biochemical responses which contribute to the physiological defence against bacterial infections and cell disruption. In fact, it has long been known that the transduction pathway underlying the chemotactic response is different from those responsible for cytotoxic functions, and several previous experiments carried out utilizing pharmacological manipulation of the signal transduction pathway have highlighted the fact that distinct mechanisms are involved in each of these neutrophil responses. This can be rationalized on the basis of the existence of at least two different functional receptor subtypes or isoforms; low doses of a full agonist (or a "pure" chemoattractant) are required to interact with a high-affinity receptor subtype (FPR) that activates the transduction pathway responsible for the chemotactic response, while the increase of the full agonist concentration - typical of infections sites - allows binding with the its low-affinity subtype (FPRLike-1), able to activate the transduction pathways responsible for superoxide anion production and lysozyme release.

The use of selective analogues (ligands able to discriminate between different biological responses) allowed us to confirm the idea that fine tuning of neutrophil activation occurs through differences in activation of a spectrum of signalling pathways. For each stimulus capable of a unique set of cellular responses, a distinctive imprint of signal protein activation may exist. Through more complete understanding of intracellular signalling, new drugs could be developed for the selective inflammatory blockade.

Keywords: inflammation, human neutrophyls, cellular responses, formyl-tripeptides, transduction pathway, protein kinases.

1. Inflammation

Inflammation is a process by which the body's white blood cells and chemical mediators protect us from infection and foreign substances. Both endogenous and exogenous factors are capable of inducing cell and tissue damage. Endogenous damaging factors include immunopathological reactions, and some neurological and genetic disorders. Exogenous factors can be divided into the mechanical (traumatic injury), physical (extremely low or high temperature, ionising irradiation, microwaves), chemical (caustic agents, poisons, venoms, genotoxic and proteotoxic compounds), nutritive (deficiency of oxygen, vitamins and basic nutrients), biological (viruses, microorganisms, protozoan and metazoan parasites) [1].

1.1. Inflammatory Mediators

The spread of the acute inflammatory response following injury is controlled by mediators released from different cells, and products of the plasma enzyme systems. Chemical mediators of the inflammatory response include histamine, lysosomal compounds, leukotrienes, 5-hydroxy-tryptamine (serotonin) and lymphokines. Product of plasma factors include: complement, the kinins, coagulation factors and the fibrinolytic system, which are all inter-related and produce various inflammatory mediators.

In the early stages of acute inflammation, oedema fluid, fibrin and neutrophil polymorphs accumulate in the extracellular spaces of the damaged tissue.

Table 1. Enzymes and other constituents of human neutrophil granules.

Constituents		Granules	
	Azurophil	Specific	Small storage
Antimicrobial	Myeloperoxidase Lysozyme Defensins BPI	Lysozyme Lactoferrin	
Neutral proteinases	Elastase Catepsin G Proteinase 3	Collagenase Complement activator	Gelatinase Plasminogen activator
Acid hydrolases	Catepsin B Catepsin D β-D-Glucoronidase α-Mannosidase Phospholipase A_2	Phospholipase A_2	Catepsin B Catepsin D β-D-Glucoronidase α-Mannosidase
Cytoplasmic Membrane receptors		CR3, CR4 fMLP receptors laminin receptors	
Others	Condroitin -4-sulphate	Cytochrome B $_{558}$ Monocyte - chemotactic factor Hystamine Vitamin B_{12}- binding protein	Cytochrome B $_{558}$

BPI = Bactericidal permeability-increasing protein
CR3 = Complement receptor 3
fMLP = *N*-formyl-Methionine-Leucine-Phenylalanine

2. Central Role of Neutrophils in Acute Inflammation

Neutrophils, also known as polymorphonuclear leukocytes (PMN), represent 50 to 60% of the total circulating leukocytes, and constitute the "first line of defence" against infectious agents or "non-self" substances which penetrate the body's physical barriers [2]. Once an inflammatory response is initiated, neutrophils are the first cells to be recruited to sites of infection or injury. Their targets include bacteria, fungi, protozoa, viruses, virally infected cells and tumour cells. Their development in the bone marrow takes about two weeks; during this period, they undergo proliferation and differentiation. The segmented neutrophil is a fully functionally active cell. It contains cytoplasmic granules and a lobulated chromatin-dense nucleus with no nucleolus. Upon release into the circulation from the bone marrow, the cells are in a non-activated state and have a half-life of only 4 to 10 h before marginating and entering tissue pools, where they survive for 1 to 2 days. Cells from the circulating and marginated pools can exchange with each other. Senescent neutrophils are thought undergo apoptosis (programmed cell death) prior to removal by macrophages. Their viability is significantly shorter in individuals suffering from infectious or acute inflammatory diseases when the tissue requirement for newly recruited neutrophils increases considerably.

The neutrophil granules are of major importance for neutrophil function. They are not produced continually, but on demand for storage purposes. Once microbial activity is initiated, human neutrophil granules allow a powerful response to be generated. On the basis of function and enzyme content, these granules can be divided into three main types: azurophil, specific and small storage granules [3] (Tab. 1). Their function is not just to provide enzymes for hydrolytic substrate degradation, as in classical lysosomes, but also to kill ingested bacteria and, finally, to secrete their contents which regulate various physiological and pathological processes, including inflammation. Individual granule populations can be characterized morphologically or biochemically using enzyme markers or other substances.

Subpopulations of neutrophils have been categorised by various criteria. These cells exist not only in resting or activated states but also in various intermediate stages. For, example, priming is a mechanism whereby resting neutrophils acquire a state of pre-activation that enables a more powerful response to be generated once microbial activity is initiated. Cells become “primed”, presumably by chemoattractants, producing a dramatic alteration in their surface characteristics. Receptors for chemoattractants increase two- to four-fold, and receptors for Fc portion of immunoglobulins (FcR) and complement component (CR)3 become mobilized from intracellular compartments, such as the specific granules. The resulting “primed” cell has increased adherence and avidity and marginates to endothelial sites.

3. Molecular Mechanisms of the Acute Cell-mediated Inflammatory Reaction

The emigration of circulating leukocytes from the blood into inflamed tissues [4] has been refined into a "three step" process comprising: (a) rolling of leukocytes along the vasculature (mediated through transient interactions between selectin proteins and their carbohydrate ligands), followed by (b) activation of both neutrophils and endothelial cells and a high affinity interaction between integrins and glycoproteins of immunoglobulin superfamily, leading ultimately to (c) extravasation (crawling along the endothelium, diapedesis, and migration into tissue) following a chemoattractant gradient (Fig. 1).

In the initial phase of an acute inflammatory response, circulating leukocytes are activated by exposure to inflammatory mediators including complement fragments C5a, cytokines such as Interleukin -1 (IL-1) , IL-8 and tumour necrosis factor-α (TNF-α) and lipopolysaccharide or classical chemoattractants such as N-formylated-L-Methionine-L-Leucine-L-Phenylalanine (fMLP) leading to their microvascular sequestration due to decreased deformability (i.e. increased cell stiffness) and increased adhesiveness of the circulating leukocytes. Endothelial cells are similarly activated, leading to enhanced expression of several adhesion molecules. Platelet activating factor (PAF) produced by endothelial cells may act on nearby neutrophils to potentiate their adhesion to the endothelium.

Fig. 1. Activation of neutrophil functions in the inflammatory process: 1) margination; 2) diapedesis; 3) chemotaxis; 4) phagocytosis.

3.1. Adhesion Molecules

At least four superfamilies of adhesion molecules participate in these events: the selectins, the integrins and certain members of the immunoglobulin superfamily [5]. The rolling step is mediated by neutrophils, L-selectine and by E- and P-selectins newly expressed on inflamed endothelial cells. Each selectin receptor molecule contains a lectin-like N-terminal domain, an epidermal growth factor-like motif, a series of consensus repeats similar to those in complement-regulatory proteins, a transmembrane domain, and a cytoplasmic tail. The lectin domain is directly involved in mediating cell-cell contact through Ca^{2+}-dependent interactions with cell-surface carbohydrates, particularly the tetrasaccharide Sialyl Lewis X antigen (SLeX).

Firm neutrophil adhesion to endothelial cells appears to involve exclusively the interaction of leukocyte integrins of the β_2 subfamily, LFA-1 (CD11a/CD18), CR3 (CD11b/CD18), and CR4 (CD11c/CD18), each of which contains the same 95kDa integrin subunit, also designated CD18 and different α-chains designated CD11a (180 kDa), CD11b (165kDa), and CD11c (150kDa). The intact CD11a/CD18 molecule is the lymphocyte function-associated antigen-1, LFA-1 and is expressed by lymphocytes (including T cells), myeloid cells (monocytes, macrophages, and granulocytes), and a variety of other cell types. Both CR3 and CR4 are expressed by myeloid cells. There are two or more ligands for LFA-1, those defined to date being intracellular adhesion molecule (ICAM)-1 (CD54) and ICAM-2 (CD102), which are members of the immunoglobulin superfamily. CR3 can bind fragments of complement components, particularly C3bi, and it mediates phagocytosis of complement-coated particles by professional phagocytes. There is additional member of the diverse immunoglobulin superfamily – the platelet-endothelial cell adhesion molecule, PECAM-1 (CD31). PECAM-1 has been identified on neutrophils, monocytes and platelets and is present in large amounts on endothelial cells where it is concentrated at the cell-cell junction.

Transmigration of neutrophils across the endothelial barrier involves interaction between leukocyte integrins and endothelial ICAM, and between glycosylated aminoglycans on the neutrophil plasma membrane and PECAM-1. Leukocytes migrate by active amoeboid movement through the walls of venules and small veins; electron microscopy shows that neutrophils can insert pseudopodia between endothelial cells, migrate through the gap thus created between the endothelial cells, and then on through the basal lamina into the vessel wall. In extravascular locations, interaction between extracellular matrix proteins and adhesion molecules, possibly by activation of cytosolic tyrosine kinases, facilitate the release of large quantities of toxic oxygen radicals and proteolytic enzymes.

3.2. Mobility and Chemotaxis

In general, cell mobility represents the integration of many processes including adhesion (integrin-dependent), lamellar protrusion (actin-dependent), de-adhesion (integrin-dependent), and contraction (actin and possibly myosin-dependent). Moving neutrophils lose their round configuration and assume a polarized morphology with an anterior lamellipodium extended in the direction of movement, a cell body that is elongated parallel to the axis of lamellar protrusion, and a knob-like tail or "uropod". The rearrangement of the cell requires a complex interaction of the cytoskeletal elements tubules impart directionally and stabilize the cell while the actin-myosin system provides the contractile forces and contributes to directionality. A spongy network of actin filaments is seen at the leading edge of the cell. It has been established that a dynamic actin network is required for the development of polarity and migration as these processes are suppressed by agents which prevent actin polymerisation, such as cytochalasin, by directly depolymerising agents, such as latrunculin, or by agents which stabilise F-actin, such as jasplakinolide. Ultrascrutural and immunofluorescence studies indicate that the leading lamellipodia ruffles are rich in acting-binding protein, myosin, and gelsolin, a 91,000-dalton calcium sensitive protein which appears to control actin filament length. The actin proteins are controlled by gelsolin, which breaks actin filaments, and acumentin, a 65,000-dalton protein that decreases actin length independently of calcium. Myosin, another calcium-dependent protein, works with actin to generate the forces of contraction. Thus calcium, and probably the calcium-binding protein, calmodulin, are thought to be critical regulators of the actin filament system, largely through the action of calcium on gelsolin. Local differences in calcium may be critical determinants of the locomotory response. The term chemotaxis refers to the movement of leukocytes (or cells in general), induced by a chemotactic stimulus. Besides chemotaxis (stimulated, directed migration), leukocytes also posses two other types of movement: random migration (undirected, spontaneous migration) and chemokinesis (stimulated, undirected migration) (Fig.2). Neutrophils are the fastest moving mammalian cells known. They migrate with an amoeboid-like movement with a speed of up to 20 μm per minute and they can perceive a chemoattractant gradient that can be as small as 1% across the dimension of the cell. Substances possessing chemotactic activity are called chemotactic factors (chemotaxins, chemoattractants). Leukocyte chemotaxis (leukotaxis) is mainly responsible for their mobilization at the inflammatory site. Both exogenous and endogenous chemoattractants participate in this event. Exogenous chemotaxins include bacterial oligopeptides of the fMLP type, lectins, denatured proteins, some lipids and lipopolysaccharides. Endogenous chemotaxins are produced by the host organism and are of humoral (complement fragment C5a, C5a-des-Arg and C3a, fibrinopeptides, kallikrein and plasminogen activator) or cellular type (from different cells – leukotriene (LT)B4, PAF, chemotactic cytokines etc.) [6,7].

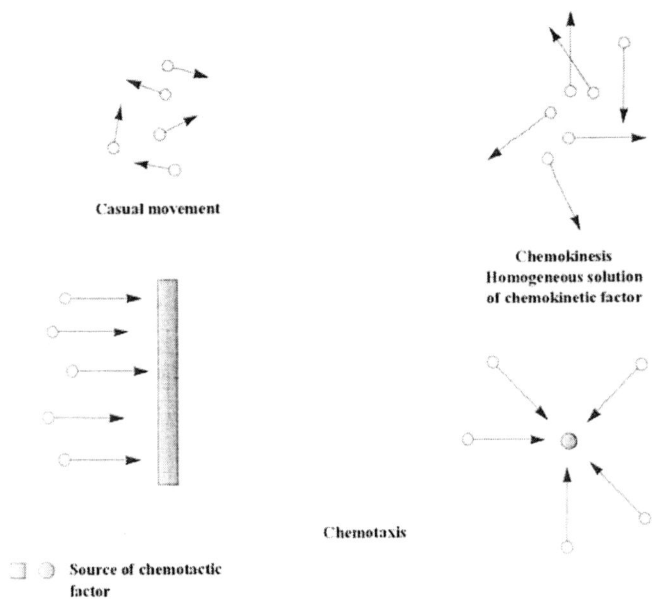

Fig. 2. Schematic representation of neutrophil movement: random movement; chemokinesis and chemotaxis. The arrows represents the velocity and the directions of neutrophils.

3.3. Phagocytosis

Once the neutrophil has arrived at the site of inflammation, it ceases to move, probably due to the lack of defined chemoattractant gradient. At the high concentrations of chemoattractants found here, the cell begins to phagocyte foreign particles Neutrophils are also able to interfere with microbial growth through alteration of the phagosome ionic environment (acidification, iron depletion). While microbicidal mechanisms of phagocytes are extremely efficient and capable of killing most microorganisms, pathogenic microorganisms have developed mechanisms to resist phagocytes. With many microbes, however, neutrophil defences are ineffective in the absence of opsonins and various agents that amplify the cytotoxic response. Opsonization is a process, in which opsonins adsorb to the surface of bacteria or other particles and facilitate their adherence to the phagocyte cytoplasmic membrane through opsonin receptors. Specific binding between the particle and phagocyte which occurs during immune phagocytosis is mediated by immunoadherent receptors. There are two types of immunoadherent receptors: FcγRIII for IgG, and CR1 and CR3 for the C3 split products C3b and C3bi, respectively. The expression of these receptors is modulated by chemoattractants and cytokines, as well as injuries and various diseases. This means that function of opsonins may be carried out by antibodies and C3bi. Specific binding between the particle and phagocyte may be also performed by lectins and lectin receptors (lectinophagocytosis), which directly recognize microbial patterns. Engagement and cross-linking of phagocytic receptors lead to rearrangement of the actin cytoskeleton, which is essential in this process, since cytochalasins and botulin toxin C2 block phagocytosis. Phagocytosis is a complex process composed of several morphological and biochemical steps such as: recognition and particle binding to the phagocyte surface and ingestion (engulfment). Simultaneous to the recognition and binding of particles, a dramatic increase in oxygen consumption (the respiratory burst) is observed. Phagocytosis is distinguished from endocytosis through the size of the particles that are internalised. Indeed, the maximum size of particles that can be taken in through endocytosis is in the submicrometer range (i.e. viruses), while particles up to the size of several micrometers can enter cells through phagocytosis (i.e. bacteria) [8,9].

3.4. Killing Mechanisms within the Phagosome

During phagocytosis, cytosolic granules (lysosomes) fuse with the invaginating plasma membrane (around the engulfed microorganism) to form a phagolysosome, into which they release their contents, thereby creating a highly toxic microenvironment. Fusion of azurophil and specific granules with the phagosome promotes oxygen independent and dependent killing [10].

- ***Oxygen independent killing***

The formation of phagolysosomes is of the utmost importance because during it two categories of cytotoxic substances, present in the preformed state in azurophil and specific granules and synthesized de novo during the respiratory burst, arrive at the same cell compartment. There is a hierarchy of granule mobilization: secretory > gelatinase > specific > azurophil, and as a consequence, azurophil granule contents are dedicated primarily towards the killing of phagocytosed microbes, while secretory granules act primarily as a reservoir for the mobilization of adhesion molecules to the neutrophil surface. This degranulation normally prevents release of the toxic components into the extracellular milieu. However, some target may be too large to be fully phagocytosed, or avoid engulfment, resulting in frustrated phagocytosis, in which no phagosome is formed. These targets may be killed extracellularly. However, tissue damage occurs when neutrophil microbicidal products are released extracellularly to such an extent that host defences (antioxidant and antiprotease screens) in the immediate vicinity are overwhelmed.

- ***Oxygen dependent killing***

Two types of free radicals are produced by neutrophils, macrophages, endothelial and other cells:

a) *reactive oxygen intermediates (ROI)* which are formed in neutrophils by the activity of nicotinamide adenine dinucleotide phosphate (NADPH) oxidase, the membrane-associated enzyme of the respiratory burst.

b) *reactive nitrogen intermediates (RNI)*, the first of which, nitric oxide, is produced by nitric oxide synthase.

a) Reactive Oxygen Intermediates (ROI)

Upon activation neutrophils and mononuclear phagocytes demonstrate increased oxygen consumption, a process known as the respiratory burst. During this, NADPH oxidase consumes oxygen (O_2) and cytosolic NADPH to generate the superoxide anion (O_2^-) and the perhydroxyl radical:

$$2O_2 + NADPH \longrightarrow 2O_2^{\cdot} + NADP^+ + H^+$$

The superoxide anion could participate directly in the killing process, or it could simply be a precursor of H_2O_2, a reaction catalysed by superoxide dismutase:

$$2O_2^{\cdot} + 2H^+ \longrightarrow O_2 + H_2O_2$$

Myeloperoxidase (MPO), contained in neutrophil azurophil granules, reacts with hydrogen peroxide to produce hypochlorous acid, and with halide ions (chloride, bromide, and iodide) to produce a number of toxic substances. The hydroxyl radical ($^{\cdot}OH$) is formed by several ways, of which decomposition of H_2O_2 catalyzed by Fe^{2+} is the most important.

The hydroxyl radical and hypochlorite are the most powerful substances involved in microbicidal and cytotoxic reactions. It appears that superoxide does not have direct toxic effects on targets but rather exerts its toxicity by penetration to important sites, where it subsequently is converted to other ROI.

NADPH oxidase [11] is an electron transport chain found in B and T lymphocytes and in the wall of the endocytic vacuole of professional phagocytes. It is thus called because NADPH is used as an electron donor to reduce oxygen to superoxide and hydrogen peroxide. NADPH oxidase is a complex enzyme composed of at least five members. While some components are membrane bound, others are stored in the cytosol. The $p21^{phox}$ and $gp91^{phox}$ subunits of a very unusual flavocytochrome b_{558}, are found in the cytoplasmic membrane. Two proteins ($p47^{phox}$, $p67^{phox}$), a quinone and a Rac-related GTP-binding protein, located in the cytosol, are thought to be the other functional components of this electron transport system. (phox stands for "phagocyte oxidase", p = protein, and gp = glycoprotein). The NADPH oxidase system is dissociated and thus inactive in dormant neutrophils. Upon activation, the cytosolic components translocate to the plasma membrane to assemble the active oxidase. The absence of, or an abnormality in, any one of these components results in chronic granulomatous disease (CGD), characterized by the absence of respiratory burst from neutrophils and monocytes [12].

ROI are also highly toxic for the cells which produce them. Therefore neutrophils have to contain large reserves of endogenous antioxidants such as glutathione and ascorbate. Their ability to maintain these antioxidants in the reduced state during phagocytosis may prevent death by oxidative suicide.

The exocytic machinery and NADPH oxidase interact to achieve maximal killing efficiency. For example, NADPH oxidase components are inserted in the phagosomal membrane through exocytotic membrane fusion, and myeloperoxidase secreted through an exocytotic mechanism enhances the efficacy of reactive species generated through NADPH oxidase activity. In addition to generating toxic products, professional phagocytes have powerful tools to alter the ionic environment within the phagosome to prevent survival of microorganisms. The insertion of an H^+-ATPase allows acidification of the phagosome to pH values which are incompatible with the growth of many bacteria. The function of Nramp1 (natural resistance associated macrophage protein 1) has not yet been elucidated, however in analogy with its homologue, Nramp2, it may function as an iron transporter [13]. As virtually all microorganisms depend on iron for growth, a transporter which is able to iron-deplete the phagosome would indeed be a powerful weapon in the arsenal of phagocytes [9].

b) Reactive Nitrogen Intermediates (RNI)

The pathway by which these are produced is an oxidative process in which short-lived nitric oxide ($NO^\cdot$) is derived from the guanidino nitrogen in the conversion of L-arginine to L-citrulline. This reaction is catalysed by $NO^\cdot$ synthase and, like the respiratory burst, it involves oxygen uptake [14,15]. Three distinct isoforms of nitric oxide synthase (NOS) representing three distinct gene products have been isolated and purified. Two of these enzymes are constantly present and termed constitutive NOS (cNOS). The third isoform is inducible NOS (iNOS). Its stimuli typically include cytokines and/or lipopolysaccharide (LPS), and, once expressed, this enzyme generates large amounts of NO. A number of cytokines are involved in the production of iNOS. Among them interferon (IFN)-γ, IL-1, IL-6, TNF-α, granulocyte-macrophage colony stimulatory factor (GM-CSF) and PAF exert the stimulatory effect, whereas suppression has been observed in the case of IL-4, IL-8, IL-10, transforming growth factor (TGF-β), platelet-derived growth factor (PDGF) and macrophage deactivating factor (MDF).

$NO^\cdot$ may react with superoxide to form the highly toxic peroxynitrite anion:

$$NO^\cdot + {}^\cdot O_2^-$$

which may be transformed in an acid milieu to peroxynitrite acid and then the hydroxyl radical:

$$ONOO^- + H^+ \rightarrow ONOOH$$
$$ONOOH \rightarrow {}^\cdot OH + NO^\cdot_2 \rightarrow NO_2- + H^+$$

Independent pathways are involved in the synthesis of ROI and RNI. Dormant neutrophils produce $NO^\cdot$ continuously, but activation arrests this pathway in favour of the oxidative burst. Thus, although the ROI and RNI pathways are independent, they may compete for common substrates such as NADPH and O_2^-, and exert other modulating effects on each other. The steady-state production of these species may dictate the anti-/pro-inflammatory balance. Microbial killing appears to be ROI dependent in normal neutrophils, but RNI may play a role in cells with deficiencies in the NADPH oxidase/MPO pathways.

Nitric oxide may also contribute to the microbicidal activity of neutrophils by reacting with ROI to form secondary cytotoxic species such as peroxynitrite.

4. *N*-FORMYLATED PEPTIDES

The tripeptide fMLP is the main chemotactic factor produced by *Escherichia coli* [16]. This compound, together with its synthetic methyl ester derivative (fMLP-OMe), is a potent chemoattractant for phagocytes, and is also generally adopted molecule as the reference model for evaluating the activity of newly synthesized analogues. In addition to its high lipophilic character, due to the presence of the three hydrophobic side chains, this compound is characterized by a pronounced conformational flexibility of the backbone, which appears to be an important feature for establishing efficient interactions with formyl-peptide receptor (FPR) (Fig. 3). For this reason, and due to the relative ease of its synthesis, both fMLP and fMLP-OMe are the most intensely studied members of the formylpeptide family [17]. Several fMLP-OMe analogs have been synthesized and utilized in different laboratories, with the aim of characterising formyl peptide interaction in phagocytes, as well as in cellular activation.

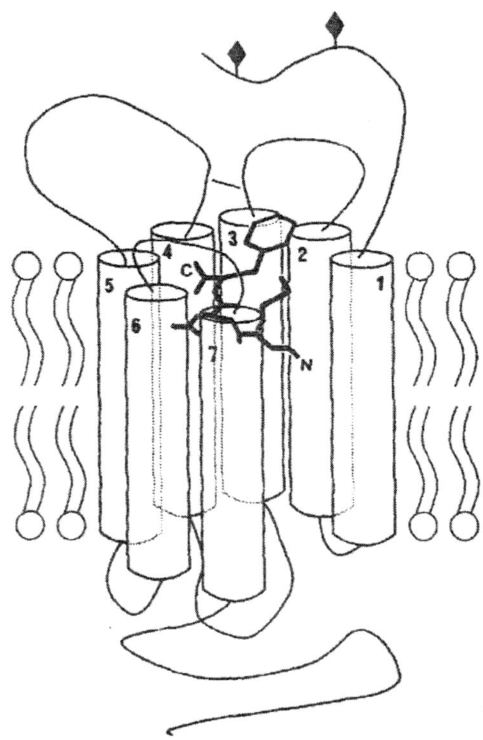

Fig. 3. A model for the fMLP-FPR interaction. The transmembrane domains of the receptor are represented by cylinders, numbered 1-7, and the loop structures are shown as *lines* connecting the *cylinders*. Two putative N-linked carbohydrate side chains at the extracellular amino terminus are depicted by *diamonds*. The skeletal model of the formyl tripeptide fMLP is drawn based on previous studies (Freer, 1982; Edmundson, 1985).

The early assumption that an *N*-formyl group was essential for optimal agonist potency, because it forms a hydrogen bond with a phenolic hydroxyl group of a tyrosine residue, has been challenged by findings that non-formylated peptides also chemoattract and activate phagocytes through the fMLP receptor(s) [18]. Structural analysis of FPR suggests that the binding pocket of this receptor is able to accommodate an amino terminal group larger than a formyl group [19,20]; this may explain the capacity of this receptor to interact with a great variety of peptide ligands. A synthetic pentapeptide, Met-Nle-Leu-Phe-Phe-OH, either N-formylated or N-acetylated, is more potent than the prototype fMLP in the induction of Ca^{2+} flux in human neutrophils [21], similarly, the pentapeptide Met-Met-Trp-Leu-Leu has been identified as moderately active FPR agonist, and its formylated form is more potent than the classical FPR agonist fMLP [22]. Amino terminal urea-substituted and carbamate-modified peptides are also potent agonists for FPR. Moreover, we have previously reported that bioactive fMLP-OMe analogues containing a N-terminal oximic or formylhydrazonic moiety are significantly active as chemoattractants [23].

Several formylpeptides, modified at position 1, have been synthesized in order to clarify the features peculiar to the methionine receptor pocket. The Met residue, however, seems to induce optimum chemotactic, as well as killing, activities due to its side-chain electron-rich sulphur atom, which allows the peptide to interact with a restricted positive arm on the receptor. Indeed, when amino acids carrying linear or branched aliphatic side chains are inserted at position 1, a more or less marked loss of biological responses is evident [24], while other substitutions produce discordant behaviour [25-28]. We previously demonstrated [29] that the substitution of Met with residues carrying a variation in electron density, and/or a shift in its position along the side chain, greatly lowers chemotactic response, together with killing mechanisms.

A great many substitutions at position 2 have been accomplished, introducing both natural and synthetic amino acids. The importance of these substitutions is linked to the ability of the peptide chain to properly locate the receptor pocket [30-33]. From the initial study onwards, the receptor pocket, in which the second residue is located, has been thought to be a hydrophobic area, and all subsequent investigations have concentrated on hydrophobic sterically hindered residues. We have demonstrated that the nature of the second residue is not vital to the biological response, which is influenced mainly by the steric hindrance of the amino acid rather than by its hydrophobic/hydrophilic nature [34,35].

Although the C-terminal aminoacid should be aromatic, Phe being the preferred choice [8,24,27] in a recent paper we demonstrated the receptor pocket in which the third residue is accommodated can also accept hydrophilic residues and is indifferent to residue steric hindrance [36].

Esterification (in particular, methylation) of the C-terminal carboxylic acid group does not alter the activity [37].

Studies of FPR have also revealed that the ligand binding pocket is large enough to accommodate a fourth amino acid residue [38,39], up to 5 amino acids in length [40].

5. THE HUMAN FORMYL-PEPTIDE RECEPTOR (FPR) FAMILY

The human formyl-peptide receptor was first defined biochemically in 1976, as a high affinity binding site on the surface of neutrophils for the prototypic for-Met-Leu-Phe. It was then cloned in 1990, by Boulay et al., from a differentiated HL-60 myeloid leukaemia-cell cDNA library [41]. Two additional human genes, designated FPRL1 (FPR-like 1) and FPRL2 (FPR-like 2), were subsequently isolated by low-stringency hybridization using FPR cDNA as a probe [42-44] and shown to cluster with FPR on human chromosome 19q13.3. The pharmacology established in transfected cells indicates that FPR is activated by picomolar to low nanomolar concentrations of fMLP, and is therefore also identified as the high-affinity fMLP receptor (K_d≈60 nM); whereas FPRL1 is considered a low-affinity fMLP receptor (K_d=1.5 μM), based on its activation only by high concentrations of fMLP (μM range) *in vitro*, and based on observations that high concentrations of fMLP could elicit only Ca^{2+} mobilization but not chemotaxis through this receptor [45,46]; however, it is unclear whether such concentrations of fMLP could be generated at sites of bacterial infection or tissue injury [42,47]. The results of K_d were well within the range of what was observed previously at low temperatures. In fact, Fabbri et al. [48], reported two fMLP-binding sites on neutrophils with K_{d1}=25 nM and K_{d2}=1.5 μM in assays performed at 4 °C. It has been generally accepted that FPRL1 shares signal transduction features with FPR, since both receptors are sensitive to pertussis toxin and possess a high degree of amino acid identity in the signalling cytoplasmic domains. However, [49] it has recently been demonstrated that PIP_2-binding peptide (PBP10) selectively inhibits FPRL1, but not FPR, -mediated superoxide production and granule mobilisation. Furthermore, the activity of this novel PBP10 sensitive pathway in neutrophils is modulated by the actin cytoskeleton network. In fact this PBP10 signalling pathway is also used by FPR, but only when the cytoskeleton network is disrupted.The cellular distribution of FPR is not restricted to neutrophils and monocytes, as previously thought. It is also expressed in hepatocytes, dendritic cells, astrocytes, microglia cells, and the tunica media of coronary arteries [50]. More recently, by using an antiserum directed against the carboxyl terminus of human FPR, Becker et al. [51] have confirmed that FPR, or an antigenically similar receptor, is widely localized in different human tissues and organs, including the thyroid, adrenals, liver, and nervous system. It is therefore speculated that FPR may have additional functions to host defence against bacterial infection, and that identification of host-derived ligand(s) may further elucidate this issue. FPR is highly conserved and is expressed in neutrophils of non-human primates, rabbits and mice. While the sequences of FPR in primates are nearly identical, rabbit and mouse FPR share 78% and 76% identity, at the amino acid level, to human FPR [18]. FPRL1 is expressed in an even greater variety of cell types in addition to phagocytic leukocytes, including hepatocytes, epithelial cells, T and microvascular endothelial cells [42,52]. FPRL2 does not bind or respond to N-formyl peptides [53] but instead shares some non-formylated chemotactic peptides identified for FPRL1. Little information is available about the expression pattern of FPRL2, except that mRNA for this receptor is present in monocytes but not in neutrophils [47]. Functional FPRL2 is also expressed in mature dendritic cells (DCs), which express reduced levels of FPR but do not appear to express FPRL1. F2L, a novel chemoattractant peptide acting specifically through FPRL2 has been recently identified. F2L is an acetylated amino-terminal peptide derived from the cleavage of the human haem-binding protein, an intracellular tetrapyrolle-binding protein. This peptide binds and activates FPRL2 in the low nanomolar range, which triggers intracellular calcium release, inhibits cAMP accumulation, and phosphorylation of extracellular signal-regulated kinase 1/2 mitogen-activated protein kinases through the G(i) class of heterotrimeric G proteins. When tested on monocytes and DCs, F2L promoted calcium mobilization and chemotaxis. Therefore, F2L appears as a new natural chemoattractant peptide for DCs and monocytes, and the first potent and specific agonist of FPRL2 [54].

5.1. Pathophysiological Role of FPRs

For ~25 years, speculation about the biological roles of the high affinity FPR was limited primarily to anti-bacterial host defence. Moreover, the discovery of novel related receptors, non-formylated peptide and lipid agonists for these receptors, and their expression in diverse tissues and cell types has indicated that this receptor family could have diverse and complex biological roles.

The capacity of FPRL1 to interact with diverse agonists suggests its potentially broad role in the anti-microbial response, inflammation process and amyloidogenic diseases. It has been reported that multiple domains are involved in FPRL1 interaction with its specific peptide agonists, with a prominent role for the sixth transmembrane domain and the third extracellular loop in mediating the chemotactic function of this receptor, while receptor residues at the base of the first and third extracellular loops appear to be involved in FPR-ligand recognition [55].

5.1.1. FPRs Regulate Anti-microbial Responses

The classic evidence supporting FPR as an anti-microbial receptor is that bacteria are the major biological source of chemotactic formyl peptides [56,57]. During an anti-bacterial response, phagocytic cells could be attracted, not only by chemotactic peptides produced by invading microorganisms, but also by host-derived anti-bacterial peptides, one of which, LL37, acts, in part, by interacting with FPRL1. LL37 is a 37 amino-acid peptide naturally cleaved from the C-terminus of the anti-microbial protein cathelicidin, which is present in human neutrophil granules and on epithelial cell surfaces. LL-37 is microbicidal, and neutralizes endotoxin by direct binding and, in addition, has chemotactic activity for human neutrophils, monocytes, and resting T cells [58]. Therefore, LL37 could act to amplify the innate immune response. FPRs have also been implicated in host responses during HIV-1 infection as several synthetic peptides corresponding to amino-acid sequences of HIV-1 envelope proteins gp41 and gp120 are chemotactic agonists for FPR and/or FPRL1 [59,60]. Whether these peptide fragments are generated *in vivo* during proteolysis of envelope proteins shed by infecting virus or not is unclear. Also, whether the interaction of FPRs with HIV-1 envelope peptides, if it occurs *in vivo*, could mobilize the innate defence, thus favouring host cellular responses. WKYMVm, a modified hexapeptide based on a sequence isolated from a random peptide library, was initially reported to be a very potent stimulant of several human leukocytic cell lines, as well as peripheral blood neutrophils. It has been found that this peptide uses both FPR and FPRL1 to stimulate phagocytes with a higher efficacy for FPRL1.

Another peptide library derived sequence, MMK-1, is a potent and very specific agonist for FPRL1 [61]. Although these peptides are synthetic peptides, and not endogenous proteins, they have been found to bear a similar sequence, and, due to their high potency in cell activation and preferential usage of FPRL1, these peptides provide valuable tools for studying leukocyte signalling via FPRL1.

5.1.2. FPRs Regulate the Inflammatory Response

Annexin I (lipocortin I), lipoxin A4 (LXA4) and the urokinase-type plasminogen activator receptor (uPAR) comprise a group of newly identified host-derived agonists for FPR and FPRL1 which have suggested novel roles for these receptors in the regulation of acute and chronic inflammation. Annexin I is a glucocorticoid-regulated protein possessing both pro- and anti-inflammatory activity, which might be mediated, in part, by FPR activation. Expressed in a variety of cell types, annexin I is particularly abundant in neutrophils. It is externalized onto the neutrophil cell surface and acts to inhibit transendothelial migration. Although both annexin I holoprotein and its N-terminal peptides (Ac1–26 and Ac9–25) are FPR agonists, at low concentrations they elicit Ca^{2+} transients through FPRs without fully activating the MAP kinase pathway [62]. This is associated with neutrophil desensitization and inhibition of transendothelial migration induced by other chemoattractants, such as the chemokine IL-8. By contrast, at high concentrations, annexin I peptides fully activate neutrophils *in vitro* and become potent pro-inflammatory stimulants. FPRL1 has also been reported to interact with a lipid metabolite, LXA4, whose most prominent biological activity is the inhibition of neutrophil-mediated pro-inflammatory responses. LXA4 suppresses the activation of FPRL1 by its peptide agonist and, in addition, inhibits the activation of neutrophils or epithelial cells by pro-inflammatory cytokines acting through receptors distinct from FPRL1 [18,63]. How LXA4 is capable of eliciting pro- versus anti-inflammatory signals through FPRL1 in different cell types is still rather puzzling. uPAR is a multifunctional protein involved in several different, but often related, biological events, such as the regulation of cell-associated proteolysis, cell adhesion, chemotaxis, cell migration, proliferation [64]. It is a three-domain (D1, D2 and D3) glycosylphosphatidylinositol (GPI)-anchored protein with a high affinity for uPA, a serine protease known for its ability to regulate fibrinolysis. However, uPA also induces leukocyte chemotaxis *in vitro* and is required for leukocyte trafficking to sites of inflammation *in vivo*. This mechanism involves indirect activation of FPRL1 through the liberation of a chemotactic peptide, $D2D3_{88–274}$, from uPAR [65]. Cells with undetectable FPRL1 or uPAR do not respond to uPA or D2D3 fragments, but transfection of FPRL1 into uPAR-deficient cells restores their responsiveness to D2D3, although not uPA, for which both uPAR and FPRL1 are required.

5.1.3. FPRL1 in Amyloidogenic Diseases

Interest in the potential role of FPRL1 in amyloidogenic diseases originated from the discovery that at least three of the polypeptides associated with these diseases are specific chemotactic agonists for FPRL1. These polypeptides are serum amyloid A (SAA), a 42 amino-acid amyloid β peptide ($A\beta_{42}$) and a peptide fragment of the aberrant human prion protein (PrP106–126). All these molecules are chemotactic and elicit a pro-inflammatory response in human leukocytes. SAA, $A\beta_{42}$ and PrP106–126 are endogenous proteins which, when aggregated, tend to precipitate and result in amyloid deposition in pathologic states such as systemic amyloidosis (SAA), Alzheimer's disease ($A\beta_{42}$) and prion disease (PrP106–126), respectively [66,67].

6. Formyl Peptide Receptor Signaling

Classical chemoattractants, fMLP, IL-8, LTB4, C3a and C5a, work through binding to cell surface receptors, and most, if not all, are classical G protein coupled receptors (GPCR), Gi α1, Gi α2, and Gi α3, exert their effects by triggering exchange of GDP to GTP binding in the α subunit and dissociation of the βγ subunits. GPCR is characterized by seven hydrophobic transmembrane segments connected by hydrophilic domains; potential N-linked glycosilation or phosphorylation sites are present in the extracellular or intracellular receptor surface, respectively [68]. Despite structural similarities, different types of chemoattractants possessing distinct structures bind to specific receptors, although cross-recognition has been observed within certain subfamilies of chemokines [69]. This high recognition specificity is crucial for directing neutrophil migration in the context of a complex chemoattractant field that is envisioned to be present during the anti-inflammatory response *in vivo*. Indeed, there is evidence that, when exposed to multiple chemoattractant sources in different temporal and spatial configurations, neutrophils can manage to discriminate and selectively migrate toward specific chemoattractants. This is accomplished, in part, through a combination of cross-desensitization of a particular receptor-mediated response and dominance exerted by certain signaling pathways. More recent studies have further suggested that dominant hierarchies of chemoattractants exist: neutrophils migrate preferentially toward end target chemoattractants, such as fMLP and C5a, emanating from the site of infection, rather than intermediary chemoattractants, such as IL-8 and LTB4 [70] encountered *en route* to sites of infectious. To date, the activation mechanisms coupling leukocyte migration and chemotactic responses to chemoattractants are not completely understood. *N*-formylated peptide-mediated responses are, by far, the best characterized, and have provided the most insights into neutrophil functions.

Following activation by fMLP, FPR undergoes rapid serine and threonine phosphorylation, and is desensitised and internalised. However, FPR internalisation can occur in the absence of desensitisation, indicating that desensitisation and internalisation are controlled by distinct mechanisms [71]. Further studies suggest that internalization of FPR and the complement C5a anaphylatoxin receptor (C5aR) [72,73] is mediated by mechanisms independent of the actions of arrestin, dynamin and clathrin, which are involved in the internalization of some other G protein-coupled receptors such as the β_2-adrenergic receptor. These receptors undergo a rapid chemoattratant-mediated phosphorylation, which is a prerequisite for the translocation of arrestins from the cytosol to the plasma membrane. After stimulation for prolonged time periods and internalisation, FPR, expressed in monocytes U937 cells, and C5aR, expressed in the rat insulinoma cell line RINm5F, persistently colocalise with arrestin 2 and arrestin 3 in endocyte vescicles. On this basis, both receptors have been classified as 'class B' receptors. By contrast, the 'class A' receptors, such as the β_2-adrenergic receptor, bind less tightly to arrestins, dissociating from them before reaching endosomal compartments, and recycle much more rapidly than the 'class B' receptors [74]. The distinct patterns of arrestin-receptor association during receptor endocytosis have suggested a possible role for arrestins in regulating GPCR traffic. Like many other G protein-coupled receptors GPCRs, human FPR has been reported to be in a state of constitutive activation. Cyclosporin H (CsH) is an inverse FPR agonist and, like Na^+ stabilises the FPR in an inactive state. The biological significance of constitutive FPR activation is not clear, but it is probably of importance *in vivo* [75].

Downstream of the heterotrimeric G-protein, a number of signalling systems are activated in human neutrophils [6]. Activation of phospholipase Cβ (PLCβ) results in hydrolysis of phosphatidylinositol 4,5-bisphosphate [Ptlns(4,5)P_2], generating diacylglycerol (DAG), which activates protein kinase C (PKC) isoforms, and inositol-1,4,5-trisphosphate (PIP_3), which releases Ca^{2+} from intracellular stores. Calcium enhancement appears to occur in two stages: an initial transient release from intracellular storage sites, followed by a more delayed influx across the plasma membrane. Increase in Ca^{2+} cytosolic levels has been postulated to be a necessary condition for neutrophil response to formylpeptides, as it is one of the earliest detectable events induced by cell exposure to fMLP. The release of Ca^{2+} from internal stores induces the opening of the store-operated Ca^{2+}-channels (SOCC) in the plasma membrane and a sustained influx of Ca^{2+} takes place [17]. Activation of the enzyme phosphoinositide 3-kinaseγ isoform (PI3Kγ) is responsible for the conversion of [Ptlns(4,5)P_2] into phosphatidilinositol- 3,4,5-trisphosphate [Ptlns(3,4,5)P_3]; this leads to the phosphorylation and activation of PKB. Moreover [Ptlns(3,4,5)P_3] regulates localisation and possibly crosslinking/stabilisation of actin filaments, although it does not appear to be directly involved in inducing actin polymerisation. Other intracellular effectors coupled to FPR signalling cascade include phospholipases A, D, mitogen-activated protein kinase (MAPK), Jun N-terminal kinases (JNK), p38 and extracellular response kinases 1 and 2 (ERK1/2), and phosphorylation mediated by tyrosine kinase occurs. Tyrosine kinases are stimulated via seven-transmembrane domain receptors, tyrosine kinase-linked receptors and integrin engagement [6].

6.1. Small Guanosine Triphosphate (GTP)ases

The small G-protein superfamily is made up of proteins that bind to guanosine triphosphate GTP and this family contains over 100 members that are generally classified, by structural similarity, into five subfamilies: Ras family GTPases (e.g. Ras, Rap and Ral), Rho family GTPases (Rho, Rac and cdc42), Arf family GTPases (Arf1–Arf6, Arl1–Arl7 and Sar), Rab family GTPases (>60 members, e.g. Rab5) and Ran family GTPases. The small GTPases are monomeric G-proteins with molecular masses over the range 20–30 kDa. The known functions of the many small G-proteins have been reviewed in detail, although many have yet to be elucidated [76-79].

Rho GTPases play a central role in all eukaryotic cells, co-ordinately controlling the organization of the actin cytoskeleton with other cellular activities such as gene transcription, cell cycle progression, and adhesion. The activation of Rho, Rac, and Cdc42 by extracellular agonists has been best characterized in quiescent Swiss 3T3 fibroblasts [80]. In these cells, lysophosphatidic acid (LPA) (a major constituent of tissue culture serum), acting via a plasma membrane receptor, activates Rho leading to the assembly of actin-myosin stress fibres and associated integrin adhesion complexes (focal adhesions). Rac can be activated by Platelet Derived Growth Factor (PDGF) or by insulin, and this leads to actin polymerization at the cell periphery causing lamellipodial extensions and membrane ruffling activity. Lamellipodia are also associated with integrin adhesion complexes, although their function is not clear; they contain many of the same constituents as classical focal adhesions, but they are much smaller. They are not required for actin polymerization, but they may be required for cell movement [81]. Bradykinin will activate Cdc42 in these cells to produce filopodia or microspikes and associated integrin complexes. There is much crosstalk within and between the Ras and Rho GTPase families. Activation of Cdc42 leads to a localized activation of Rac, hence filopodia are intimately associated with lamellipodia; in fact in these cells it is difficult to see filopodia unless Rac activity is first inhibited. Quiescent Swiss cells have no detectable stress fibres, and activation of Rac under these conditions leads to weak and delayed activation of Rho producing a few weak stress fibres. However, activation of Rac in cells growing in serum, which already have very pronounced and abundant stress fibres, leads to a loss of stress fibres. Finally, in addition to its well characterized effects on the ERK/MAP kinase pathway, oncogenic Ras is a potent activator of Rac. In neutrophils, the small GTPases of the Rho subfamily are active regulators of cell shape change and motility [82,83]. Many of the results connecting these small GTPases to particular cytoskeletal proteins have been derived from studies using neutrophil extracts, since mature native neutrophils have not been successfully microinjected or transfected. The paradigm, derived from the now-classic studies of the Hall lab, is that individual small GTPases lead to specific functional responses within the actin cytoskeleton. Cdc-42h is thought to induce fine finger-like projections called filopodia, and to activate the small GTPase, Rac. Rac activation induces lamellae formation, a broad, thin, actin-rich, organelle-excluding region most often found at the front of motile cells. In cytoplasmic extracts of neutrophils, active CDC42h stimulates actin filament assembly [84], and Rac stimulates actin assembly in a permeabilised neutrophil model [85].

Ras family GTPases regulate cell signalling events which lead to alterations in gene transcription. The mechanism by which GPCRs engage the Ras/MAPK signalling cascade varies between different cell types and also depends on which GPCR activates the cascade. There are two primary mechanisms by which GPCRs activate Ras/MAPK signalling: (i) via signals initiated by classical heterotrimeric G-protein effectors, such as the second-messenger-dependent protein kinases, and (ii) via the transactivation of receptor tyrosine kinases. In particular, activation of the Gi-coupled chemoattractant fMLP receptor, leads to both Ras-dependent and Ras-independent activation of the Ras family GTPase, Ral. Ral is required for initiation of a number of fMLP receptor-stimulated neutrophil responses, including the cytoskeletal reorganization required for chemotaxis [76,86].

Rab and Arf family GTPases control the formation, fusion and movement of vesicular traffic between different membrane compartments of the cell Arf6 is involved in regulating both recycling of endosomal vesicles and receptor-mediated endocytosis. The activity of Arf6 is regulated by two Arf GEFs, ARNO (ADP ribosylation factor nucleotide-binding site opener) and EFA6. Several recent studies have linked Arf6 and ARNO to the regulation of GPCR signalling, desensitization and endocytosis. The effects of Arf6 and ARNO on GPCR activity involve both the direct association of Arf6 with the receptors themselves, as well as the association of both Arf6 and ARNO with β-arrestin [76]. The 60 different Rab GTPases constitute the largest and most diverse group of Ras-like GTPases and, although these proteins are ubiquitous and highly conserved in their structure, each Rab has a distinct intracellular localization. Multiple Rab GTPases, such as Rab1, Rab4, Rab5, Rab7 and Rab11, have been identified to regulate ER (endoplasmic reticulum)–Golgi transport as well as the endocytosis and trafficking of GPCRs between early, late and recycling endosomes and lysosomes [87,88].

Ran GTPases are some of the most highly conserved proteins in eukaryotes; they are essential for cell viability in all organisms tested and regulate both microtubule organization and nucleo-cytoplasmic protein transport. Alterations in the Ran GTPase cycle produce a very pleiotropic phenotype. Whether this reflects functions of Ran in multiple cellular processes or merely its pivotal role in nucleo-cytoplasmic transport has been debated for some time. The recent demonstration that Ran functions in mitotic spindle and nuclear envelope formation independently of its role in nucleo-cytoplasmic transport argues for the former explanation. However, in contrast to the role of Ran in nucleo-cytoplasmic transport, the molecular mechanisms of these additional functions of Ran are poorly understood [89]. All of these GTPases function as molecular switches that control eukaryotic cell function by cycling between two inter-convertible forms, a GDP-bound 'inactive' form and a GTP-bound 'active' form.

6.2. MAP Kinase Cascade

The mitogen-activated protein (MAP) kinase cascades are some of the most intensely studied groups of signal transduction pathways. These signalling pathways are present in all eukaryotes and have been implicated in many physiological processes, including cell growth, differentiation, oncogenic transformation, immune responses, apoptosis [90] and also in migration [6]. MAP kinase cascades are organized by a core signalling module consisting of three protein kinases: a MAP kinase kinase kinase (MKKK), a MAP kinase kinase (MKK) and a MAP kinase. Signals are transmitted through the module by sequential phosphorylation and activation of these components. The MAPKKK is in turn activated by interaction with a family of small GTPases and/or other protein kinases connecting the MAPK module to the cell surface receptor or external stimuli. Distinct signalling modules activate different MAP kinases. The activated MAP kinase phosphorylates appropriate substrates, including other protein kinases and transcription factors, to achieve the desired response to extracellular stimulation. Three are the major groups of MAP kinase cascades in humans which lead to altered gene expression: extracellular signal–regulated kinases (ERK1/2), c-Jun *N*-terminal kinases (JNKs, also termed SAPKs) and p38 MAPKs.

6.2.1. The ERK1/2 Cascade

ERK1 and ERK2 are proteins of 43 and 41 kDa that are nearly 85% identical overall, with much greater identity in the core regions involved in binding substrates.The two phospho-acceptor sites, tyrosine and threonine, which are phosphorylated to activate the kinases, are separated by a glutamate residue in both ERK1 and ERK2 to give the motif TEY in the activation loop. ERK1 and ERK2 are activated by a pair of closely related MEKs, MEK1 and MEK2. Both of these MEKs have been shown to fully activate ERK1/2 *in vitro*. The MEKs are activated by c-Raf, the MAPKKK of this signalling pathway, which is in turn regulated by growth factor receptors and tyrosine kinases activating through Ras [91]. Upon translocation to the nucleus, ERKs are responsible for the phosphorylation of multiple substrates, depending on the initial stimulus. These include activators of transcription including p90 RSK S6 kinase [92], MAPK-activated protein kinase-1, MAPKAP-K1, phospholipase A2 and MSK (serine/threonine protein kinases), as well as transcription factors (Elk-1, Ets 1, Sap1a, m-Myc), signal transducers and activators of transcription (STAT) proteins such as Stat3, adapter proteins such as Sos, growth factor receptors such as epidermal growth factor (EGF), and the oestrogen receptors.

6.2.2. The JNK Cascade

The JNK signalling pathway responds to a variety of stress signals including heat shock, osmotic stress, pro-inflammatory cytokines, ischemia and UV exposure [93] JNK protein kinases are encoded by the three genes JNK-1, JNK-2, and JNK-3, which are alternatively spliced to form the JNK isoforms. The JNKs are activated by dual phosphorylation at the Thr-Pro-Tyr motif by JNKK1 and JNKK2, also known as MAPK kinase 4 (MKK4) and MKK7. Upstream of the MKKs are their MAPKKKs, which include MEKKs 1-4, ASK, and a member of the mixed-lineage kinases. In turn, these are activated by GTP-binding proteins from the Rho family (Racs, Rhos, the Cdc42s). JNKs were originally identified as the major kinases responsible for the phosphorylation of c-Jun, leading to increased activity of the AP-1 (activator protein-1) transcription factor [94] ; other nuclear transcription factors are now also known to be targets, including ATF-2, Elk-1, Myc, Smad3, tumour suppressor p53, NFAT4, DPC4 and MADD, a cell death domain protein [95].

6.2.3. The p38 Cascade

Enzymes in the p38 MAPK module are subject to dual phosphorylation at the Thr-Gly-Tyr motif, and are generally activated by environmental stresses, including heat, osmotic and oxidative stresses, ionizing radiation and ischemia-induced vasoactive stresses, as well as inflammatory cytokines and tumour necrosis factor (TNF) receptor signalling. Murine p38 was first identified in 1994 as a kinase activated in response to bacterial lipopolysaccharide (LPS). To date, four p38 MAP kinase isoforms have been identified, which share about 60% homology, and two of these isoforms (p38α, p38β) are ubiquitously expressed. p38γ is predominantly expressed in skeletal muscle, whereas p38δ gene expression is found in the lungs, kidneys, testes, pancreas, and small intestine. In most inflammatory cells, p38α is the major isoform activated. Extracellular stimuli of the p38 MAP kinase pathway include a variety of cytokines (IL-1, IL-2, IL-7, IL-17, IL-18, TGF-β, and TNF-α) and a number of pathogens that activate p38 through the different Toll receptors, including LPS, staphylococcal peptidoglycan, staphylococcal enterotoxin B, echovirus 1, and herpes simplex virus 1. The main biological response of p38 activation involves the production and activation of inflammatory mediators to initiate leucocyte recruitment and activation. The upstream kinases acting on p38 include MKKs 3 and 6. These upstream kinases have preferential effects on different p38 isoforms [96], which are in turn activated by MEKKs, MLKs and ASK1. GTPases are responsible for the transmission of stress stimuli to the MAPKKKs of this pathway, including the Racs, the Rhos and the Cdc42s. The p38 isoforms defined to date vary, based on their substrate specificity. The α and β isoforms of p38 are responsible for the activation of heat shock proteins (hsps) and the MAPK-activated protein (MAPKAP)-2. The γ and δ isoforms of p38 activate ATF2. Other transcription factors affected by the p38 family include Stat1, Max/Myc complexes, MEF-2A/C, Elk-1 and CREB through the activation of MSK1. Therefore, the p38 subfamily is also involved in affecting cell motility, transcription and chromatin remodelling [97].

6.3. Protein Kinase C

PKC is a family of isotypes classified into three groups on the basis of their structure and mode of activation: conventional PKCs (cPKCs; PKC—α, –βI, –βII –γ), novel PKCs (nPKCs; KC–δ, –ε, –η –θ) and atypical PKCs (aPKCs; PKC–ζ –λ) [98]. All members have a conserved kinase core carboxyl terminal to a regulatory moiety in common. This regulatory moiety contains two key functionalities: an auto-inhibitory sequence (pseudosubstrate) and one or two membrane-targeting modules (C1 and C2 domains and, in the case of protein kinase D, PH domain). The distinguishing feature of novel PKCs, as compared with traditional cPKCs, is that they lack a conserved region of homology found in cPKCs known as the C2 domain, which allows them to function in a calcium-independent manner. The atypical aPKC isoforms have altered C1 domains and are not DAG sensitive; regulation occurs in part through the N-terminal PB1 domain. A new serine-threonine kinase with homology to PKC has recently been described and is known as PKC m (or PKD), which may form a distinct PKC subgroup [99]. In resting cells or in the absence of lipid hydrolysis, PKCs are localized primarily in the cytosol. Activation leads to hydrolysis of phosphatidylinositol 4,5-bisphosphate (PIP2), generating DAG and *myo*-inositol 1,4,5-trisphosphate (IP3). DAG binds to the C1 domain and increases the affinity of PKC for membrane phospholipids. Upon activation, the kinase translocates from the soluble to the particulate compartment. As a consequence, not only is the residence time of PKC on the membrane increased, but also the conformational changes associated with phospholipid binding displaces the pseudosubstrate moiety from the catalytic domain, thus enabling PKC to phosphorylate protein substrates. Increasing evidence exists to suggest that individual PKC isoforms can translocate to subcellular locations other than the plasma membrane, including other membrane vesicles, nuclear structures, and cytoskeletal components. The subcellular location of a specific isoform may directly control the potential of that isoform to perform distinct functions, as the targeting of PKCs to discrete subcellular compartments would restrict their access to potential substrates.

Activation of PKC in a variety of different cell types leads to changes in the cell cytoskeleton, including lymphocyte surface receptor capping smooth muscle contraction and actin rearrangement in T cells and neutrophils [100,101]. Given the ubiquitous expression of PKC and the diversity of cytoskeletons in different cell types, it is not surprising that PKC has been shown to phosphorylate or be associated with a wide range of cytoskeletal components.

The structure and function of focal adhesions is believed to be regulated by protein phosphorylation. PKCα and PKCδ have been shown to be present in focal adhesions in REF52 cells and Swiss 3T3 cells [102], respectively. PKC has been shown to phosphorylate proteins localised in focal adhesions, including talin, vinculin, and integrins [103]. PKC-mediated phosphorylation of integrins has been implicated in changes in their ligand-binding affinity. Jaken *et al.* [104] suggested that PKC α–mediated phosphorylation of focal adhesion proteins may be responsible for smooth muscle contraction.

The relationship between PKC and membrane-cytoskeletal cross-linking proteins is important in the context of a number of cellular functions. The best-characterised cross-linker is the ubiquitously expressed myrostylated alanine-rich C kinase substrate (MARCKS) protein, which has been shown to be a direct substrate for PKC both in vitro and in vivo. The MARCKS protein bridges the plasma membrane and the actin cytoskeleton. When MARCKS is phosphorylated by PKC, it is released from the plasma membrane into the cytoplasm. A specific association between PKCα and MARCKS has been illustrated and related to the process of zymosan phagocytosis in macrophages [105]. The intracellular distribution of PKC isoforms suggests specific isoform-related biological functions, but this specialisation has only been partly explored. Interdependence between isozymes exists and may be important. PKC can be used by many receptor types to regulate the activation of MAP kinases, whose phosphorylation has an impact on processes in the cytoplasm, the nucleus, the cytoskeleton and the membrane. Upon activation, the phosphorylated proteins then move to the nucleus, where they catalyse phosphorylations and activation of specific target transcription factors.

6.4. cAMP

Adenosine 3',5'-cyclic monophosphate (cAMP) synthesis has also been demonstrated in neutrophils in response to fMLP. The mechanism responsible for this increase of cAMP levels is still the subject of debate. No direct coupling between FPR and adenylyl cyclase could in fact be demonstrated in neutrophils membranes, and an inhibitory action of fMLP on cAMP phosphodiesterase activity has never been clearly observed [106]. Nevertheless, cAMP elevation by the formylpeptide is counteracted by adenylyl cyclase inhibitors [107], suggesting that the nucleotide synthesizing enzyme is a target for the peptide's action. Neutrophil cAMP enhancement, in the presence of formylpeptides, seems to be a consequence of PLC stimulation. This hierarchy is suggested by the demonstration that the PLC blocker, U-73122, inhibits the nucleotide increase in both basal and fMLP activated conditions; in contrast, the adenylyl cyclase inhibitor MDL 12330A, at doses able to significantly reduce basal or fMLP-enhanced cAMP levels, is completely ineffective on neutrophil PLC activity [108]. Several experiments carried out utilizing drugs [109] or pharmacological manipulation of the signal transduction pathway [110] indicate that distinct mechanisms are responsible for the different neutrophil responses. It has been demonstrated that each physiological function shows different requirements for receptor occupancy. O_2^- production needs continuous occupancy of almost 100% of the receptor to achieve and maintain an optimal response, whereas exocytosis of secretory granules and chemotaxis do not [8]. Moreover, it has been demonstrated, not only by our team, that the transduction pathway underlying chemotactic response is different from those responsible for O_2^- production or lysozyme release [111-113].

Consequently, the availability of ligands capable of eliciting only a single response, or a limited set of responses, is of particular interest in order to obtain information about the biochemical pathways involved.

7. FMLP-OMe Analogs Trigger a Specific Signalling Pathways in the Physiological Functions of Human Neutrophils

In 1990 we started a research program based on the synthesis of new fMLP-OMe analogs in order to: i) obtain more effective and selective ligands; ii) gain information on the structural features and conformational preferences of chemotactic formyltripeptides; iii) identify the transduction pathways known to be responsible for different biochemical responses which contribute to the physiological defence against bacterial infections and cell disruption; and iv) clarify whether signalling requirements are identical or specific for each physiological function. First, we identified two fMLP-OMe analogues, whose structures, together with that of the parent fMLP-OMe, are depicted in Fig. 4.

The first analogue to be identified was for-Thp-Leu-Ain-OMe [Thp^1,Ain^3], characterised by the presence of an unnatural α-amino acid, in which the native Met and Phe external residues have been replaced by 4-amino-tetrahydrothiopyran-4-carboxylic acid (Thp) and 2-aminoindane-2-carboxylic acid (Ain). The α,α- disubstituted amino acids have demonstrated great versatility as peptide constrained due to the limited values of their ϕ and torsion angles when inserted into peptide backbones. Both new residues are achiral and, when compared to the parent amino acid, present a CH_2 bridge joining to the α-carbon atom to the side chain. Thus, in terms of conformational consequence, two different types of constraint have been introduced: the first concerning the Met and Phe side chains, which are blocked in a well-defined spatial orientation with respect to the backbone; the latter affecting the backbone itself, which adopts a local folded conformation of the helical type at the level of the α-α-bisubstituted residues (Fig. 5). The [Thp^1,Ain^3] analogue adopts a conformation characterised by helical folding centred at each of the three residues; the central Leu has right helical handedness, as opposed to those of the two adjacent achiral residues [26,114], (Fig. 6). This conformation presents strong similarities with that adopted in the crystal form by fMLP-OMe [115], and resembles the conformation of fMLP bound to the Bence-Jones dimmer, an immunoglobulin species which possesses binding properties similar to those of neutrophil receptors [116], (Fig. 7). In Fig. 8 the similarity between the two conformations adopted in the crystals by fMLP-OMe and [Thp^1,Ain^3] are depicted. The rigid fitting could be improved if the aromatic rings were on the same side [26].

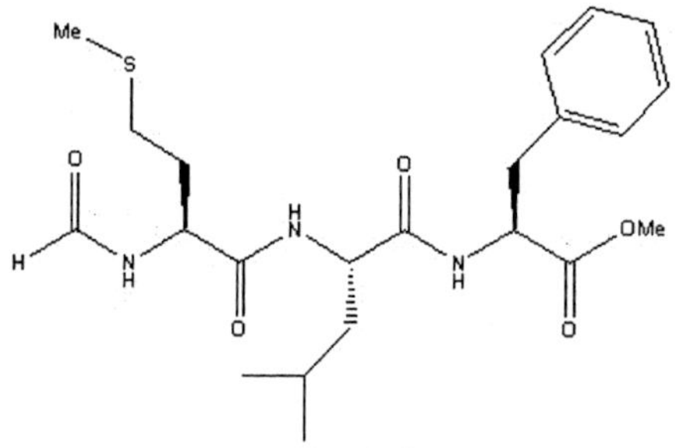

fMLP-OMe

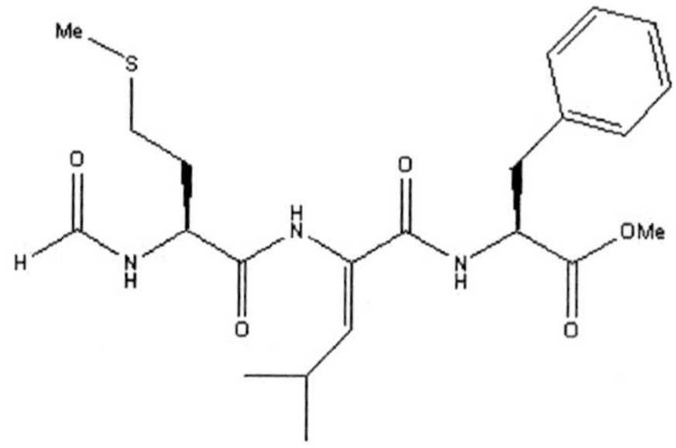

[ΔzLeu²]

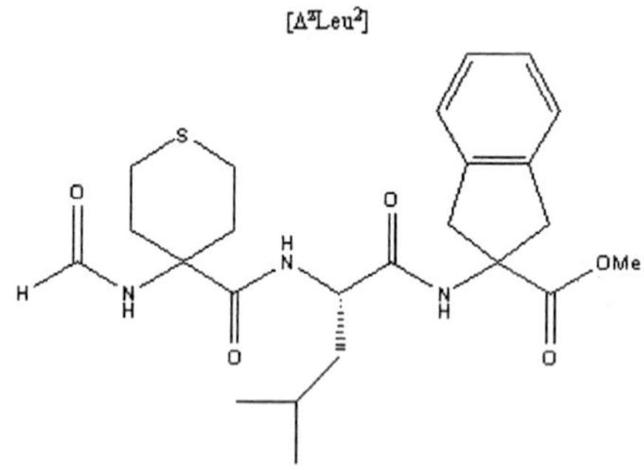

[Thp¹Ain³]

Fig. 4. Structures of the fMLP-OMe and of its synthetic analogues.

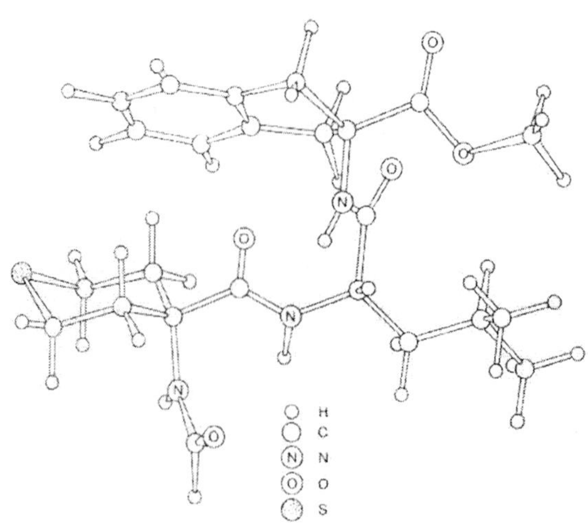

Fig 5. A perspective view of the conformation adopted by [Thp[1],Ain[3]] in the crystal (Torrini, 1991).

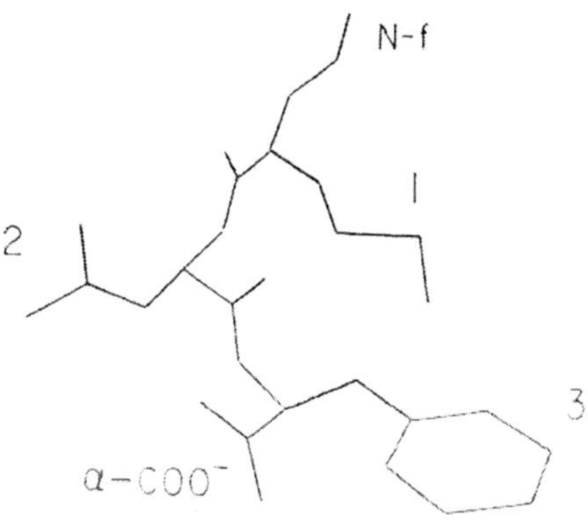

Fig. 6. Skeletal model of a prototype tripeptide to illustrate the overall wedge shape assumed during binding to the Mcg dimer. The N-formyl and α-carboxyl groups are labeled N-f and α-COO$^-$ and the three side chains are represented by numbers (1: methionine; 2: leucine; and 3: phenylalanine) (Edmunson, 1985).

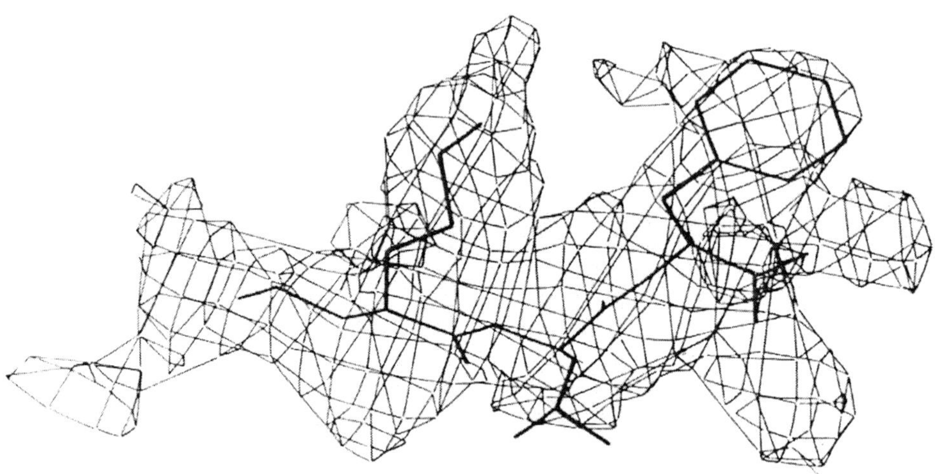

Fig. 7. Photograph of a computer graphics display of "cage" electron density corresponding to the chemotactic peptide, fMLP, as it bound to the Mcg Bence-Jones dimmer. The superimposed skeletal model of the peptide was fitted to the cage density by interactive computer graphics. The N-formyl group is on the far left and the phenylalanine side chain is on the upper right (Edmunson, 1985).

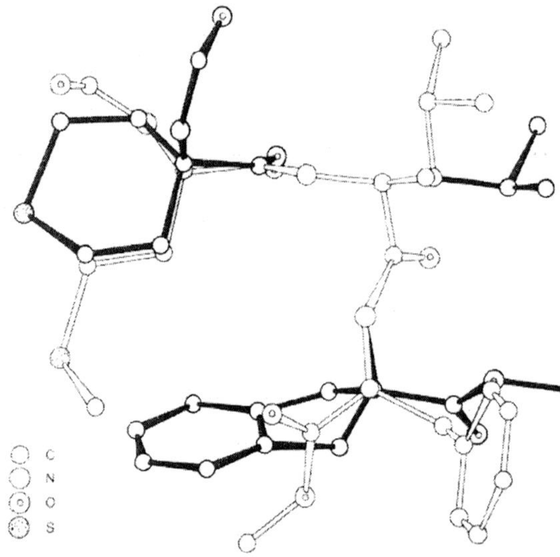

Fig 8. A rigid molecular fitting between the crystal conformations of fMLP-OMe and [Thp^1,Ain^3] analogue. The bonds of [Thp^1,Ain^3] are shown filled where exact superimposition does not occur (Torrini, 1991).

The second analogue identified was for-Met-Δ^zLeu-Phe-OMe [Δ^zLeu^2], in which the central Leu has been replaced by (*Z*)-2,3-didehydroleucine (Δ^zLeu) [117]. This α,β-unsaturated residue induces a type-II β-turn conformation in which the new residue occupies the (i + 2) position of the ten-membered pseudo-cyclic system. A weak intra-molecular H-bond occurs between the formyl C=O and the Phe NH groups (Fig. 9). This conformation differs from that adopted by the parent tripeptide, fMLP-OMe, both in the crystal form and in solution, and corresponds to the intra-molecularly H-bonded β-bend conformations adopted by the conformationally-restricted analogues of the general formula for-Met-Xaa-Phe-OMe, where Xaa respresents an α,α-dialkylated glycine residue [118]. The Met and Δ^zLeu side chains of this analogue are close together, which leaves the aromatic side chain of the C-terminal residue on the opposite side. This spatial arrangement differs from that found by Edumdson [116], and that previously highlighted by us in the case of [Thp^1, Ain^3], as described above [26].

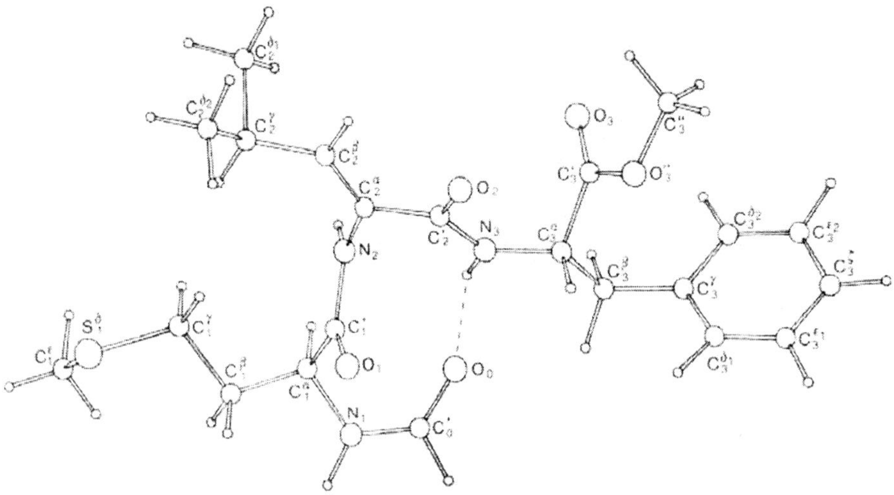

Fig 9. A perspective view of the peptide [Δ^ZLeu^2] crystal conformation (Pagani Zecchini, 1993).

Although these synthetic residues maintain close structural similarity with the corresponding natural amino acids, they can determine a significant reduction of peptide backbone flexibility inducing, at the same time, the adoption of preferred conformations. The resulting "conformationally constrained" residues were expected to show selective activity so they were used as useful tools to collect information on the nature of the bioactive, receptor-bound conformations.

In accordance with the above considerations, it has been previously found that fMLP-OMe, which has an identical behaviour to that of fMLP, is able to induce a full response by neutrophils ("full" agonist), its effects ranging from chemotaxis to superoxide anion production and lysozyme release [26]. The conformational flexibility could be responsible for the lack of selective action shown by these ligands. Conversely, the two constrained analogues exhibit a different and selective behaviour. In particular, the [Thp^1,Ain^3] analogue was found to be able only to induce the chemotactic response; the Thp group proves to be essential for inducing the switch from a "full" agonist to a "pure" chemoattractant [119]. With regard to this point, Haines et al. [120] postulated that chemotactic substances can be divided into "classical" and "pure" chemoattractants. The former group comprises fMLP and other agents (included C5a and IL-8), which evoke a number of physiological responses by neutrophils, causing an increase in Ca^{2+} at the intracellular level. Pure chemoattractants, such as substance P and the transforming growth factor β1 [121], are molecules that induce chemotaxis, but not oxidative burst and granular enzyme release; a common feature of these substances seems to be their inability to influence Ca^{2+} concentration. In addition, it has been reported that the murine S100 protein, CP-10, also belongs to the pure chemoattractant group, in that it possesses a potent chemotactic activity for neutrophils and monocytes/macrophages, but fails to enhance Ca^{2+} concentration in human or mouse neutrophils [122]. The requirement of Ca^{2+} enhancement for chemotaxis has further been questioned following the report by Laffafian and Hallet [123], who demonstrated that Ca^{2+} intracellular levels, measured in human neutrophils moving towards a source of the classical chemoattractant fMLP, do not significantly change. They observed an abrupt rise in the cation concentration only when cells reach the highest peptide concentration zone, when chemotaxis stops. We previously demonstrated that [Thp^1,Ain^3] belongs to the "pure" chemoattractants as it is a potent chemotactic agent for human neutrophils, but induces neither degranulation nor O_2^- production [26]. Moreover, the peptide does not alter the intracellular Ca^{2+} concentration. The inability of this peptide to increase the cytoplasmatic Ca^{2+} concentration could be related to its poor efficacy in enhancing cAMP level, as we have previously reported [106,119]. Moreover we previously demonstrated the inability of "pure" chemoattratactants to significantly alter the neutrophil PLC activity, measured in terms of inositol phosphate productions. In fact, PLC activation is one of the earliest events to follow fMLP-receptor interaction in the neutrophil plasmamembrane: as a consequence, an increase of diacylglycerol and inositol phosphate, as well as of Ca^{2+} cytosolic levels, is evoked [108].

[$\Delta^z Leu^2$] analogue, which is conformationally constrained at the central position, is only able to elicit O_2^- production and degranulation [118]. An activation of specific transduction pathways seems to underlie the different biological activities exhibited by the above described constrained analogues. In fact [$\Delta^z Leu^2$], in the presence of extracellular Ca^{2+}, similarly to fMLP and fMLP-OMe, induces a considerable enhancement of PLC activity, as well as of Ca^{2+} and cAMP intracellular levels [106,108,119]. The two constrained analogues also show different affinities towards the FPR. In particular, the full agonist fMLP-OMe and the constrained analogue [$\Delta^z Leu^2$] are both effective in displacing the labelled peptide from its binding sites, while [Thp^1,Ain^3] is much less efficacious [48]. All these described data are summarised in Tab.2. In the absence of calcium in the extracellular medium, neutrophil chemotaxis appear insensitive, while superoxide anion release and $[Ca^{2+}]_i$ are present, but reduced (Tab.3) [119].

The role played by PKC isoforms (α, β_1, β_2, ξ) and MAPKs (p38, ERK1/2 and JNK) in the signal transduction pathway leading to chemotaxis, superoxide anion (O_2^-) production and degranulation triggered by the classical peptide fMLP-OMe, the conformationally constrained analogue [$\Delta^z Leu^2$] and the "pure" chemoattractant [Thp^1,Ain^3] has been studied in the presence or absence of extracellular Ca^{2+} in order to discover the pattern of second messengers that specifically align each peptide [112,113]. Under our experimental conditions, the presence of extracellular Ca^{2+}, both in fMLP-OMe and [$\Delta^z Leu^2$] stimulated neutrophils contributes to the rise in $[Ca^{2+}]_i$, leading to membrane-association of PKC α, β_1, β_2 and ζ isoforms and phosphorylation of p38 and ERK1/2 MAPKs. Unlike a number of previous studies, we observed not only cytosolic PKC ζ, but also a small percentage of PKC ζ translocation [124,125] and membrane localisation of p38 MAPK. Since activation of p38 MAPK has been reported to be indispensable for chemotaxis [126], the presence of the pp38 form by [$\Delta^z Leu^2$] stimulation apparently contrasts with the peptide's inability to trigger chemotaxis. However, recent data reporting that p38 MAPK regulates exocytosis of intracellular granules [127] and phosphorylation of different NADPH oxidase components [124], or proteins inducing oxidase activation, provides an explanation of our experimental data.

Table 2. Chemotactic activity, superoxide anion production, cAMP levels, intracellular calcium, phospholipase C activity and binding affinity of fMLP-OMe and its derivatives.

Peptides	Chemotactic Index	O_2^- nmoles	cAMP %[a]	$\Delta[Ca^{++}]_i$ level[a] %[b]	PLC %[c]	IC_{50} nM
fMLP-OMe	1.15 ±0.08 (10^{-9} M)	48 ±3 (10^{-7} M)	250 ±3 (10^{-6} M)	450 ±25 (10^{-6} M)	245 ±18 (10^{-6} M)	60 ±3
[Thp[1],Ain[3]]	1.47 ±0.09 (10^{-9} M)	2 ±0.09 (10^{-6} M)	7 ±0.5 (10^{-6} M)	ineffective	ineffective	> 10.000
[Δ^zLeu[2]]	0.16 ±0.02 (10^{-7} M)	59 ±5 (10^{-6} M)	350 ±4 (10^{-6} M)	200 ±15 (10^{-6} M)	183 ±15 (10^{-6} M)	550 ±60

Values indicate the efficacy (activity at the optimal peptide concentrations).
Optimal concentration values in parentheses.
Data are means ± standard errors of the mean (SEM) of four separate determinations.
[a]Basal cAMP level was 1.9 ± 0.2 pmol/1x10^6 cells
[b]Basal Ca^{2+}level was 92.4 ± 4 nM
[c]Basal level of inositol phosphate production was 1890 ±142 dpm/10^7 cells

Table 3. Effect of Ca^{+2} chelator EGTA in the extracellular buffer on fMLP-OMe or [Δ^zLeu[2]]-induced chemotaxis, O_2^- production and $[Ca^{2+}]_i$ levels in neutrophils.

Peptides	Chemotactic Index	O_2^- nmoles	$\Delta\ [Ca^{2+}]_i$ level[a] %
fMLP-OMe	0.98 ±0.08 (10^{-9} M)	20 ±3 (10^{-7} M)	216 ±13[b] (10^{-6} M)
[Δ^zLeu[2]]	0.15 ±0.02 (10^{-7} M)	25 ±5 (10^{-6} M)	170 ±12[b] (10^{-6} M)

Data are means ± standard errors of the mean (SEM) of at least six separate determinations.
[a] Basal level was 92.4 ± 4 nM
[b] P<0.001 versus control conditions

In the absence of extracellular Ca^{2+}, the two formylpeptides assume different roles: fMLP-OMe stimulation selectively triggers the translocation of the PKC β1 isoform and the phosphorylation of p38 MAPK, whereas [Δ^zLeu[2]] induces the translocation of PKC α and β_2 isoforms but failed to phosphorylate MAPKs. The rate of translocation in both peptides was strongly reduced when compared to normal conditions. These latter findings suggest that the p38 active-form plays a central role in regulating neutrophil chemotaxis, and also that PKC β_1 translocation can to contribute to the same physiological function. The selective translocation of different PKC isoforms by the two peptides can be explained by the presence in neutrophils of at least two fMLP-receptor isoforms [128], which may activate separate second messenger pathways, and hence specific biological responses.

The formylpeptide fMLP-OMe, at the molar concentration 10^{-9} M (optimal concentration to induce chemotaxis), and the "pure" analogue [Thp[1],Ain[3]] selectively trigger the translocation of the PKC β_1 isoform. Cellular functional assays using the specific PKC inhibitor, GF109203X indicated that the activation of PKC was indispensable for both fMLP-OMe- and [Thp^1,Ain^3]-induced chemotaxis of human neutrophils. PKC is considered an important regulator of cytoskeletal functions, and it has previously been associated with intermediate filament proteins, membrane-cytoskeletal cross-linking proteins, components of the actin filaments and microtubules, as well as with β-integrin vesicle trafficking [129] and therefore the link between PKC β and β_2-integrins may not be coincidental. The genes encoding for PKC β_1 isoform, for leukocyte adhesion receptor (CD43), and for CD11a, CD11b and CD11c occur in a cluster on human chromosome 16, suggesting that they could be functionally linked [98]. PKC has also been shown to phosphorylate proteins localised in specialised regions including talin, vinculin and integrins (focal adhesions). Within these regions, several components of the cytoskeleton are concentrated, together with a number of signalling proteins. As previously demonstrated [119], neutrophil chemotaxis is almost insensitive to any variation of Ca^{2+} concentration. This was confirmed by our experiments, which showed PKC β_1 translocation and p38 MAPK phosphorylation in neutrophils triggered by fMLP-OMe and [Thp^1,Ain^3] in either the presence or absence of extracellular Ca^{2+}. Since the increase in intracellular Ca^{2+} is not important for chemotaxis, but is necessary for the activation of conventional PKC (α, β_1, β_2, γ), the PKC β_1 translocation showed by both peptides could be misleading. However, it has been observed that localised Ca^{2+} signalling spikes are present both in the absence of extracellular Ca^{2+} and in the presence of trans-membrane blocking Ni^{2+} , thereby demonstrating that the presence of localized Ca^{2+} signalling was due to release from Ca^{2+} stores, and that there is no requirement for trans-membrane influx of extracellular Ca^{2+} [130]. Fluctuations of localised intracellular Ca^{2+} could explain the translocation of PKC β_1 by both formyl peptides observed in our study.

Functional chemotactic experiments using MAPK inhibitors revealed that fMLP-OMe- and [Thp^1,Ain^3]-induced chemotaxis of human neutrophils was reduced by the p38 MAPK inhibitor SB20358, but not by the p44/42 MAPK inhibitor PD98059. In addition, Western blotting analysis confirmed that exposure of neutrophils to formylpeptides induced phosphorylation and activation of p38 MAPK, but not of p44/42 MAPK. These observations strongly suggest that the p38 MAP kinase-mediated signalling pathway plays a central role in regulating neutrophil chemotaxis. Schematic signalling pathways of chemotaxis in response to stimulation of human neutrophil with formyl-peptides are proposed in Fig.10.

All these data reinforce the idea that calcium and cAMP intracellular levels, PLC activity together with selected PKC isoforms and specific MAPK phosphorylations lead to transduction pathways in specific neutrophil functions. For each stimulus capable of a unique set of cellular responses, a distinctive imprint of signal protein activation may exist. Through more complete understanding of intracellular signalling, new drugs could be developed for the selective inflammatory blockade.

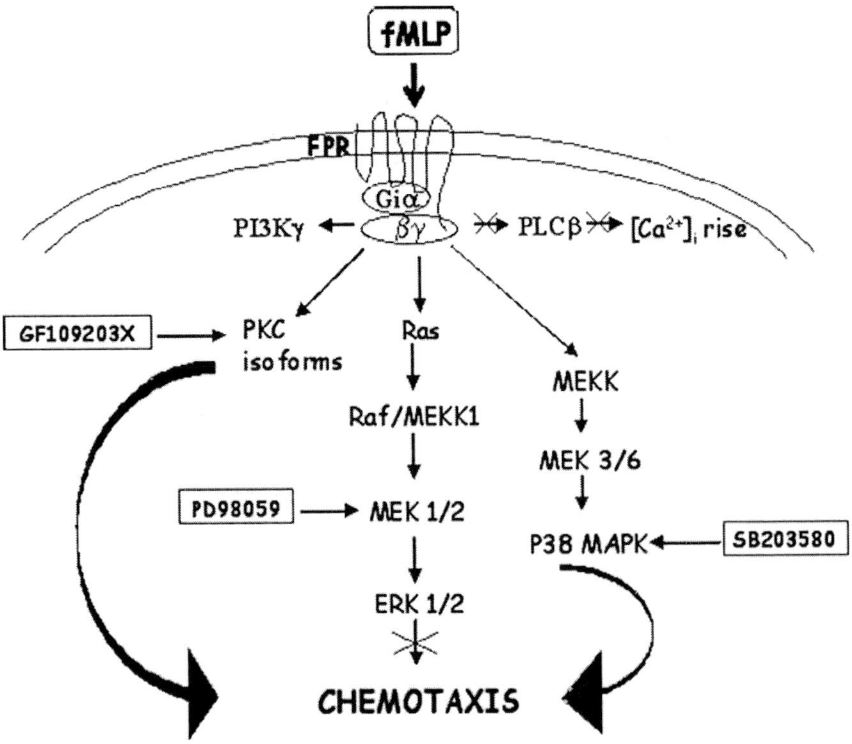

Fig 10. Schematic signalling pathways of chemotaxis. Upon formyl-peptide binding, trimeric G-proteins are uncoupled from FPR and a series of signal transduction events ensue that results in chemotactic activation.

Acknowledgements

This work was supported by Grants from University of Ferrara, E. and E. Rulfo for Medical Genetics, Parma, Italy and Fondazione Cassa di Risparmio of Ferrara, Italy. We are grateful to Anna Forster for the English revision of the text.

References

[1] Rankin, J. A. (2004). Biological mediators of acute inflammation. *AACN Clin. Issues, 15*, 3-17.

[2] Kobayashi, S. D., Voyich, J. M. & DeLeo, F. R. (2003). Regulation of the neutrophil-mediated inflammatory response to infection. *Microbes. Infect., 5*, 1337-1344.

[3] Faurschou, M., & Borregaard, N. (2003). Neutrophil granules and secretory vesicles in inflammation. *Microbes Infect., 5*, 1317-1327.

[4] Liu, Y., Shaw, S. K., Ma, S., Yang, L., Luscinskas, F. W. & Parkos, C. A. (2004). Regulation of leukocyte transmigration: cell surface interactions and signaling events. *Immunol. J., 172*, 7-13.

[5] Witko-Sarsat, V., Rieu, P., Descamps-Latshaba, B., Lesavre, P. & Halbwachs-Mecarelli, L. (2000). Neutrophils: molecules, functions and pathophysiological aspects. *Lab. Invest., 80*, 617-653.

[6] Niggli, V. (2003). Signalling to migration in neutrophils: importance of localised pathways. *Int. J. Biochem. & Cell Biol., 35,* 1619-1638.

[7] Gallin, J. I. Inflammation. In: William EP, editor. *Fundamental Immunol.,* New York: Raven Press; 1993; 1015-1032.

[8] Prossnitz, E. R. (1997). The N-formyl peptide receptor: A model for the study of chemoattractant receptor structure and function. *Pharmacol. Ther., 74*, 73-102.

[9] Krause, K. H. (2000). Professional phagocytes: predators and prey of microorganisms. *Schweiz Med. Wochenschr., 130*, 97-100.

[10] Hampton, M. B., Kettle, A. J. & Winterbourn, C. (1998). Inside the neutrophils phagosome: oxidants, mieloperoxidase, and bacterial killing. *Blood, 92*, 3007-3017.

[11] Babior, B. M. (2004). NADPH oxidase. *Curr. Opin. Immunol. 16,* 42-47.

[12] Smith, R. M. & Cornutte, J. T. (1991). Molecular basis of chronic granulomatous disease. *Blood, 77*, 673-686.

[13] Fleming, M. D. & Andrews, N. C. (1998). Mammalian iron transport: an unexpected link between metal homeostasis and host defence. *J. Lab. Clin. Med., 132*, 464-468.

[14] Xia, Y. & Zweier, J. L. (1997). Superoxide and peroxynitrite generation from inducible nitric oxide synthase in macrophages. *Proc. Natl. Acad. Sci. U.S.A. 94*, 6954-6958.

[15] Cedergre, J., Follin, P., Forslund, T., Lindmark, M., Sundqvist, T. & Skogh, T. (2003). Inducible nitric oxide synthase (NOS II) is constitutive in human neutrophils. *Apmis, 111*, 963-967.

[16] Marasco, W. A., Phan, S. H., Krutzsch, H, Showell, H. J., Feltner, D. E., Nairn, R., Becker, E. L. & Ward, P. A. (1984). Purification and identification of formyl-methionyl-leucyl-phenylalanine as the major peptide neutrophil chemotactic factor produced by Escherichia coli. *J. Biol. Chem., 259*, 5430-5439.

[17] Dalpiaz, A., Spisani, S., Biondi, C., Fabbri, E., Nalli, M. & Ferretti, M. E. (2003). Studies on human neutrophil biological functions by means of formylpeptide receptor agonists and antagonists. *Curr. Drug Targets – Immune & Metabolic Disorders, 3*, 33-42.

[18] Le, Y., Li, B., Gong. W., Shen, W., HU, J., Dunlop, N.M., Oppenheim, J.J. & Wang, J.M. (2000). Novel pathophysiological role of classical chemotactic peptide receptors and their communications with chemokine rceptors. *Immunological Rev., 177,* 185-194.

[19] Derian, C. K., Solomon, H. F., Higgins, J. D. Beblavy, M. J., Santulli, R. J., Bridger, G. J., Pike, M. C., Kroo, D. J., & Fiscman, A. J. (1996). Selective inhibition of N-formylpeptide-induced neutrophils activation by carbamante-modified pepetide analogues. *Biochemistry, 35*, 1265-1269.

[20] Higgins, J. D., Bridger, G. J., Derian, C. K., Beblavy, M. J., Hernandez, P. E., Gaul, F. E. Abrams, M. J., Pike, M.& Solomon, H. F. (1996). N-terminus urea-substituted chemotactic peptides: new potent agonists and antagonists towards the neutrophils fMLF receptor. *J. Med. Chem., 39*, 1013-1015.

[21] Gao, J. L., Becker, E.L., Freer, R. J., Muthukumaraswamy, N. and Murphy, P. M. (1994). A high potency nonformylated pepetide agonist for the phagocyte N-formylpeptide chemotactic receptor. *J. Exp. Med,. 180*, 2191-2197.

[22] Chen, J., Bernstein, H.S., Chen, M., Wang, L., Ishii, M., Turck, C.W. & Coughlin, S. R. (1995). Tethered ligand library for discovery of peptide agonists. *J. Biol. Chem., 270*, 23398-23401.

[23] Torrini, I., Paglialunga Paradisi, M., Lucente, G., Mastropietro, G. & Spisani, S. (2000). Bioactive fMLF-OMe analogs containing a N-terminal oximic or formylhydrazonic moiety. *J. Pept. Res., 55*, 102-109.

[24] Freer, R. J., Day, A. R., Radding, J. A., Schiffmann, E., Aswanikumar, S., Showell, H. J. & Becker, E. L. (1980). Further studies on the structural requirements for synthetic peptide chemoattractants. *Biochemistry, 19*, 2404-2410.

[25] Cavicchioni, G., Fraulini, A., Turchetti, M., Varani, K., Falzarano, S., Pavan, B. & Spisani, S. (2005). Biological activity of for-Met-Leu-Phe-OMe analogs: Relevant substitutions specifically trigger killing mechanisms in human neutrophils. *Eur. J. Pharmacol., 512,* 1-8.

[26] Torrini, I., Pagani Zecchini, G., Paglialunga Paradisi, M., Lucente, G., Gavuzzo, E., Mazza, F., Pochetti, G., Spisani, S. & Giuliani, A. L. (1991). Synthesis and properties of chemotactic peptide analogs. HCO-Met-Leu-Phe-OMe analogs containing cyclic α,α-disubstituted amino acids as Met and Phe mimicking residues. *Int. J. Pept. Prot. Res., 38*, 495-504.

[27] Toniolo, C., Bonora, G. M., Showell, H., Freer, R.J. & Becker, E. L. (1984). Structural requirements for formyl homooligopeptide chemoattractants. *Biochemistry 23*, 698-704.

[28] Torrini, I., Pagani Zecchini, G., Paglialunga Paradisi, M., Lucente, G., Gavuzzo, E., Mazza, F., Pochetti, G., Traniello, S. & Spisani, S. (1994). Modified chemotactic peptides: Synthesis, crystal conformation, and activity of For-HSE(Me)-Leu-Phe-OMe. *Biopolymers, 34,* 1-9.

[29] Cavicchioni, G., Monesi, L. G., Ferretti, M. E., Fabbri, E., Rizzuti, O. & Spisani, S. (1998). fMLP-OMe analogs sustituted at the methionine residue: an insight into the receptor properties. *Arch. Pharm. Pharm. Med. Chem., 331,* 368-370.

[30] Spisani, S., Traniello, S., Cavicchioni, G., Formaggio, F., Crisma, M. & Toniolo, C. (2002). Probing structural requirements of fMLP receptor: on the size of the hydrophobic pocket corresponding to residue 2 of the tripeptide. *J. Pept. Res., 7*, 56-65.

[31] Toniolo, C., Crisma, M., Valle, G., Bonora, G. M., Polinelli, S., Becker, E. L., Freer, R. J., Sudhanand, R., Balani Rao, P., Balaram, P. & Sukumar, M. (1989). Conformationally restricted for-Mer-tripeptide chemoattractants: a three dimensional structure-activity study of analogs incorporatine a Cα,α-dialkylated Gly at position 2. *Pept. Res., 2,* 275-281.

[32] Dentino, A. R., Antony Raj, P., Bhandary, K. K., Wilson, M. E. & Levine, M. J. (1991). Role of peptide backbone conformation on biological activity of chemotactic peptides. *J. Biol. Chem., 266*, 18460-18468.

[33] Formaggio, F., Pantano, M., Crisma, M., Toniolo, C., Boesten, W. H. J., Schoemaker, H. E., Kamphuis, J. & Becker, E. L. (1993). Backbone modified formyl-methionyl tripeptide chemoattractant. *Bioor. Med. Chem. Lett., 3*, 953-956.

[34] Cavicchioni, G. & Spisani, S. (2001). A hydrophilic residue at position 2 can improbe specific biological responses in fMLP-OMe analogs. *J. Pept. Res., 58*, 257-262.

[35] Cavicchioni, G., Turchetti, M. & Spisani, S. (2002). Biological variation responses in fMLP-OMe analogs, introducing bulky protecting groups on the side chain of hydrophilic residues at position 2. *J. Pept. Res., 60*, 223-231.

[36] Spisani, S., Turchetti, M., Varani, K., Falzarano, S. & Cavicchioni, G. (2003). Hydrophilic residue at position 3 highlight unforeseen features of the fMLP receptor pocket. *Eur. J. Pharmacol., 469*, 13-19.

[37] Spisani, S.,Cavalletti, T., Gavioli, R., Scatturin, A., Vertuani, G. & Traniello, S. (1986). Response of human neutrophils to formyl-peptide modified at the terminal amino and carboxyl groups. *Inflammation, 10,* 363-369.

[38] Cavicchioni, G., Varani, K., Niccoli, S., Rizzati, O. & Spisani, S. (1999). An evaluation of fMLP pocket dimensions and features using formyltetrapeptides. *J. Pept. Res., 54*, 336-343.

[39] Dalpiaz, A., Scatturin, A., Vertuani, G., Pecoraro, R., Borea, P. A., Varani, K., Traniello, S. & Spisani, S. (2001). Met-Ile-Phe-Leu derivatives: full and partial agonists of human neutrophil formylpeptide receptors. *Eur. J. Pharmacol., 411*, 327-333.

[40] Prossnitz, E.R., Gilbert, T. L., Chiang, S., Campbell, J. J., Qin, S., Newman, W., Sklar, L. A. & Ye, R. D. (1999). Multiple Activation Steps of the *N*-Formyl Peptide Receptor. *Biochemistry, 38,* 2240-2247.

[41] Boulay, F., Tardif, M., Brouchon, L. & Vignais, P. (1990). Synthesis and use of a novel N-formyl peptide derivative to isolate a human N-formyl peptide receptor cDNA. *Biochem. Biophys. Res. Commun., 16*, 1103-1109.

[42] Ye, R. D,, Cavanag, S. L., Quehenbergher, O., Prossnitz, E. R. & Cohrane, C. G. (1992). Isolation of a cDNA that encodes a novel granulocyte N-formyl-peptide receptor. *Biochem. Biophys. Res. Commun., 184,* 582-589.

[43] Murphy, P. M., Ozcelik, T., Kenney, R. T., Tiffany, H. I., McDermott, D. & Francke, U. (1992). A structural homologue of the N-formyl-peptide receptor. Characterisation and chromosome mapping of a peptide chemoattractant receptor family. *J. Biol. Chem., 267*, 7637-76443.

[44] Bao, L., Gerard, N. P., Eddy, R., Jr., Shows, T. B. & Gerard, C. (1992). Mapping of genes for the human C5a receptor (C5Ar), human FMLP receptor (FPR) and two FMLP receptor omologue orphan receptors (FPRH1, FPRH2) to chromosome 19. *Genomics, 13*,437-440.

[45] Le, Y., Oppenheim, J. J. & Wang, J. M. (2001). Pleiotropic roles of formyl peptide receptors. *Cytokine and growth factor Rev., 12,* 91-105.

[46] Paclet, M. H., Davis, C., Kotsonis, P., Godovac-Zimmermann, J., Segal, A. W. & Dekker, L. V. (2004). *N*-Formyl peptide receptor subtypes in human neutrophils activate L-plastin phosphorylation through different signal transduction intermediates. *Biochem. J., 377*, 469-477.

[47] Durstin, M., Gao, J. L., Tiffany, H. L., McDermott, D. & Murphy, P. M. (1994). Differerential expression of members of the N-formylpeptide receptor gene cluster in human phagocytes. *Biochem. Biophys. Res. Commun., 201*, 174-179.

[48] Fabbri, E., Spisani, S., Barbin, L., Biondi, C., Buzzi, M., Traniello, S., Zecchini, G. P. & Ferretti, M. E. (2000). Studies on fMLP–receptor interaction and signal transduction pathway by means of fMLP-OMe selective analogues. *Cell. Signal., 12*, 391-398.

[49] Fu, H., Björkman, L., Janmey, P., Karlsson, A., Karlsson, J., Movitz, C. & Dahlgren, C. (2004). The two neutrophil members of the formylpeptide receptor family activate the

NADPH-oxidase through signals that differ in sensitivity to a gelsolin derived phosphoinositide-binding peptide. *BMC Cell. Biology*, *5*, 1471-1478.

[50] Le, Y., Hu, J., Gong, W., Shen, W. Li, B., Dunlop, N. M., Halverson, D.O., Blair, D. G. & Wang, G. M. (2000). Expression of functional formyl peptide receptors by human astrocytoma cell lines. *J. Neuroimmunol.*, *111*, 102-108.

[51] Becker, E. L., Forouhar, F. A., Grunnet, M. L., Boulay, F., Tardif, M., Bormann, B. J., Sodja, D., Ye, R. D., Woska, J. R. & Murphy, P. M. (1998). Broad immunocytochemical localization of the formylpeptide receptor in human organs, tissues and cells. *Cell. Tissue Res. 292,* 129-135.

[52] Le, Y, Murphy, M. P. & Wang, M. (2002). Formyl-peptide receptors revised. *TRENDS Immunol., 23,* 541-548.

[53] Haviland, D. L., Borel, A. C., Fleisher, D. T., Haviland J. C. & Wetsel, R. A. (1993). Structure, 5'-flanking sequence, and chromosome location of the human N-formyl peptide receptor gene. A single-copy gene comprised of two exons on chromosome 19q.13.3 that yelds two distinct transcripts by alternative polyadenylation. *Biochemistry, 32*, 4168-4174.

[54] Migeotte, I., Riboldi, E., Franssen, J. D., Gregoire, F., Loison, C., Wittamer, V., Detheux, M., Robberecht, P., Costagliola, S., Vassart, G., Sozzani, S., Parmentier, M. & Communi, D. (2005). Identification and characterization of an endogenous chemotactic ligand specific for FPRL2. *J. Exp. Med., 201*, 83-93.

[55] Le, Y., Ye, R. D., Gong, W., Li, J., Iribarren, P. & Wang, J. M. (2005). Identification of functional domains in the formyl peptide receptor-like 1 for agonist-induced cell chemotaxis. *FEBS J.*, *272*, 769-782.

[56] Goldstein, I. M., (1992). Complement: biologically active products; In; Gallin, J. L., Goldstein, I. M. & Snyderman, R. Ed. Inflammation Basic Principles and Clinical Correlates. 2nd. New York: Raven Press, 55-74.

[57] Hwang, S. B. (1990). Specific receptors of platelet-activation factor, receptor heterogeneity, and signal transduction mechanisms. *J. Lipid Med., 2,* 123-158.

[58] Yang, D., Chen, Q:, Schmidt, A. P., Anderson, G. M., Wang, J. M., Wooters, J., Oppenheim, J. J. & Chertov, O. (2000). LL-37, the neutrophil granule- and epithelial cell-derived cathelicidin, utilises formyl peptide receptor-like 1 (FPRL-1) as a receptor to chemoattractant human peripheral blood neutrophils, monocytes and T cells. *J. Exp. Med., 192,* 1069-1074.

[59] Su, S. B., Gao, G. L., Gong, W. H., Dunlop, N. M., Murphy, P. M., Oppenheim, J. J. & Wang, J. M. (1999). T21/DP107, a synthetic leucine zipper-like domain of the HIV-1 envelope gp41, attracts and activates human phagocytes by using G-protein-coupled formyl peptide receptors. *J. Immunol., 162,* 5924-5930.

[60] Le, Y., Jiang, S., Hu, J., Gong, W, Su, S., Dunlop, N. M., Shen, W., Li, B. & Ming Wang, J. (2000). N36 a synthetic N-terminal heptad repeat domain of the HIV-1 envelope protein, is an activator of human phagocytes. *Clin. Immunol., 96*, 236-242.

[61] Le, Y., Gong, W., Li, B., Dunlop, N. M., Shen, W., Su, S. B., Ye, R. D. & Wang, J. M. (1999). Utilisation of two seven-transmembrane, G protein-coupled, formyl peptide receptor-like 1 and formyl peptide receptor, by the synthetic hexapeptide WKKYMVm for human phagocyte activation. *J. Immunol., 163*, 6777-6784.

[62] Walther, A., Riehermann, K. & Gerke, V. (2000). A novel ligand of the formyl peptide receptor: Anexin I regulates neutrophil extravasation by interacting with the FPR. *Mol. Cell., 5*, 831-840.

[63] Chiang, N., Fierro, I. M., Gronert, K. & Serhan, C. N. (2000). Activation of lipoxin A4 receptors by aspirin-triggered lipoxins and select peptides evokes ligand-specific responses in inflammation. *J. Exp. Med., 191*, 1197-1207.

[64] Blasi, F. & Carmeliet, P. (2002). uPAR: A versatile signalling orchestrator. *Nature Rew., 3,* 932-943

[65] Resnati, M., Pallavicini, I.,. Wang, J. M , Oppenheim, J., Serhan, C. N. Romano, M. & Blasi, F. (2002). The fibrinolytic receptor for urokinase activates the G protein-coupled chemotactic receptor FPRL1/LXA4R. *Proc. Natl. Acad. Sci. U. S. A., 99*, 1359-1364.

[66] Le, Y., Gong, W., Tiffany, H. L., Tumanov, A., Nedospasov, S., Shen, W. Dunlop, N. M., Gao, Ji-L., Murphy, P. M., Oppenheim, J. J. & Wang, J. M. (2001). Amyloid β_{42} Activates a G-Protein-Coupled Chemoattractant Receptor, FPR-Like-1. J. *Neurosci., 21*, 1-5.

[67] Su, S. B., Gong, W., Gao, J. L., Shen, W., Murphy, P. M., Oppenheim, J. J. & Wang, J. M. (1999). A Seven-transmembrane, G Protein-coupled Receptor, FPRL1, Mediates the Chemotactic Activity of Serum Amyloid A for Human Phagocytic Cells. *J. Exp. Med., 189*, 395-402.

[68] Liu, Y., Shaw, S. K., Ma, S., Yang, L., Luscinskas, F. W. & Parkos, C. A. (2004). Regulation of leukocyte transmigration: cell surphace interactions and signalling events. *J. Immunol. 23*, 1767-1773.

[69] D'Ambrosio, D., Panina-Bordignon, P. & Sinigaglia, F. (2003). Chemokine receptors in inflammation: an overview. J. Immunol. *Method, 273*, 3 -13.

[70] Heir, B., Tavener, S., Raharjo, E. & Kubes. P. (2002). An intracellular signaling hierarchy determines direction of migration in opposing chemotactic gradients. *J. Cel. Biol., 159*, 91-102.

[71] Gilbert, T. L., Bennett, T. A., Maestas, D. C., Cimino, D. F. & Prossnitz, E. R. (2001). Internalization of the human N-formyl peptide and C5a chemoattractant receptors occurs via clathrin-independent mechanisms. *Biochemistry, 40*, 3467-3475.

[72] Bennett, T. A., Maestas, D. C., Prossnitz, E. R. (2000). Arrestin binding to the G protein-coupled *N*-formyl peptide receptor is regulated by the conserved ADRYB sequence. *J. Biol. Chem., 32*, 24590-24594.

[73] Boulay, F. & Rabiet, M. J. (2005). The chemoattractant receptors FPR and C5aR: same functions different fates. *Traffic, 6*, 83-86.

[74] Oakley, R. H., Laporte, S. A., Holt, H. A., Caron, M. G. & Barak, L. S. (2000). Differential affinities of visual arrestin, beta arrestin1, and beta arrestin2 for G-protein-coupled receptors delineate two major classes of receptors. *J. Biol. Chem., 275*, 17201-17210.

[75] Wenzel-Seifert, K., Hurt, C. M. & Seifert, R. (1998). High constitutive activity of the human formyl peptide receptor. *J. Biol. Chem., 273*, 24181-24189.

[76] Bhattacharya, M., Babwah, A.V. &. Ferguson, S. S. G. (2004). Small GTP-binding protein-coupled receptors. *Biochem. Soc. Transact., 32,* 1040-1044.

[77] Takai, Y., Sasaki, T. & Matozaki, T. (2001). Small GTP-binding proteins. *Physiol. Rev., 81*, 153-208.

[78] Pasqualato, S., Renault, L. & Cherfils, J. (2002). Arf, Arl, Arp and Sar proteins: a family of GTP-binding proteins with a structural device for 'front–back' communication. *EMBO Reports, 11*, 1035-1041.

[79] Feig, L. A. (2003). Ral-GTPases: approaching their 15 minutes of fame. *Trends Cell. Biol., 13*, 419-425.

[80] Mackay, D. J. & Hall, A. (1998). Rho GTPases. *J. Biol. Chem., 273*, 20685-20688.

[81] Machesky, L. M., & Hall, A. (1997). Role of actin polymerization and adhesion to extracellular Matrix in Rac- and Rho-induced cytoskeletal reorganization *J. Cell. Biol.,* 138, 913-926.

[82] Katanaev V. L. (2001). Signal transduction in neutrophil chemotaxis. *Biochemistry (Mosc), 66*, 351-368.

[83] van Nieuw Amerongen, G. P. & van Hinsbergh, V. W. (2001). Cytoskeletal effects of rho-like small guanine nucleotide-binding proteins in the vascular system. *Arterioscler. Thromb. Vasc. Biol., 21*, 300-311.

[84] Zigmond S. H., Joyce, M., Yang, C., Brown, K., Huang, M. & Pring, M. (1998). Mechanism of Cdc42-induced actin polymerization in neutrophil extracts. *J. Cell. Biol., 142*, 1001-1012.

[85] Glogauer, M., Hartwig, J. & Stossel, T. (2000). Two pathways through Cdc42 couple the N-formyl receptor to actin nucleation in permeabilized human neutrophils. *J. Cell. Biol., 150*,785-796.

[86] Nobes, C. D. & Hall, A. (1995). Rho, rac, and cdc42 GTPases regulate the assembly of multimolecular focal complexes associated with actin stress fibers, lamellipodia, and filopodia. *Cell, 81*, 53-62.

[87] Leung, T., Chen, X. Q., Manser, E. & Lim, L. (1996). The p160 RhoA-binding kinase ROK alpha is a member of a kinase family and is involved in the reorganization of the cytoskeleton *Mol. Cell. Biol., 16*, 5313-5327.

[88] Ishizaki, T., Naito, M., Fujisawa, K., Maekawa, M., Watanabe, N., Saito, Y. & Narumiya, S. (1997). p160ROCK, a Rho-associated coiled-coil forming protein kinase, works downstream of Rho and induces focal adhesions. *FEBS Lett., 404*, 118-124.

[89] Kunzler M., Hurt, E. (2001). Targeting of Ran: variation on a common theme? *J. Cell. Sci., 114*, 3233-3241.

[90] Schaeffer, H. J. & Weber, M. J. (1999). Mitogen-Activated Protein Kinases: Specific Messages from Ubiquitous Messengers. *Mol. Cell. Biol., 19*, 2435-2444.

[91] Moodie, S. A. & Wolfman, A. (1994). The 3Rs of life: Ras, Raf and growth regulation. *Trends Genet. 10*, 44-48.

[92] Frodin, M. & Gammeltoft, S. (1999). Role and regulation of 90 kDa ribosomal S6 kinase (RSK) in signal transduction. *Mol. Cell. Endocrinol., 151*, 65-77.

[93] Irving, E. A,. Bamford, M. (2002). Role of mitogen- and stress-activated kinases in ischemic injury. *J. Cereb. Blood Flow Metab., 22*, 631-647.

[94] Shaulian, E. & Karin, M. (2002). AP-1 as a regulator of cell life and death. *Nat. Cell. Biol., 4*, 131-136.

[95] Ali, A., Hoeflich, K. P. & Woodgett, J. R. (2001). Glycogen synthase kinase-3: properties, functions, and regulation. *Chem. Rev., 101*, 2527-2540.

[96] Chan-Hui, P. Y. & Weaver, R. (1998). Human mitogen-activated protein kinase kinase kinase mediates the stress-induced activation of mitogen-activated protein kinase cascades. *Biochem. J., 336*, 599-609.

[97] Kyriakis, J. M. & Avruch, J. (2001). Mammalian mitogen-activated protein kinase signal transduction pathways activated by stress and inflammation. *Physiol. Rev., 8*, 807-869.

[98] Dekker, L. V. & Parker, P. J. PKC isozymes and myeloid cell differentiation. In: Parker P J & Dekker LV, editors. *Protein kinase C* , R.G. Austin, Texas, U.S.A: Landes Biosciences; 1997; 121-129.

[99] Parker, P. J. & Rust, J. M. (2004). PKC at a glance *J. Cell Sci., 117*, 131-132.

[100] Keenan, C. & Kellehe, D. (1998). Protein Kinase C and the Cytoskeleton *Cell. Signal., 10*, 225–232.

[101] Downey, G. P., Chan, C. K., Lea, P., Takai, A. & Grinstein, S. (1992). Phorbol ester-induced actin assembly in neutrophils: role of protein kinase C. *J. Cell. Biol., 116*, 695-706.

[102] Barry, S. T. & Critchley, D. R. (1994). The RhoA-dependent assembly of focal adhesions in Swiss 3T3 cells is associated with increased tyrosine phosphorylation and the recruitment of both pp125FAK and protein kinase C-delta to focal adhesions. *J. Cell Sci., 107*, 2033-6532.

[103] Gimond, C., De Melker, A., Aumailley, M. & Sonnenberg, A. (1995). The cytoplasmic domain of alpha 6A integrin subunit is an in vitro substrate for protein kinase C. *Exp. Cell Res. 216*, 232-235.

[104] Jaken, S., Leach, K. & Klauck, T. (1989). Association of type 3 protein kinase C with focal contacts in rat embryo fibroblasts. *J. Cell Biol., 109*, 697-704.

[105] Allen, L. A. H. & Aderem, A. (1995). A role for MARCKS, the alpha isozyme of protein kinase C and myosin I in zymosan phagocytosis by macrophages *J. Exp. Med., 182*, 829-840.

[106] Ferretti, M. E., Spisani, S., Pareschi, M. C., Buzzi, M., Cavallaro, R., Traniello, S., Reali, E., Torrini, I., Paglialunga Pardisi, M., Pagani Zecchini, G., & Biondi, C. (1994). Two new formylated peptides able to activate chemotaxis and respirtory burst selectively as tools for studying human neutrophil responses. *Cell. Signal., 6*, 91-101.

[107] Spisani, S., Pareschi, M. C., Buzzi, M., Colamussi, M. L., Biondi, C., Traniello, S., Pagani Zecchini, G., Paglialunga Pardisi, M., Torrini, I. & Ferretti, M. E. (1996). Effects of cyclic AMP level reduction on human neutrophil responses to formylated peptides. *Cell. Signal., 8*, 269-277.

[108] Ferretti, M. E., Nalli, M., Biondi, C., Colamussi, M. L., Pavan, B., Traniello, S. & Spisani, S. (2001). Modulation of neutrophil phospholipase C activity and cyclic AMP levels by fMLP-OMe analogues. *Cell. Signal., 13*, 233-240.

[109] Elferink, J. G. & de Koster, B. M. (1993). Differential stimulation of neutrophil functions by pyrithioxine. *Int. J. Immunopharmacol., 15*, 641-646.

[110] Mocsai, A., Banfi, B., Kapus, A., Farkas, G., Geiszt, M., Buday, L., Farago, A. & Ligeti, E. (1997). Differential effects of tyrosine kinase inhibitors and an inhibitor of the mitogen-activated protein kinase cascade on degranulation and superoxide production of human neutrophil granulocytes. *Biochem. Pharmacol., 54*, 781-789.

[111] Sarndahl, E., Bokoch, G. M., Boulay, F., Stendahl, O. & Andersson, T. (1996). Direct or C5a-induced activation of heterotrimeric Gi2 proteins in human neutrophils is associated with interaction between formyl peptide receptors and the cytoskeleton. *J. Biol. Chem., 271*, 15267-15271.

[112] Selvatici, R., Falzarano, S., Traniello, S., Pagani Zecchini' G. & Spisani, S. (2003). Formylpeptides trigger selective molecular pathways that are required in the physiological functions of human neutrophils. *Cell. Signal., 15*, 377-383.

[113] Spisani, S. Falzarano, S., Traniello, S., Nalli, M. & Selvatici, R. (2005). A "pure" chemoattractant formylpeptide analogue triggers a specific signalling pathway in human neutrophil chemotaxis. *FEBS J., 272*, 883-891.

[114] Spisani, S., Traniello, S., Giuliani, A. L., Torrini, I., Pagani Zecchini, G., Paglialunga Paradisi, M., Gavuzzo, E., Mazza, F., Pochetti, G. & Lucente, G. (1992). New chemotactic peptide analogs with high biological activity for human neutrophils. *Biochem. Int., 26*, 1125-1135.

[115] Gavuzzo, E., Mazza, F., Pochetti, G. & Scatturin, A. (1989). Crystal structure, conformation, and potential energy calculations of the chemotactic peptide N-formyl-L-Met-L-Leu-L-Phe-OMe. *Int. J. Pept. Protein Res., 34*, 409-415.

[116] Edmundson, A. B. & Ely, K. R. (1985). Binding of N-formylated chemotactic peptides in crystals of the Mcg light chain dimer: similarities with neutrophil receptors. *Mol. Immunol., 22*, 463-475.

[117] Pagani Zecchini, G., Paglialunga Paradisi, M., Torrini, I., Lucente, G., Gavuzzo, E., Mazza, F., Pochetti, G. & Spisani, S. (1991). for-Met-Δ^{Z}Leu-Phe-OMe: a new active analog of chemotactic N-formylpeptides with β-turn crystal conformation. *Tetrahedron Lett., 32*, 4375-4376.

[118] Pagani Zecchini, G., Paglialunga Paradisi, M., Torrini, I., Lucente, G., Gavuzzo, E., Mazza, F., Pochetti, G., Paci, M., Sette, M., Di Nola, A., Veglia, G., Traniello, S. & Spisani, S. (1993). Synthesis, conformation, and activity of HCO-Met-Δ^{Z}Leu-Phe-OMe, an active analogue of chemotactic N- formyltripeptides. *Biopolymers, 33*, 437-451.

[119] Fabbri, E., Spisani, S., Biondi, C., Barbin, L., Colamussi, M. L., Cariani, A., Traniello, S., Torrini, I. & Ferretti, M. E. (1997). Two for-Met-Leu-Phe-OMe analogues trigger selective neutrophil responses. A differential effect on cytosolic free Ca^{2+}. *Biochim. Biophys. Acta, 1359*, 233-240.

[120] Haines, K. A., Kolasinski, S. L., Cronstein, B. N., Reibman, J., Gold, L. I. & Weissmann, G. (1993). Chemoattraction of neutrophils by substance P and transforming growth factor-beta 1 is inadequately explained by current models of lipid remodeling. *J. Immunol., 151*, 1491-1499.

[121] Reibman, J., Meixler, S., Lee, T. C., Gold, L. I., Cronstein, B. N., Haines, K. A., Kolasinski, S..L. & Weissmann, G. (1991). Transforming growth factor beta 1, a potent chemoattractant for human neutrophils, bypasses classic signal-transduction pathways. *Pro. Natl. Acad. Sci. U. S. A., 88*, 6805-6809.

[122] Cornish, C. J., Devery, J. M., Poronnik, P., Lackmann, M., Cook, D. I. & Geczy, C. L. (1996). S100 protein CP-10 stimulates myeloid cell chemotaxis without activation. *J. Cell. Physiol., 166*, 427-437.

[123] Laffafian, I. & Hallett, M. B. (1995). Does cytosolic free Ca^{2+} signal neutrophil chemotaxis in response to formylated chemotactic peptide? *J. Cell. Sci,. 108*, 3199-3205.

[124] Dang, P. M. C., Fontayne, A., Hakim, J., El Benna, J. & Périanin, A. (2001). Protein kinase C zeta phosphorylates a subset of selective sites of the NADPH oxidase

component p47phox and participates in formyl peptide-mediated neutrophil respiratory burst. *J. Immunol., 166*, 1206-1213.

[125] Laudanna C., Mochly-Rosen D., Liron T., Costantin G. & Butcher E. C. (1998). Evidence of zeta protein kinase C involvement in polymorphonuclear neutrophil integrin-dependent adhesion and chemotaxis. *J. Biol. Chem., 273*, 30306-30315.

[126] Zu Y. L., Qi J., Gilchrist A., Fernandez G.A., Vazquez-Abad D., Kreutzer D.L., Huang Chi-K. and Sha'afi R.I. (1998). p38 mitogen-activated protein kinase activation is required for human neutrophil function triggered by TNF-α or FMLP stimulation. *J. Immunol., 160*, 1982-1989.

[127] Ward, R. A., Nakamura, M. & McLeish K. R. (2000). Priming of the neutrophil respiratory burst involves p38 mitogen-activated protein kinase-dependent exocytosis of flavocytochrome b558-containing granules. *J. Biol. Chem., 275*, 36713-36719.

[128] Ye R.D. & Boulay F. Structure and function of leukocyte chemoattractant receptors. In: August JT, Anders MW, Murad F. & Coyle J., Eds. *Advances in Pharmacology*. San Diego USA: Academic Press; 1997; 39, 221-288,

[129] Keenan, C. & Kelleher, D. (1998). Protein kinase C and the cytoskeleton. *Cell. Signal., 10*, 225-232.

[130] Pettit, E. J. & Hallett, M. B. (1996). Localised and global cytosolic Ca^{2+} changes in neutrophils during engagement of Cd11b/CD18 integrin visualised using confocal laser scanning reconstruction. *J. Cell. Sci., 109*, 1689-1694.

In: Trends in Cellular Signaling
Editor: Dave E. Caplin, pp. 41-78

ISBN 1-59454-633-9

Chapter II

Transcription of Genes in Response to Activated cAMP/Protein Kinase A Signalling Pathway: There is More to it than CREB

Mona Johannessen* and Ugo Moens
Department of Microbiology and Virology, Institute of Medical Biology, N-9037 Tromsø, Norway.

Abstract

The general accepted paradigm for cAMP-induced transcription of genes presages that cAMP activates protein kinase A, which in turn phosphorylates the cAMP responsive element-binding protein CREB at the serine-133 residue. Phosphorylated CREB, bound to cAMP responsive element(s) in the promoter of such genes, will subsequently recruit the co-activator CBP or its paralog p300. These co-activators enhance CREB-mediated transcription through linking CREB to the basal RNA polymerase II transcription complex and through their intrinsic acetyltransferase activity, which enables to formation of an open chromatin structure. However, CREB-mediated transcription of cAMP-responsive genes remains enigmatic in multiple ways. Despite the ubiquitous expression of protein kinase A, CREB and CBP, many genes display a tissue-specific responsiveness to cAMP. In fact, genes with cAMP responsive element(s) in their promoters can be up-regulated, refractory, or even repressed by increased cAMP levels. This review will discuss several mechanisms that contribute to the outcome of the responsiveness of CREB target genes to increased cAMP levels.

Keywords: CREB, CREB-interacting proteins, chromatin, TATA box, acetylation, methylation, mRNA stability

* Correspondence concerning this article should be addressed to Mona Johannessen, Department of Microbiology and Virology, Institute of Medical Biology, N-9037 Tromsø, Norway.monaj@fagmed.uit.no.

Introduction

The second messenger adenosine 3′,5′cyclic monophosphate (cAMP) is generated in cells in response to numerous stimuli. Increased intracellular cAMP levels activate the cAMP-dependent protein kinase or protein kinase (PKA). One of the nuclear substrates of PKA is the transcription factor CREB (cAMP response element-binding protein). CREB was originally described as a protein that binds to an eight base-pair (bp) element in the somatostatin promoter. Screening promoters of other cAMP-inducible genes revealed the consensus sequence TGACGTCA, referred to as the cAMP response element (CRE). PKA was shown to phosphorylate CREB exclusively at serine-133 *in vivo*, a modification that strongly enhances the transcriptional activity of CREB. This phosphorylation event enables CREB to recruit the coactivator CBP or its paralog p300. CBP and p300 possess histone acetyltransferase activity, and have been suggested to mediate target gene activation, in part, by acetylating and disrupting promoter-bound nucleosomes. Moreover, they can link CREB to the basal transcription machinery as CBP was found to indirectly contact RNA polymerase II by the bridging protein RNA helicase A (reviewed in Mayr and Montminy, 2002). Furthermore, acetylation of CREB by CBP modulates CREB-mediated gene expression (Lu et al., 2003). Cessation of CREB-dependent transcription is obtained through dephosphorylation of CREB. Both protein phosphatase PP1 and PP2A have been shown to dephosphorylates CREB *in vivo* (Wadzinski et al; 1993; Alberts et al., 1994). The contemporary model for CREB-dependent transcription is summarized in Figure 1.

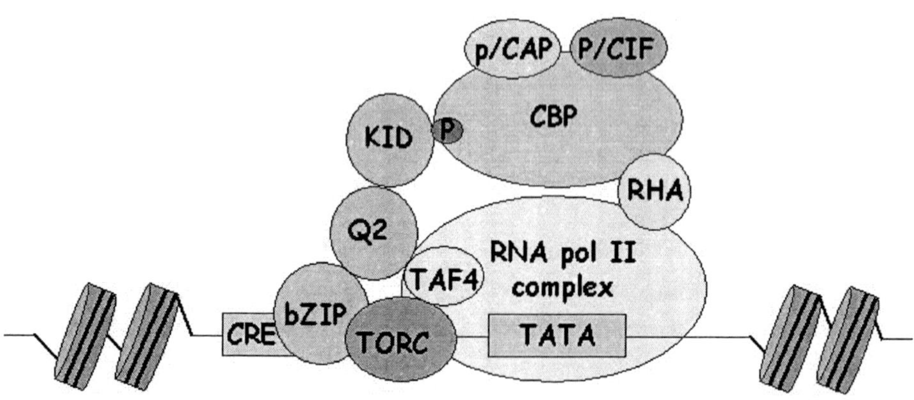

Figure 1. General model for signal-induced CREB-mediated transcription. CREB is bound as a homodimer to its cognate sequence (the cAMP response element, CRE) in the promoter of cAMP-responsive genes. For the sake of simplicity, CREB is visualized as a monomer. CREB consists of distinct functional domains, including the basic leucine zipper motif (bZIP, a DNA-binding and dimerization domain), the basal transactivator domain Q2, and the stimulus-inducible domain KID, which can be phosphorylated by PKA on serine-133 in response to stimuli that elevate intracellular cAMP concentrations. The Q2 domain interacts with the general transcription factor TAF4, while bZIP can bind the co-activator TORC. TORC augments the association of CREB with TAF4. The phosphorylated KID of CREB will recruit the co-activator CBP (or its paralog p300). CBP/p300 increase CREB-mediated transcription through linking CREB to the basal RNA polymerase II transcription complex via e.g. RNA helicase A (RHA) and through their intrinsic acetyltransferase activity, which enables to formation of an open chromatin structure. Other co-activators like p/CAP and p/CIF with histone acetyltransferase activity can enhance transcription. See text for details.

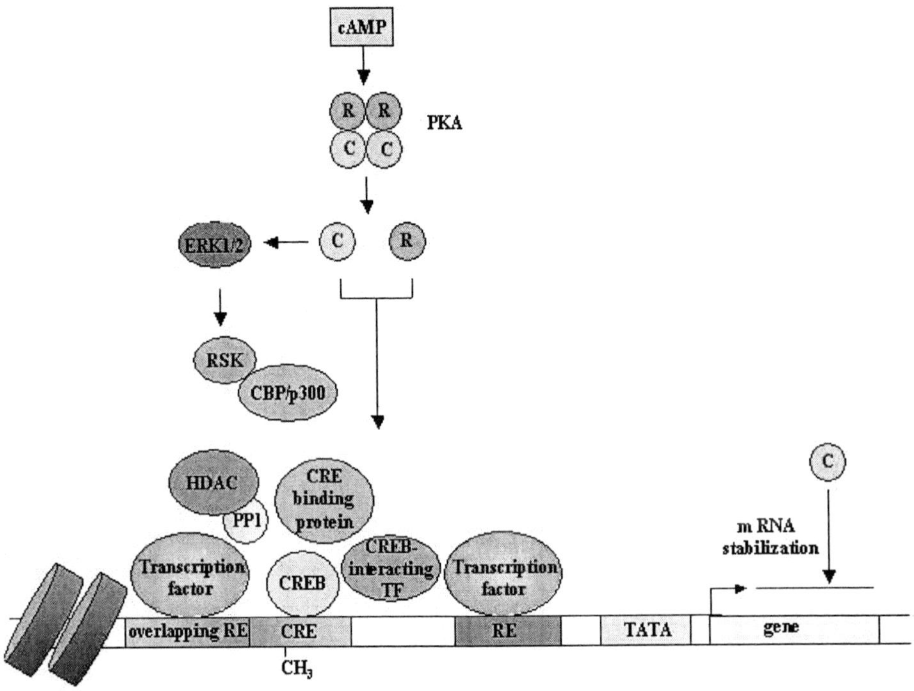

Figure 2. Multiple control levels of CREB-mediated transcription through the cAMP/PKA signalling pathway. Increased cAMP levels bind the regulatory subunits (R) of the PKA holoenzyme. As a result, the catalytic subunits (C) dissociate and can enter the nucleus and phosphorylate CREB at serine-133. The regulatory subunits may also interact with CREB and affect the transcriptional potentials of CREB. However, activated PKA can also signal through the mitogen-activated protein kinases ERK1/2 and $p90^{RSK}$. $p90^{RSK}$ can sequester CBP, thereby inhibiting CREB-dependent transcription. CREB binds its cognate sequence (cAMP response element, CRE) as a homodimer or as a heterodimer. These heterodimers can act as repressors or activators of transcription. Regulatory elements (RE) can overlap with the CRE motif and transcription factors binding to such site can prevent CREB to bind to CRE. Alternatively, other transcription factors may cooperate with CREB to induce full cAMP-response. Finally, PKA may also control the stability of transcripts. See text for further details.

Although PKA-mediated CREB phosphorylation at serine-133 and subsequent recruitment of CBP/p300 is generally accepted as the model for cAMP-induced activation of cAMP-responsive genes, this model is oversimplified. An incoherent picture of cAMP/CREB-regulated transcription emerges from the results obtained from studies aimed at identifying new cAMP-responsive genes and novel CREB target genes. The methodological approaches included: *i.* microarray analyses that compared the altered expression pattern of control cells with cells exposed to cAMP elevating agents or cell lines expressing a dominant mutant form of CREB (the non-phosphorable alanine-133 mutant or the non-DNA binding K-CREB mutant) or an activated form of CREB (VP16-CREB chimera protein); *ii.* the use of CRE decoy oligonucleotides to trap CREB, which will affect expression of CREB target genes; *iii.* Chromatin immunoprecipitation (ChIP) with anti-CREB or anti-phosphoCREB antibodies to determine sites in the genome occupied by CREB. All these studies have identified genes with or without CRE motifs, but responsive to cAMP. Furthermore, genes containing CRE were up- or down-regulated by cAMP, or not affected albeit CREB was bound to the promoter. These data illustrate that the presence of CRE motifs in the promoter of a putative CREB target gene not always can predict the outcome of the transcriptional response of this gene in cells with increased intracellular cAMP concentrations (Lee et al., 2000; Popovici et al., 2000; Park et al., 2001; Cho-Chung et al., 2002; Park et al., 2002; Reusch and Klemm, 2002; Yan et al., 2002; Fass et al., 2003; Tierney et al., 2003; Impey et al., 2004; Kudo et al., 2004; Bailey et al., 2005; Zhang et al., 2005). Thus cAMP-induced activation of CREB-directed transcription through recruitment of CBP remains enigmatic in multiple ways. Some of the unsolved questions in CREB-directed transcription are illustrated in the next section, while mechanisms that contribute to the complexity and diversity of cAMP-responses for gene transcription form the focus of this review and are illustrated in Figure 2.

Enigmas of cAMP-Induced Regulation of CREB-Dependent Transcription

A plethora of stimuli can trigger the phosphorylation of CREB at serine-133. Distinct stimuli induce CREB phosphorylation with comparable kinetics and stoichiometry, yet not all of them increase CREB-dependent transcription. The mechanism of signal discrimination by CREB is unknown, but it has been proposed that one or more cofactors may bind CREB and prevent activation in the absence of the appropriate stimulus. The discrepancy between CREB phosphorylation and CREB activation has been discussed recently and is beyond the scope of this review (Mayr and Montminy, 2001; Johannessen et al., 2004).

Another puzzling observation is that not all CRE-containing promoters are activated upon increased cAMP levels generated by forskolin, an activator of the adenylate cyclase enzymes that produce cAMP. Some are insensitive, while others may in fact be repressed. A recent study demonstrated that CREB occupies ~4,000 genes and the profiles for CREB occupancy were very similar for forskolin-treated cells (phosphoCREB) and in different human tissues. However, less than 2% responded to forskolin, indicating that the majority of phosphoCREB-occupied genes are not induced by cAMP (Zhang et al., 2005). Another example comes from transient transfection studies in PC12 cells with reporter plasmids containing fragments of the *bcl*-2 promoter that encompass the CRE motif. These studies showed that dibutyryl cAMP provoked a robust CREB phosphorylation, yet failed to induce the promoter. However, the isolated *bcl*-2 CRE cloned upstream of the Herpes simplex virus thymidine kinase promoter rendered the promoter cAMP responsive. The authors proposed that other elements in the *bcl*-2 promoter inhibited any response to cAMP (Freeland et al., 2001). Expression levels of genes on chromosome 22 were compared by microarray analysis of untreated and forskolin-treated human JEG-3 choriocarcinoma (placental) cells. The authors found that the expression of genes was up-regulated, reduced, or unaffected after 1.5 h stimulation with 10 μM forskolin. Not all of the genes whose expression was altered contained binding sites for CREB. This study clearly shows that the outcome of cAMP-induced transcription is not only governed by the presence of a CRE or binding of CREB and that repression of gene expression by cAMP can occur in the presence or absence of CRE motifs (Euskirchen et al., 2004).

Cyclic AMP-induced gene expression can be cell-specific despite the ubiquitous expression pattern of PKA, CREB and CBP. Many examples have been reported in the literature and a few will be discussed here. The COX-2 promoter contains a CRE motif, but CREB does not play a positive regulatory role in the expression of COX-2 in RAW264.7 macrophages, NIH3T3, mast, and osteoblast cells, while this promoter is cAMP responsive in other cell types (Wadleigh et al., 2000 and references therein). The promoter of the BRCA1, a gene associated with inherited breast cancer, forms an example of a CRE-containing promoter that is refractory to cAMP. The promoter contains a CRE-like motif (TTACGTCA) at position –176 to –169 (with +1 the start site of transcription), but lacks a TATA box. Electrophoretic mobility shift assay (EMSA) with nuclear extracts of MCF-7 and T-47D breast cancer cells confirmed binding of nuclear proteins to the CRE motif and supershift was obtained with anti-CREB antibodies, suggesting that CREB can bind to this site *in vitro*. Transient transfection studies in MCF-7, T-47D, and HepG2 cells with a luciferase reporter plasmid containing the BRCA-1 promoter established that mutations in the CRE that prevented CREB binding *in vitro* strongly reduced the basal promoter activity. Administration of forskolin failed to activate the native, as well as the mutated promoter, despite robust phosphorylation of CREB in treated cells. In fact, both promoters were repressed in forskolin-exposed HepG2 cells (Atlas et al., 2001). Opposite to MCF-7 and T-47D nuclear extracts, extracts from HeLa cells did not bind the BRCA1 CRE motif in EMSA and the BRCA1 promoter fragment encompassing the CRE was not induced or repressed by forskolin in HeLa cells. *In vivo* binding of CREB to the BRCA1 promoter was not investigated in this study (Suen and Goss; 1999). The lack of CREB binding to the BRCA1 promoter and the presence of high phosphoCREB levels in non-forskolin treated HeLa cells (Meijer et al., 2000), may explain why the BRCA-1 promoter is refractory to forskolin treatment in these cells.

Other promoters that are induced by cAMP in a cell-specific mode are the somatostatin, CYP1B1, typeII deiodinase and different cyclin promoters. The CRE-containing somatostatin promoter did not provide cAMP responsiveness in the pancreatic islet cell line Tu6, while a strong response was observed in PC12 cells stably transfected with the somatostatin gene (Leonard et al., 1992). The cytochrome P450 superfamily (CYP) encodes constitutive and inducible enzymes responsible for the catalytic hydroxylation of a multitude of compounds, and is therefore also designated as drug detoxification enzymes (McFadyen et al., 2004). The *CYP1B1* promoter is insensitive to cAMP stimulation in HEK293 cells, while cAMP-responsiveness is observed in Y-1 mouse adrenal cells (Zheng and Jefcoate, 2005). Expression of the type II deiodinase, an enzyme involved in maintaining appropriate intracellular concentrations of tri-iodothyronine, is induced by cAMP in a cell-specific mode. While a vigorous response is observed in e.g. thyroid, glial, neuroblastoma, and osteablast cells, cAMP had only a modest effect on type II deiodinase gene expression in pituitary tumour cells (Kim et al., 1998; Canettieri et al., 2004 and references therein; Morimura et al., 2005). The expression of the genes for cyclins A, E, D1 and D2 is induced by cAMP in some cell types, but repressed in others. In both cases, cAMP responses seem to involve the CRE in the promoters of these genes, indicating that CRE can mediate opposite regulation in responses to increased cAMP levels. For example, elevated cAMP concentrations reduced cyclin D1 mRNA and protein levels in Hs68 cells, bovine tracheal myocytes, human diploid fibroblasts, and astrocytic cell line. Surprisingly, forskolin triggered CREB phosphorylation in bovine tracheal myocytes and binding of CREB to cyclinD1 CRE in EMSA, yet no transactivation of the promoter activity was measured (Musa et al., 1999). On the other hand, cyclins D1 and E mRNA levels increased in cAMP-treated FRTL5 cells, while they were repressed in the astrocycte cell line C.LT.T.1.1 after exposure to forskolin (Won et al., 1992; Gagelin et al., 1999 and references therein).

Table 1. Mechanisms involved in diversification of cAMP-induced regulation of CREB-mediated transcription.

<u>The role of PKA in differential regulation of CREB-dependent transcription</u>
• type I PKA has higher affinity for cAMP than type II PKA
• cAMP-induced nuclear translocation of the catalytic subunits
• catalytic Cα isoform possess stronger transcriptional activation properties than Cγ
• RIα and RIIβ regulatory subunit display distinct properties to activate or repress CRE-mediated transcription
<u>The role of substained cAMP-induced activation of the MEK/ERK signalling pathway</u>
• sequestration of CBP by $pp90^{RSK}$
<u>The role of CREB coregulators</u>
• cAMP-dependent upregulation of the expression of coregulator (e.g. FHL3)
• cAMP-dependent translocation of coactivator (e.g. TORC2)
• cAMP/PKA induced phosphorylation enhances activity of coactivator (e.g. CBP)
• coactivator recruited by CREB coactivator (e.g. SRCAP associates with CBP)
<u>The role of CRE-binding and CREB-interacting transcription factors</u>
• repressor competes with CREB for binding to CRE (e.g. ICER)
• repressor forms heterodimer with CREB (e.g. ATF-1)
• repressor binds to CREB (e.g. DREAM)
• activator interacts with CREB (e.g. Ets-2)
<u>The role of general transcription factors</u>
• splice variant of general transcription factor with antagonistic effect on CREB-dependent transcription (e.g. TAF4 Δ3)
<u>The role of chromatin structure and DNA elements</u>
• cAMP-induced chromatin remodelling
• cAMP-regulated activity of histone deacetylases
• cAMP/PKA-induced phosphorylation of HDAC8
• methylation of CRE prevents CREB binding
• TATA box enhances cAMP responsiveness of the promoter
• position of the CRE in the promoter determines its functionality
• sequencing adjacent to or overlapping with CRE affect CREB binding
• cAMP response unit: cooperativity between multiple regulatory elements to obtain maximal cAMP responsiveness (e.g. CREB and C/EBP binding motifs)
<u>cAMP-induced stabilization/destabilization of mRNA</u>

All the examples above demonstrate a cell-dependent response of a CRE containing promoter to cAMP. In addition, different promoters may have different sensitivity for cAMP in a cell-specific mode. For example, isolated hepatocytes and H4-II-E-C3 hepatoma cells display similar sensitivity to cAMP at the level of PKA activation, yet a 100-fold higher concentration of Cpt-AMP was required to induce tyrosine aminotransferase (TAT) mRNA levels in H4-II-E-C3 cells (500 μM) compared to hepatocytes (5 μM). Surprisingly, similar stoichiometric profiles of CREB phosphorylation were obtained in both cell types. The mechanism for this differential sensitivity to cAMP remains elusive, but may be attributed to cell specific differences in the expression of proteins that interact with CREB functions (Pickering et al., 1998). In the next part, we will focus on some of the mechanisms that can be involved in the outcome of gene expression in response to cAMP. These distinct additional factors that determine cAMP-induced CREB-mediated transcription are summarized in Table 1.

Mechanisms that Contribute to the Diversity in cAMP-Induced CREB-Dependent Transcription

A. The Role of PKA in cAMP-Induced Regulation of CREB-Dependent Transcription

A signal transduction pathway consists of a relay of events that links the uptake of a signal to the conversion of the signal into a cellular response. Cyclic AMP is generated by many stimuli and this second messenger acts mainly by activating the cAMP-dependent protein kinase/protein kinase A (PKA), which subsequently can phosphorylate the nuclear substrate CREB. The outcome of CREB-mediated transcription elicited by increased cAMP can therefore be controlled by components of the cAMP/PKA signalling pathway. The PKA holoenzyme is a heterodimer composed of two regulatory and two catalytic subunits of which different isoforms exist. Both the regulatory (RIα, RIβ, RIIα, RIIβ) and the catalytic (Cα, Cβ, Cγ) subunits are differentially expressed, and are able to form different PKA isoforms. PKA type I (RI-containing PKA) has a 2- to 8-fold higher affinity for cAMP than PKA type II (with RII subunits) because the R isoforms exhibit different cAMP binding affinities. The composition of the PKA holoenzyme determines the threshold for the cAMP-induced dissociation of the catalytic subunits from the regulatory subunits and hence activation of PKA and can therefore contribute to cell-specific responses of CREB to elevated cAMP levels (reviewed in Skålhegg and Taskén, 2000; Taskén and Aandahl, 2004). In addition, the different catalytic and regulatoric subunit may also modulate CREB-mediated transcription as will be discussed in the following section.

1. The Role of PKA Catalytic Subunit Isoforms in Diversifying CREB-Mediated Transcription in Response to cAMP

Elegant studies in rat pheochromocytoma PC12 cells have demonstrated that transcriptional activation via cAMP can be rate limited by nuclear entry of the catalytic subunit of PKA. Nuclear translocation of the C subunit is a passive process that plateaus after 15 to 30 min, coinciding with peak levels of CREB serine-133 phosphorylation and target gene activation. Forskolin treatment induced 70% of total cellular PKA activity within 30 min of stimulation and this effect persisted throughout the 2 hours time course of stimulation (Hagiwara et al., 1993; Michael et al., 2000). Similar studies in other cell types are lacking, but it can be imagined that disturbance in nuclear translocation of the catalytic subunit may explain the unresponsiveness to cAMP observed in certain cell types. Indeed, we found that the cAMP analogue dibutyryl cAMP failed to cause nuclear translocation of PKA, to phosphorylate and activate CREB, and to stimulate the cAMP-responsive c-*fos* promoter in NIH3T3 cells (Seterens et al., 1999).

Differential expression of PKA isoforms may contribute to the specificity in the cAMP/PKA signalling pathway (Skålhegg and Taskén, 2000; Taskén and Aandahl, 2004). To test this on CREB-mediated transcription, Morris and colleagues examined the effect of different catalytic subunit isoforms of PKA on CRE-driven reporter plasmids by transient transfection studies in different cell lines. The Cα subunit of PKA was more efficient in activating the α-glycoprotein hormone (GHα) promoter (two CREs) than Cγ in four distinct mammalian cells. In COS-7 and HEK 293 cells Cα transactivated this promoter 20- to 50-fold better than Cγ, while in mouse adrenal cells Y1 and Kin8 cells Cα was 6.5-fold, respectively 1.4-fold stronger activator than Cγ (Morris et al., 2002). Thus the observation that Cα and Cγ differentially regulate the cAMP-responsive GHα promoter provides a mean to diversify expression of cAMP-regulated genes in a cell-specific way and during development and differentiation.

Another example demonstrating that particular PKA isoforms may contribute to the specificity in the cAMP/PKA signalling pathway was observed in NG108-15 cells. This cell-line, which is a hybrid cell line between gliom and neuroblastoma cells, was shown to contain type I PKA (CβRIβ) primarily in cytosol and type II PKA (CαRIIβ) in the particulate and nuclear fractions. By using specific inhibitors for either PKA type I or PKA type II, Constantinescu et al. observed that forskolin-induced nuclear translocation of the catalytic subunits was only blocked by type II PKA inhibitor, indicating a nuclear redistribution of Cα, but not Cβ. Accordingly, forskolin-triggered phosphorylation of CREB depended only on PKA type II. Remarkably, induction of CREB-mediated transcription by forskolin required both PKA isoforms, suggesting that both PKA types are involved in mediating maximal CREB-driven transcription in response to cAMP in these cells (Constantinescu et al., 2002). The reason for differential requirement of PKA type I and type II for CREB phosphorylation and CREB activation remains unknown.

2. The Role of PKA Regulatory Subunit Isoforms in Diversifying CREB-Mediated Transcription in Response to cAMP

The R subunits are differential expressed in several cells and tissues, and at various differentiation and developmental stages. In mice for example, RIα is predominantly found in heart and central nervous system, while RIβ is mainly present in brain and spinal cord. RIIα is the major isoform in heart, but is also present in brain. RIIβ is also found in brain, but it is predominantly expressed in liver and fat tissue (reviewed in Skålhegg and Taskén, 2000). The difference in cAMP response of cAMP-regulated genes may be explained by the opposite effect of the distinct R subunits on CRE-dependent transcription and the different expression levels in various cells. Paolillo and colleagues compared cAMP signal transmission to the nucleus in cortical neurons and in cerebellar granule cells. Cortical neurons possess relatively high levels RIIβ, which are paralleled by the presence of the specific RII-anchoring protein AKAP150, while cerebellar granule cells express RIα and RIIα (Ventra et al., 1996). In cortical cells, but not in granule cells, the catalytic subunit of PKA translocated in response to cAMP stimuli, and CREB became phosphorylated, while almost no phosphorylated CREB was detected in stimulated cerebellar granule cells. Microinjection of cDNA of RIIβ and AKAP75 increased cAMP-induced transcription in cerebellar granule cells, whereas forskolin-induced CREB phosphorylation was significantly reduced in cortical cells treated with the RIIβ antisense oligonucleotide. These data illustrate the need for RIIβ in cAMP-dependent transcription (Paolillo et al., 1999). The necessity for RIIβ in cAMP-mediated CREB phosphorylation is consistent with findings that overexpression of RIIβ restored cAMP-dependent transcription in the cAMP-unresponsive mutant pheochromocytoma cell line A126-1B2 (Tortora et al., 1990). In another study, the ability of cAMP to transactivate somatostatin-CAT was compared in DT cells, DT cells expressing wild-type RIIβ (referred to as DTRIIβ cells) and DT cells expressing a RIIβ mutant unable to autophosphorylate serin-114 (designated as RIIβ-P cells). DTIIβ cells displayed positive effects on the CRE-mediated transcription in the absence of cAMP agonist, indicating that the precence of RIIβ is sufficient for increased CREB-mediated transcription. Stimulation with 8-Cl-cAMP and forskolin further increased transcription of the reporter gene under control of the somatostatin promoter. On the other hand, the cells expressing inactive RIIβ showed lower basal transactivation of somatostatin-CAT, and the activity remained low even after forskolin stimulation. The authors suggested that dominant negative activity of RIIβ-P is due to the trapping of wild-type RIIβ in inactive heterodimers (Srivastava et al., 1998). The observation that DTRIIβ cells displayed constitutive positive effects on CREB-mediated transcription may be due to intrincic transcriptional activity as studies in COS cells showed that GAL4-RIIβ was able to transactivate expression of a reporter plasmid, although to a lesser extent than GAL4-CREB (Srivastava et al., 1998). The finding that RIIβ possesses intrinsic transcriptional activity is further emphasized by experimental data from another group. Overexpression of RIIβ upregulated CD154 mRNA in activated T cells in a phospho-serine-114 dependent mode and increased CRE-directed transcription of a reporter gene in the nonhematopoietic KB and NIH3T3 cells (Elliot et al., 2004).

All the examples above underscore a positive role of RIIβ in CREB mediated transcription. However, RIIβ has also been shown to exert a negative role in cAMP-induced CREB-mediated transcription of the IL-2 promoter in activated T cells. Elliot et al. demonstrated that cAMP-induced activation of PKA resulted in phosphorylation of RIIβ at serine-114, subsequent increased nuclear translocation of this R-subunit and repression of the IL-2 promoter. A non-phosphorable RIIβ alanine-114 mutant did not down-regulate IL-2 promoter activity in activated T cells, illustrating the importance of serine-114 phoshorylation. RIIβ also inhibited the transcriptional activity of GAL4-CREB fusion protein and repressed CRE-directed reporter gene expression, as well as expression of the cAMP-responsive c-*fos* gene. The authors demonstrated that RIIβ physically interacted with CREB, resulting in repression of CREB-dependent transcription from the IL-2 promoter. This interaction required the N-terminal 135 amino-acid residues of RIIβ, while the domain of CREB necessary for interaction has not been mapped. The exact mechanism by which RIIβ interferes with CREB-dependent transcription is not completely understood, but RIIβ did not bind to, nor prevented CREB binding to the c-*fos* CRE *in vitro*, and inhibition of CREB activity by RIIβ was not associated with a decrease in serine-133 phosphorylation of CREB (Elliot et al., 2003). Others have, however, reported that RIIβ can bind to a consensus CRE *in vitro* (Fax et al., 2001; Srivastava et al., 1998). The work of Elliot et al. further revealed that inhibition of protein phosphatase-1 increased nuclear phosphoRIIβ levels. Thus PP1 may have opposite effects on CREB-mediated transcription: PP1 dephosphorylates CREB at serine-133 and this will decrease the transcriptional activity of CREB (Alberts et al., 1994), whereas PP1-mediated dephosphorylation of phosphoRIIβ will prevent nuclear translocation of RIIβ. Non-phosphorylated RIIβ residing in the cytoplasm will not affect CREB-mediated transcription (Elliot et al., 2003; Elliot et al., 2004).

The RIα subunit may also affect transcription of cAMP-responsive genes. Overexpression of the RIα subunit caused a 3- to 4-fold increase in IL-2 production in activated T cells. The involvement of the CRE motif in the IL-2 promoter in RIα-induced activation was not investigated, but RIα neither binds CRE nor CREB (Elliot et al., 2003; Elliot et al., 2004). The mechanism by which RIα affects transcriptions remains therefore obscure.

In conclusion, both the catalytic and the regulatory subunits of PKA can alter CREB-mediated transcription. The cell-type specific expression and the different affinity for cAMP by the different isoforms of PKA will influence the response of cAMP-responsive genes to a particular stimulus in a defined cell.

3. Sustained Activation of the ERK Signalling Pathway by cAMP

Increased cAMP levels not only activate PKA, but may also lead to activation of various other signalling pathways that can converge to CREB. For example, the cAMP/PKA can interfere with the mitogen-activated protein kinase MEK/ERK pathway (Delghandi et al., 2005 and references therein). Although interaction between these signalling pathways is beyond the scope of this review, one example illustrates that the cross-talk between the cAMP/PKA and MEK/ERK pathways may affect CREB-dependent transcription. Persistent activation of the ERK pathway by treatment of human hepatoma Hep3B cells with compound 5, an inhibitor of protein phosphatase Cdc25A, resulted in down-regulation of CREB activity and suppression of CRE-mediated transcription of cyclin D1 and bcl-2 genes due to increased recruitment of CBP to $pp90^{RSK}$ (Wang et al., 2003). A similar mechanism may occur in cells where increased cAMP levels can endure ERK activation. Increased ERK1/2 phosphorylation levels were monitored in mouse adrenocortical Y1 cells after 30 min stimulation with forskolin (1 μM), and these levels kept increasing for at least 6 hours after stimulation (Gyles et al., 2001). The authors did not monitor the expression levels of CREB target genes, but a similar scenario as in compound 5-treated Hep3B cells can be imagined, i.e. persistent ERK activation abolishes forskolin-induced CREB-mediated transcription because $pp90^{RSK}$ sequesters CBP and renders it unavailable for phosphorylated CREB.

B. Regulation at the Level of CREB-Interacting Proteins

1. Regulation of the Activity of Co-Regulators for CREB by cAMP

Co-activators can modulate the transcriptional activity of CREB after cAMP stimulation. The four-and a half-LIM only proteins, a family of proteins containing zinc finger protein-

protein interaction motifs, are expressed in a tissue-specific manner. The FHL2 and FHL3 members of this family can stimulate the transcriptional activity of CREB in a phosphoserine-133 independent mode (Fimia et al., 2000). It was recently shown that cAMP elevating agents induced nNOS expression in A673 cells. Inhibition of CREB phosphorylation did not abrogate cAMP-induced expression of nNOS, suggesting that CREB bypasses in part the classical requirement for phosphorylation and association with CBP. In fact, the authors found the FHL1-3 proteins to be expressed in A673 cells, and FHL3 expression itself was upregulated by increased cAMP levels (Boissel et al., 2004). Thus, the increase in nNOS expression after cAMP stimulation may result from elevated levels of FHL3, which will augment CREB its transcriptional activity independent of the phosphorylation status of serine-133. Consequently, cAMP-induced coactivator synthesis represents an extra regulation level of cAMP/CREB-mediated transcription.

Recently, a new family of cell- and tissue-specific expressed CREB coactivator, known as transducers of regulated CREB activity (TORC), were identified. The TORC family consists of three highly related proteins, TORC1-3. TORCs bind the bZip domain of CREB and interact with the general transcription factor TAF4. Overexpression of TORC enhanced the association of CREB with TAF4. TORC proteins potentiate CREB activity via a phosphoserine-133-independent mechanism and this seems to be promoter context dependent, at least in HEK293 cells where only TATA containing reporters responded to TORC1 and TORC2 (Conkright et al., 2003b; Iourgenko et al., 2003). The subcellular localization of the different TORCs varies and depends on both cell-type and external stimuli. TORC1 and TORC2 resided in the cytoplasm of resting HIT, MIN6 and ATYB1 cells, whereas TORC2 was mainly nuclear, while TORC1 was cytoplasmic in resting HeLa cells. TORC3 localized to the nucleus in resting ATYB1 and HeLa cells (Bittinger et al., 2004; Screaton et al., 2004). In order to assess the effect of TORC1 localization on gene expression, HeLa cells and HeLa cells stably expressing a TORC1-EGFP fusion protein were compared in their ability to induce expression of the luciferase reporter gene under control of CRE. Resting HeLa cells and the stably transfected HeLa cells displayed similar levels of CRE-driven luciferase, whereas treatment with leptomycin B, an inhibitor of nuclear export, resulted in a significant induction only in the HeLa::TORC1-EGFP cells. This observation indicates that nuclear accumulation of TORC1-EGFP was sufficient to activate CRE-driven gene expression. In fact, costimulation with cAMP- and Ca^{2+}-elevating agents induced TORC2 translocation to the nucleus in MIN6 and HIT cells, resulting in increased CREB mediated transcription of NR4A2 (nuclear receptor superfamily 4, group A, member 2) mRNA and Evx-1-luciferase reporter, respectively. Forskolin alone was sufficient to provoke nuclear redistribution of TORC2 in ATYB1 cells, suggesting that the forskolin-induced translocation of TORC2 in ATYB1 cells may then lead to increased CREB mediated transcription (Bittinger et al., 2004). These findings illustrate a cell- and signal-specific dependence for subcellular redistribution of the TORC proteins. Hence, signal-regulated translocation of cytoplasmatic CREB coactivators to the nucleus can modulate CRE-mediated gene transcription in a cell-specific fashion. CBP, another co-activator for CREB, contains a consensus PKA site at serine-1772, suggesting that cAMP-induced phosphorylation by PKA may modify the activity of CBP and therefore contribute to cAMP-regulated transcription of CREB (Kwok et al., 1994). Indeed, the intrinsic transcriptional activity of a GAL4-CBP fusion proteins was increased in PC12 and F9 teratocarcinoma cells stimulated with cAMP analogues (Chrivia et al., 1993; Liu et al., 1998). Moreover, PKA was able to phosphorylate CBP *in vitro*, while dibutyryl cAMP stimulation of PC12 cells resulted in phosphorylation of CBP (Liu et al., 1999; Janknecht et al., 1996; it-Si-Ali et al., 1999). Xu et al showed that phosphorylation of CBP at serine-1772 was crucial for induction of Pit-1 by forskolin, indicating that phosphoCBP may increase Pit-1-mediated transcription (Xu et al., 1998). Still the role of CBP phosphorylation by PKA in transcription regulation remains controversial as mutating the PKA consensus site did not affect either basal- or PKA-stimulated transcriptional potentials of GAL4-CBP in PC12 cells (Kwok et al., 1994), whereas transfection studies with the constitutive CBP-binding $CREB_{DIEDML}$ mutant showed increased transactivity in the presence of co-transfected PKA, suggesting additional phosphorylation sites on CBP or other CREB regulators (Cardinaux et al., 2000). Serine-436 has been reported to be a growth factor-dependent phosphoacceptor site in CBP, but did not seem to be involved in 8-bromo-cAMP stimulated CRE-mediated transcription in CV-1 cells (Zanger et al., 2001). Whether PKA contributes to gene activation of cAMP responsive genes through phosphorylating CBP or proteins that regulate the activity of CBP is not known. A putative target that can interfere with CBP/CREB-dependent transcription is the arginine methyltransferase CARM1. Post-translationally methylation of CBP by CARM1 disrupts the interaction between CBP and CREB and result in an inhibition of CREB signalling as demonstrated by the observation that overexpressing of CARM1 into 293T cells reduced forskolin-dependent somatostatin-CRE reporter and endogenous somatostatin gene expression (Xu et al., 2001). It remains unknown whether cAMP/PKA signalling pathway can affect the expression or the activity of CARM1, but by doing so, it could modulate CREB-mediated transcription.

The group of Montminy showed that forskolin induced comparable levels of phosphoCREB on the cAMP-responsive NR4A2 and unresponsive CDC37 genes by ChIP assay of HEK293T cells. However, forskolin induced recruitment of CBP to the NR4A2 promoter, but not the CDC37 promoter. The authors suggested that the ability of cAMP to activate selective CREB target gene subsets is reflected at the level of CBP recruitment and other cofactors (Zhang et al., 2005). A putative candidate recruited by CBP is SNF2-related CBP activator protein SRCAP. SRCAP is an ATPase-dependent chromatin remodelling factor that interacts with CBP and increases the transactivation function of CBP. SRCAP enhances transcription of the CRE-responsive PEPCK, somatostatin, and enkephalin genes in a phosphoCREB-dependent mode. Whether all CREB-modulated promoters utilize SRCAP or only a subset has not been investigated (Monroy et al., 2001).

2. Interference with CREB-Dependent Transcription by CRE-Binding Transcription Factors that also Associate with CREB

Several other transcription factors have been isolated that can bind directly to CRE motifs. Some of these CRE-binding proteins, including the CREM variants, C/EBP, CREB2, ATF2, E4BP4 can also heterodimerize with CREB and can therefore interfere with CREB-mediated transcription by competing for the CRE site and/or by forming transcriptional complexes with CREB. Some examples will be discussed in the next section.

CREM Members

The CREMα, β and γ isoforms can bind as homodimers or heterodimers with CREB to CRE sequences and attenuate CREB-dependent transcription. For example, cAMP-induced activation of the c-*fos* promoter is blocked by CREM isoforms. This CREM-mediated repression required an intact CRE motif around position –60 (+1 is defined as start of transcription site) in the c-*fos* promoter (Foulkes et al 1991). CREM repression mediated through CRE motifs has also been reported for the IL-2 promoter and the promoter of immunoglobulin light κ chain (Pongubala and Atchison, 1995; Tenbrock and Tsokos, 2004). A special class of CREM isoforms are derived from four splice variants, which are generated by alternative splicing of a primary transcript synthesised from the P2 promoter in the *CREM* gene. These alternative spliced mRNAs are translated into repressors known as induced cAMP early repressors (ICER), which basically consist only of the bZip DNA binding and dimerization domain. The ICER promoter (P2 promoter in the *CREM* gene) contains four cAMP-autoregulatory elements and transcription of ICER is cAMP-inducible and shows kinetics of early response genes with a rapid increase, usually within minutes, followed by a progressive decline to basal transcript levels by a few hours. ICER proteins can either heterodimerize with CREB or bind CRE motifs as homodimers. Thus, ICER works as a powerful repressor of cAMP-induced transcription (reviewed in De Cesare and Sassone-Corsi, 2000). Indeed, cAMP-induced ICER expression can block CREB-dependent transcription of CREB target genes. For example, treatment of WEHI7.2 cells with forskolin resulted in rapid induction of c-*fos* and Jun B. However, the expression of these transcription factors was promptly attenuated after the following ICER induction (Mao et al., 1998). The cAMP mediated regulation of ICER itself seems to be cell-specific. Transcription of a reporter gene under control of the ICER promoter was more responsive to cAMP in WEHI7.2 cells than in JEG-3 cells. A truncated ICER promoter containing the two most proximal CRE motifs was cAMP-inducible in both cell lines, while a promoter containing the two distal CRE motifs was only stimulated in JEG-3 cells. Gel mobility shift analysis revealed, however, that the proteins bound to the distal CREs were not CREB. The authors suggested the presence of cell type-specific factors to explain the distinct sensitivity of the ICER promoter in these two cell lines (Mao et al., 1998). Another group reported that CRE-mediated luciferase expression was repressed after prolonged cAMP stimulation in JEG3, CHO, and GH3 cells, but not COS-7 cells. The time point of luciferase repression coincided with increased expression levels of ICER, suggesting that ICER binds the CRE and prevents expression of the luciferase gene. COS-7 cells did not express ICER and prolonged cAMP stimulation did not result in reduction of luciferase expression. These findings underscore an inhibitory role of ICER in cAMP-mediated induction of CRE-containing promoters. The authors also demonstrated that cAMP-induced ICER expression was time- and concentration-dependent. Therefore, transcription repression of a CRE-containing promoter by ICER depends on the duration and the amplitude of elevated cAMP levels induced by a particular stimulus and the cell-type (Kemp et al., 2002).

ATF-1

Another example where the ability of CREB to mediate cAMP response is modulated by heretodimerization is exemplified by the presence/absence of ATF-1 in F9 cells. The somatostatin promoter was not activated in response to cAMP in undifferentiated F9 cells, while a 8-fold induction was observed in differentiated F9 cells. The authors showed that the transcriptional activity of CREB was down-regulated by heterodimer formation with ATF-1. Undifferentiated F9 cells have an excess of ATF-1, while the ATF-1 levels decrease approximately 10-fold upon differentiation of F9 cells. Therefore, CREB is mainly found as CREB:ATF-1 heterodimers in undifferentiated cells, while CREB homodimers are predominantly present in differentiated F9 cells. How exactly ATF-1, when associated with CREB, inhibits CREB-dependent transcription remains unknown. Although ATF-1 is highly homologous to CREB, CREB contains three regions with no or little homology to ATF-1. ATF-1 lacks the Q1 domain, and shows low homology to the KID domains and the Q2 regions of CREB. These differences may be responsible for the inhibitory properties of ATF-1. Alternatively, the authors suggested that ATF-1 might recruit a specific repressor (Ellis et al., 1995).

E4BP4 (NFIL3)

A protein array used to investigate interactions between bZIP transcription factors revealed that CREB binds with E4BP4, also known as NFIL3 (Newman and Keating, 2003). E4BP4, which can also bind to CRE, acts as a repressor in most cases studied but it activates the IL-3 promoter in resting T cells. The transcriptional repressor domain of E4BP4 spans residues 298-363 and was shown to bind the repressor Dr1 (Cowell, 2002; Ozkurt and Tetradis, 2003). Overexpression of E4BP4 in U937 repressed LPS+TPA+dbcAMP-induced activation of the IL-1β promoter activity and this repression was mediated through the CRE site. Phosphorylation of E4BP4 was shown to enhance its binding to the CRE site *in vitro*. Interestingly, E4BP4 contains 5 putative PKA sites (Chen et al., 1995), and stimulation of primary mouse osteoblasts with forskolin or with 8-Br-cAMP increased E4BP4 mRNA levels (Ozkurt et al., 2004). Therefore, a mechanism can be imagined where increased intracellular cAMP levels leads to activation of PKA and phosphorylation of E4BP4, as well as increased E4BP4 mRNA levels. Phosphorylated E4BP4 homodimers or phosphoE4BP4:phoshoCREB heterodimers may compete with phoshoCREB for binding to CRE motifs and may then repress the expression of genes. Support for this model comes from studies with the CRE-containing COX-2 promoter. The primary mouse osteoblasts and osteoblastic cell lines MC3T3-E1 and ROS17/2.8 have low basal levels of E4BP4 mRNA. Parathyroid hormone increases the intracellular levels of cAMP, leading to activation of PKA and increase in the mRNA levels of E4BP4 in both cell lines. Moreover, ectopic expression of E4BP4 repressed PTH-induced activation of the COX-2 promoter in mouse osteoblasts, suggesting that E4BP4 can interfere with cAMP-responsiveness of CREB target genes (Ozkurt and Tetradis, 2003).

C/EBP Isoforms

CCAAT/enhancer binding proteins (C/EBPs) are a family of leucine zipper, transcription factors that bind to their cognate DNA sequence as homodimers and heterodimers. C/EBP can bind to CRE motifs, as well as physically associate with CREB. Six C/EBP genes have so far been identified, and they are designated *C/EBPα*, *C/EBPβ*, *C/EBPγ*, *C/EBPδ*, *C/EBPε*, and *C/EBPζ* (also known as C/EBP homologous protein-10, *CHOP* or *GADD153*). The expression profile of the different C/EBP genes is cell-specific. In general, C/EBPα is expressed at high levels in the adipose tissue, liver, intestine, lung, adrenal gland, peripheral-blood mononuclear cells and placenta. In the liver and in adipose tissue, the highest levels of mRNA are present in the terminally differentiated cells. Constitutive expression of C/EBPβ is particularly high in the liver, intestine, lung, adipose tissue, spleen, kidney and myelomonocytic cells, whereas that for C/EBPδ occurs in the adipose tissue, lung and intestine (reviewed in Ramji and Foka P, 2000). C/EBPβ can interact with the Q1 domain of CREB and this interaction stimulates CREB-mediated transcription of some CREB responsive promoters. Moreover, forskolin induced the C/EBPβ levels 2-fold in RAW264 cells (Caivano et al., 2001). Thus, forskolin may have a dual mechanism on stimulating CREB-dependent transcription. This cAMP elevating agent can both provoke CREB phosphorylation and can increase the levels of C/EBPβ, which favours the formation of stronger transcriptional active C/EBPβ:CREB hetereodimers. On the other hand, C/EBPβ is also implicated in attenuation of CREB mediated transcription (Chen et al., 2003). Studies in mouse granulosa GRMO2 cells have suggested a biphasic repression of cAMP-induced activation of the inhibin α subunit promoter involving ICER and C/EBPβ. ICER, whose transcription is rapidly upregulated by phosphoCREB in forskolin-treated cells, mediates an early repression by competing for the atypical CRE motif in the inhibin α subunit promoter. Autorepression of the ICER gene will gradually reduce the protein levels of ICER and the repression is released. Subsequently, the second inhibition mechanism, depending on increased levels of C/EBPβ, becomes operational. This repression also requires the atypical CRE in the promoter and it was shown that C/EBPβ efficiently competes with CREB for binding to this CRE *in vitro* (Burkart et al., 2005).

A role for particular C/EBP isoforms has also been reported in the cell-specific activation of CRE-containing promoters upon increased intracellular cAMP levels. Two CRE motifs and three C/EBP binding sites in the IL-10 promoter are required for full cAMP responsiveness in differentiated monocytic cells. Treatment with cAMP elevating agents resulted in an increase in IL-10 mRNA levels in THP-1 cells, but not in HL-60 cells. HL-60 cells contain significantly less endogenous C/EBPα compared with THP-1 cells, which may explain the different responses in these cell lines. Indeed, co-expression of C/EBPα in HL-60 restored IL-10 promoter activity in response to cAMP (Brenner et al., 2003).

The link between cAMP and C/EBP is intriguing and is operational at different levels. As mentioned above, the expression of C/EBP is stimulated by cAMP. Moreover, cAMP enhances the translocation of C/EBP from the cytoplasm to the nucleus. Finally, there is compelling evidence that C/EBPα and C/EBPβ contain a cAMP-inducible transactivation domain in addition to a constitutive active transactivation domain. In contrast to CREB, this cAMP-inducible domain lacks a PKA phosphorylation site. The mechanism by which cAMP activates these domains remains unclear. Considering the multiple cAMP-dependent mechanisms to regulate C/EBP, it is not unforeseen that several cAMP-responsive promoters with CRE motifs also contain C/EBP motifs, which are required for full cAMP responsiveness (reviewed by Wilson and Roesler, 2002).

3. Interference with CREB-Dependent Transcription by CRE-Binding Transcription Factors that Cannot Heterodimerize with CREB

CREB-2 (ATF4)

The transcriptional repressors CREM, ICER, ATF-1 and E4BP4 are all able to form heterodimers with CREB. They can interfere with cAMP-induced gene expression by their intrinsic repression activity when occupying CRE motifs as homodimers or as heterodimers with CREB. Other CRE-binding proteins, however, do not form heterodimers with CREB, but they can compete for the CRE motifs and prevent CREB binding. One of these is CREB2 (=ATF-4), which has been reported to activate or repress transcription through CRE motifs (reviewed in Hai and Hartman, 2001). For example, CREB2 homodimers repress transcription from the CRE-containing enkephalin, tyrosine hydroxylase, chromogranin B and secretogranin II promoters (Karpinski et al., 1992; Jungling et al., 1994; Cibelli et al., 1996; Lim et al., 2000). Accordingly, Cibelli and co-authors found that the secretogranin II promoter was cAMP inducible in HIT cells, but overexpression of CREB-2 impaired basal and cAMP inducible transcription of secretogranin II (Cibelli et al., 1996). On the other hand, CREB2 also possesses a strong activator domain as shown by studies with CREB2-CREB fusion proteins. Transient transfection studies were performed in murine noradrenergic locus coeruleus-like CATH.a cells. Cells were co-transfected with expression plasmids encoding the DNA binding and dimerization domain (bZip) of CREB fused to the constitutive transactivation domain of CREB2, and luciferase reporter plasmids containing multiple CRE motifs of the c-*fos* or the secretogranin promoter linked to a minimal HIV TATA box plus adenovirus late promoter initiator element, or the tyrosine hydroxylase promoter that contains a single CRE copy. The CREB2:CREB hybrid induced all three promoters (Thiel et al., 2005). In summary, CREB2 is able to repress CREB mediated transcription in some promoters, despite the presence of a constitutive transactivation domain. The exact mechanisms for this repression or activation potentials of CREB2 remains elusive. The competition for crucial co-activators may explain the antagonistic activity of CREB2 on CRE-dependent transcription. CREB2 binds CBP, TATA-binding protein, TFIIB and TFIIF. In cells with high levels of CREB2 compared to CREB, CREB2 will occupy most of the available CRE sites. The recruitment of e.g. CBP may then boost CREB2-mediated transcription. However, in cell lines where both CREB2 and CREB compete and occupy a fraction of CRE sites each, CREB2 may deplete crucial co-activators, resulting in repression of CREB-mediated transcription of the particular promoter.

4. Interference with CREB-Dependent Transcription by Proteins that fail to bind Directly to CRE but that can Interact with CREB

Several proteins that do not bind CRE motifs can interact directly with CREB. These proteins can either repress or stimulate basal or/and cAMP-induced CREB-mediated transcription. Some examples are discussed here.

ZPK(=DLK=MUK)

Zipper protein kinase (ZPK), a nuclear leucine zipper protein with serine/threonine kinase activity, is able to heterodimerize with CREB. Overexpression of ZPK was shown to inhibit PKA induced transcriptional activation of CREB in NIH3T3 and NT2 cells (Reddy et al., 1999). The mechanism by which ZPK interferes with PKA-induced CREB activation has not been solved, so far.

YY1

The Yin-Yang 1 (YY1) transcription factor can act as an activator or a repressor, hence its name. YY1 repressed c-*fos* transcription by direct interaction with CREB/ATF. In the case of c-*fos* promoter there is also a binding sites for YY1 within proximity of the ATF/ CREB site (Zhou et al., 1995). YY1 was also shown to repress the human RNA methyltransferase like 1 (RNMTL1) gene whose promoter does not seem to contain a cognate YY1 site. Transient transfection studies with deletion mutants of the RNMTL1promoter upstream of the luciferase reporter gene revealed that YY1 repression required CREB/ATF1 (Xu et al, 2002). Whether the interaction between CREB and YY1 is affected by cAMP inducing agents was not studied in these reports.

Ets-2

The mammalian family of Ets transcription factors consist of approximately 30 different members. Ets-1, the first member to be identified, is the cellular homologe of the v-ets oncogene in the avian erythroblastosis virus E26. Most of the Ets members are downstream nuclear targets of the Ras-MAP kinase signalling pathways, and they play important roles in cell development, cell differentiation, cell proliferation, apoptosis and tissue remodelling. (Sharrocks, 2001; Oikawa, 2004). CREB and Ets-2 form a complex that can be coimmunoprecipitated from choriocarcinoma cells, and the association between the two transcription factors is increased by PKA activation. The interaction between these proteins was necessary for optimal cAMP-induced stimulation of the human chorionic gonatotropin α (hCGα) promoter. The hCGα promoter contains CRE sites and binding sites for Ets-2 proximal to the CRE motifs. Ectopic expression of Ets-2 is sufficient to increase hCGα promoter-dependent reporter gene expression and, in the presence of activated PKA pathway, enables maximal stimulation of the hCGα promoter. Deletion of either the CRE or the Ets-2 binding sites virtually eliminated responsiveness to the combined effects of Ets-2 and 8-bromo-cAMP, while elimination of endogenous Ets-2 protein reduced both basal and cAMP stimulated hCGα expression (Silver et al., 1987; Ghosh et al., 2005). This example illustrates that specific promoters may be cAMP responsive, but cooperation with another transcription factor is required to obtain efficient cAMP responsiveness.

DREAM

Expression of the human prodynorphin gene is regulated through its downstream regulatory element (DRE) with the GAGTCAAGG sequence, which specifically binds the transcriptional repressor, DRE-antagonist modulator (DREAM). DREAM is expressed in a tissue-specific way, with e.g. high protein levels in brain tissue, but low levels in the liver. DREAM is a Ca^{2+}-binding protein and both its DNA binding and its repressor function are prevented by Ca^{2+}. In addition to the prodynorphin promoter, DREAM can also repress transcription from the early response gene c-*fos* through the GAGTCAACG DRE motif (Carrion et al., 1999). Recently, it was shown that DREAM could interact directly with the KID domain of CREB *in vivo* and blocked CREB binding to CRE *in vitro*. Phosphorylation of serine-133 in the KID did not impair the interaction with DREAM, suggesting that DREAM may affect recruitment of CBP/p300 by phosphoCREB, thereby disturbing CREB-dependent transcription (Ledo et al., 2002). The outcome of cAMP-responsiveness of CREB-target genes may therefore depend on the levels of a coactivator (CBP) and a repressor (DREAM) that compete for CREB.

5. Splice Variants of the General Transcription Factors

The general transcription factor TAF4 (formerly known as $hTAF_{II}130/135$) interacts with the Q2 domain of CREB (reviewed in Johannessen et al., 2004). In a recent report, Brunkhorst et al. cloned five novel truncated forms of TAF4 from a neonatal Balb/c whole brain cDNA library. One of these variants, mTAF4Δ3, was tested for its effect on cAMP-induced CREB-mediated transcription. Overexpression of this isoform in COS-7 cells, a cell line with low levels of endogenous TAF4, decreased cAMP-stimulated GAL4-CREB dependent expression by more than 60%. The mechanism of inhibition remains unknown, but mTAF 4 has been reported to homodimerize. Hence, heterodimers of full-length TAF4 and truncated variants may inhibit the activity of full-length TAF4 (Brunkhorst et al., 2004). The deletion in mTAF4Δ3 variant removes exon 8. This exon encodes a Q/P-rich domain, which corresponds to the Q4 region of hTAF4 (Tanese et al., 1996). Deletion of the Q4 domain reduced the interaction with CREB by 20% in a yeast two-hybrid assay, suggesting that this domain is important in CREB-dependent transcription (Saluja et al., 1998). The expression pattern of the different TAF 4 isoforms in different cell types has not been studied so far, but it can be imagined that cells expressing mTAFΔ3 may have reduced basal and cAMP-induced CREB-mediated transcription.

C. Regulation of CREB-Mediated Transcription at the DNA Level

1. Modulation of Chromatin Structure by cAMP

Transcriptional activity of the DNA has been strongly correlated with the state of the chromatin and chromatin remodelling is believed to constitute an integral component in the process of targeted gene activation by extracellular stimuli. Histone modifications such as acetylation, phosphorylation and methylation can affect the transcriptional state of chromatin. Combined phosphorylation of serine-10 and serine-28 of histone H3 is linked to chromosome condensation, while phosphorylation of serine-10 alone corresponds to transcription activation. Acetylation of lysine 14 and methylation of lysine 9 are influenced by serine-10 phosphorylation. Therefore, the histone modification code influences the transcriptional state of the chromatin (Servillo et al., 2002; Peterson and Daniel, 2004). Changes in chromatin structure induced by the cAMP/PKA signalling pathway can contribute to the cAMP-responsiveness of target genes. For example activation of the cAMP/PKA pathway in rat striatal neurons or in rat striatum triggers phosphorylation of serine-10 and acetylation of lysine-14 of histone H3, as well as acetylation of lysine-4 of histone H4. These signal-induced modifications were not randomly distributed in chromatin, but enriched in chromatin around specific genes and were associated with transcriptional activation (Li et al., 2004). Although the effect of cAMP-induced histone modification on CREB target genes was not investigated in this study, it is likely that this mechanism contributes to differential expression of cAMP-responsive genes (Cheung et al., 2000; Clayton et al., 2000). Other groups have shown that elevated cAMP concentrations resulted in increased histone acetylation over the promoter region of cAMP-responsive genes. For example, treatment of Y-1 cells with cAMP analogues stimulated histone H3 acetylation ~4-fold in the *CYP1B1* promoter region, while histone H4 acetylation on the human chorionic gonadotropin α promoter increased 2- to 3-fold in forskolin exposed JEG-3 cells (Canettieri et al., 2003; Zheng and Jefcoate, 2005). Forskolin treatment of NIH3T3 cells with integrated rat somatostatin gene provoked a moderate (1.5- to 2-fold), but reproducible increase in histone H4 acetylation over the promoter (Michael et al., 2000; Asahara et al., 2001).

Work with the mouse mammary tumour virus promoter further illustrates a role for cAMP in stimulus-induced chromatin remodelling. Stably replicating template containing this promoter in highly organized chromatin was repressed in the presence of cAMP, while transiently transfected template with a disorganized structure was activated. Increased cAMP levels resulted in a decrease in acetylation of histones H3 and H4, declines phosphoserine-10 levels of H3, and a reduction in methylation of lysines 4 and 9 of histone H3. Moreover, the kinetics of the decline in histone H4 acetylation correlated well with those of cAMP-induced transcriptional repression (Mulholland et al., 2003).

The examples above illustrate the role of cAMP in signal induced remodelling of chromatin. Opposite, chromatin may also determine the cAMP responses on transcription of genes. The group of Schütz found that the rat tyrosine aminotransferase (TAT) promoter was refractory to cAMP in the hepatoma HTC cells, while the promoter was inducible in the hepatoma cell line 7AD-7. DnaseI-hypersensitivity mapping revealed the lack of an hypersensitivity site over the CRE region in HTC cells, while such a site was present when analyzing the TAT promoter in 7AD-7 cells, suggesting an accessible CRE motif in the latter cells (Nitsch et al., 1990). Other studies confirmed that the ability of CREB to bind DNA depends on the chromatin structure. For example, CREB was unable to bind the c-*fos* CRE motif *in vitro* when the promoter was incorporated in a nucleosome, while concomitant incubation of naked DNA with CREB and histones during the nucleosome reconstitution process allowed CREB to bind (Schild-Poulter et al., 1996).

The chromatin may also influence the duration of cAMP induced CREB phoshorylation. Michael and co-workers observed that forskolin induced a transient phosphorylation of endogenous CREB with maximal phosphoCREB levels 30 min after stimulation, remaining elevating for 2 hrs, and returning to basic levels after 4 hrs. The phosphorylation kinetics of CREB coincided with the transcription rate of somatostatin. Ectopic expressed non-CRE binding GAL4-CREB fusion protein, however, remained phosphorylated on Ser-133 for at least 4 hrs after forskolin stimulation. This observation suggests that chromatin somehow can modulate the phosphorylation state of CREB and thus affect the transcriptional activity of this transcription factor (Michael et al., 2000).

2. The Role of Histone Deacetylases in CREB-Controlled Transcription

The mechanisms of PKA-induced acetylation of histones can be mediated by inhibition of histone deacetylases (HDAC). The cAMP signalling pathway triggers phosphorylation of HDAC8 at serine-39 by PKA, resulting in decreased enzymatic activity. This in turn leads to hyperacetylation of histones H3 and H4, an event that generally correlates with transcriptional activation (Lee et al., 2004).

Several transcription factors can interact with different HDAC (reviewed in Legube and Trouche, 2003). Acetylation of CREB has been shown to stimulate the transcriptional activity (Lu et al., 2003) and CREB is a substrate for HDAC1 (see further). Whether CREB is a substrate for HDAC8 is not known, but if so HDAC8-mediated deacetylation of CREB may repress the activity of CREB. Thus cAMP signal-responsive inhibition of HDAC8 by phosphorylation may potentiate CREB mediated transcription by a dual mechanism: hyperacetylation of CREB and of the histones H3 and H4.

HDACs can also be implicated in controlling cAMP-induced transcription by forming a complex with transcription factor(s) and phosphatase(s). Canettieri *et al.* showed that PP1, which has been demonstrated to dephosphorylate CREB *in vivo*, was recruited by HDAC associated with chromatin. HDAC1 interacted directly with CREB and the interaction was blocked by forskolin exposure. This implies that the PP1-HDAC1 complex is responsible for silencing CREB activity or involved in attenuation (Canettieri et al., 2003). Thus promoters where increased cAMP can displace HDAC-PP1 complexes may be more active as CREB phosphorylation will be prolonged compared to promoters where HDAC-PP1 is not displaced. The role of HDAC in cAMP-induced transcription is strengthened by studies with the HDAC inhibitor trichostatin (TSA). Treatment of cells with the HDAC inhibitor trichostatin TSA was shown to enhance cAMP-induced activation of CRE reporter genes (Asahara et al., 2001; Lu et al., 2003; Fass et al., 2003). However, Fass and co-workers found that TSA augmented cAMP-triggered activation of the c-*fos* and *nur-77* genes, while TSA blocked cAMP-induced activation of the *ICER* and *Nor-1* genes in PC12 cells. This suggests that the role of HDAC in cAMP-induced transcription may be promoter dependent. The authors demonstrated that TSA counteracted forskolin-induced recruitment of both RNA polymerase II and TFIIB to the *ICER* promoter, while TSA alone induced the binding of RNA polymerase II and TFIIB to the *nur-77* promoter and enhanced binding of RNA polymerase II to the c-*fos* promoter. These examples illustrate that HDAC activity can regulate the formation of the preinitiation complex at a subset of CREB target promoters. In accordance with this, the authors observed that CREB target genes inhibited by TSA had significantly slower cAMP-induced activation kinetics. The promoters of CREB target genes that either bind or are prepared to bind the preinitiation complex under basal conditions may not require HDAC for activation and thus can be transcribed rapidly. How can HDAC activity contribute to activation of *ICER* and *nor-1* transcription by cAMP in PC12 cells? The authors proposed several mechanisms. Acetylated histone lysines can bind repressors with bromodomains, so that deacetylation will remove these repressors. Histone lysine that is acetylated prematurely might cause recruitment of bromodomain-containg activators in an improper order. Alternatively, acetylation of histone lysine might prevent another modification (e.g. methylation) required for activation. Finally, HDAC can deacetylate a transcription factor. Acetylation of this factor might block transcription, while deacetylation might release this repression function (Fass et al., 2003).

3. Methylation of CRE Abolishes Binding of CREB

Transient transfection studies with cAMP-responsive promoters have shown that CpG methylation of the CRE results in loss of cAMP-inducibility, probably because methylation prevents CREB binding as demonstrated in gel mobility shift assays (Iguchi-Ariga and Schaffner, 1989; Moens et al., 1993). Methylation of CpG within the consensus CRE-binding site in the promoter of the *Pdha-2* gene (a gene encoding the testis-specific E1α subunit of the pyruvate dehydrogenase complex) prevented binding of CREB *in vitro*. The effect of methylation of the CRE-motif on cAMP-inducibility of the promoter was, however, not addressed (Iannello et al., 2000). Methylation of the major CRE in the c-*fos* promoter negatively interfered with cAMP-induced expression of this gene, confirming that methylation of the CRE abolishes cAMP-responsiveness (Moens et al., 1993). The *in vivo* role for methylation in regulation of CREB target genes was confirmed by two observations in the work of Montminy and his group. The authors examined CREB occupancy on methylated CREs in 47 different genes by ChIP assay. None of the methylated CREs bound CREB. Moreover, they observed that unmethylated CRE in the homeobox gene CDX4, a gene whose expression is restricted to early development, was constrained to embryonic stem cells, while this CRE was methylated in 293T, HepG2, Jurkat, liver, muscle, MACF10a, and peripheral blood mononuclear cell cells (Zhang et al., 2005).

4. Role of the TATA Box

An early study by Kanei-Ishii and Ishii hinted that the TATA box could be of importance, but not absolutely required in the cAMP responsiveness of CRE-containing promoters. The authors cloned the consensus CRE element of rat somatostatin promoter (sequences –57/-32) into three different reporter plasmids containing the chloramphenicol acetyltransferase gene coupled to the SV40 promoter encompassing 23 bp of the 72 bp enhancer, the 21 bp repeats and the TATA box, or a 550 bp fragment of the human c-Ha-ras promoter, or a 1350 bp fragment of the chicken β-actin promoter, respectively. Transient transfection studies with these plasmids revealed that the promoter activity of the CRE-SV40 construct was inducible by 8-Br-cAMP in PC12, NIH3T3, and CV-1 cells, while no activity was measured in HeLa, KB, A431, and L cells. The construct with the c-Ha-ras promoter was constitutive active in all cell lines tested, but was refractory to increased cAMP levels. The CRE-β-actin promoter displayed no activity in PC12, NIH3T3, CV-1, and L-cells, while a constitutive, but not cAMP-inducible activity was detected in HeLa, KB, and A431 cells (Kanei-Ishii and Ishii, 1989). Both the SV40 and the chicken β-actin promoter contain a TATA-like sequence, while the c-Ha-ras promoter lacks a TATA-box (Mathis and Chambon, 1981; Kost et al., 1983; Ishii et al., 1985). The TATA box-containing SV40 promoter, but not the TATA-less c-Ha-ras promoter, fused to a CRE motif was able to mediate cAMP responses in some cells. The fact that the TATA-containing chicken β-actin promoter fused to a CRE was not inducible illustrates that the TATA box is not absolutely required for all promoters to render them cAMP responsive. Another example is the cyclin D1 promoter, which is cAMP-inducible, but lacks a TATA box (Motokura and Arnold, 1993).

More profound evidence for a contributing role of the TATA box in cAMP-induced transcription comes from work by the laboratory of Montminy. In an effort to identify putative CREB genes, this group performed comparative genome-wide analysis for conserved CRE motifs in the 10 kbp regions upstream of the start codon in human and mouse genes. Their analysis revealed 1,349 mouse sites and 1,663 human sites. Many of the conserved CREs occurred within 1000 bp of the ATG initiation codon. Only ~15% of the CREB target genes contained TATA box motifs. ChIP analyses were performed on 8 different target genes (5 TATA-less and 3 TATA-containing genes) to verify CREB binding *in vivo*. CREB occupancy on these promoters was comparable and was unaffected by stimulation with forskolin. This indicates that the presence of a TATA box does not affect CREB binding and supports the notion that cAMP does not regulate CREB's binding to consensus CRE motifs. The TATA box seemed, however, important in mediating cAMP-responsiveness to the promoter, as tested with 16 different promoters by transient transfection studies in two different cell lines (293T and JEG3). All TATA-containing promoters were induced on average 15-fold in both cell types, while TATA-less promoters were only marginally activated (1- to 3-fold). Insertion of a TATA motif in a TATA-less promoter rendered the promoter cAMP responsive. The requirement of a proximal TATA box for activation via CREB led the authors to examine whether recruitment of TFIID complexes by CREB might be impaired on promoters lacking a TATA sequence. ChIP assays with anti-TBP antibodies revealed that cAMP elevating agents stimulated recruitment of TBP to the TATA-containing NR4A2 (nuclear receptor superfamily 4, group A, member 2) promoter, but had no effect on TBP occupancy over the TATA-less cell cycle division protein 37 (CDC37) gene. Thus the TATA element imposes important constraints on co-activator recruitment and transcriptional activation via CREB in response to cAMP. Later, transient transfection studies showed that TORC1 and TORC2 activated TATA-containing but not TATA-less promoters, suggesting that augmentation of CREB activity by TORC is promoter context dependent. (Conkright et al., 2003b). Recently, the same group confirmed the importance of the TATA box. Their results revealed that CREB regulates approximately 4,000 genes in the human genome, and about 1/3 of the CRE-containing genes also possessed TATA boxes. Most genes that were up-regulated by forskolin or down-regulated by a dominant negative CREB mutant in HEK293T cells contained functional CRE and TATA motifs. These genes exhibited a much higher tendency to be cAMP responsive than CRE-containing genes lacking a TATA box (Zhang et al., 2005). In another study, the CRE consensus TGACTGTCA sequence or the sequence TGATGTCA was detected in 2.4% of 13,010 human promoter sequences spanning nucleotides –2500/+500 relative to the transcription start site that were examined. The TATA canonical sequence TATAAAA was present in 2.6% of the investigated promoters, while two variant sequences (TATATAD and TATAAGD) were represented in 2% of the promoters. A positive correlation between the presence of a TATA box and a CRE motif was calculated (FitzGerald et al., 2004). The promoter of the type II deiodinase gene is cAMP responsive and contains a functional consensus CRE that binds CREB that accounts for the cAMP response. In addition, this promoter contains two TATA motifs, a canonical TATAAAAG and an atypical TTTAAAA. Canettieri and co-workers showed that mutation of the canonical TATA box almost completely abolished the response to cAMP in JEG3 cells, while mutation of the atypical TATA box had no effect. These findings indicate that the functional CRE-related TATA box in these cells is the canonical TATA sequence (Canettieri et al., 2004). Based on these analyses, CRE motifs seem to favour a canonical TATAAAA TATA box. In this context, it is interesting to point out that the general transcription factor TAF4, which can interact with the Q2 domain of CREB, preferentially activated promoters with a canonical TATAAA motif (Johannessen et al., 2003). The preference of CREB for a canonical TATA motif may therefore reside in the presence of TAF4 at such a TATA sequence.

5. Position of the CRE in the Promoter

The reason that not all CRE-containing genes are inducible to cAMP elevating agents can also be ascribed to the position of the CRE in the promoter. Most known functional CRE are located within the proximal 170 bp of the upstream region (Conkright et al., 2003a). In agreement with this observation, the rat tyrosine hydroxylase (TH) promoter possesses a CRE at –45 to –38 bp upstream of the transcription initiation site. This CRE motif binds CREB and is crucial for both basal and cAMP-inducible transcription in different cell lines. Transient transfection studies in human neuroblastoma SK-N-BE(2)C cells with a luciferase reporter plasmid containing the rat TH promoter showed that the transcriptional activation by the CRE is distance-dependent. The CRE remains functional in either orientation when positioned within 200 bp upstream of the transcription initiation site, but a single CRE copy placed at either -503, -773, -2400, or +1600 did not support basal promoter activity and did not mediate transcriptional induction upon forskolin treatment. Two CREs in opposite orientation at position –168 resulted in a doubling of the basal promoter activity, but the cAMP-inducibility of this promoter was as the native promoter. However, three CRE copies at –773 and four copies at –2400 restored cAMP-inducibility to the level of the wild-type promoter. Inserting 5 (half a helix turn) or 10 (complete helix turn) bp between the CRE and the TATA box (TTTAAA sequence) increased basal TH promoter activity approximately 3-fold, but the cAMP-inducibility of these mutated promoters was comparable with that of the native promoter. These studies demonstrate that the action of CRE in the TH promoter is distance-dependent, but that spacing between the TATA box and the CRE seems not to be crucial for mediating cAMP-responsiveness to the promoter (Tinti et al., 1997). The proximal position for a CRE to be functional seems not be an absolute prerequisite for all cAMP-responsive genes as e.g. a functional CRE is located at –3650 in the tyrosine aminotransferase promoter (Boshart et al., 1990).

6. CRE Flanking and Overlapping Sequences

In vitro CREB-binding studies revealed that the binding-site specificity of CREB to CRE appears to require 18-20 bp, since the 5 or so bases flanking the 8 bp core sequence have been shown to dictate, in some cases, the permissivity of transcriptional activation (Deutsch et al. 1988). The sequence context of the CRE motif may explain the variations in the relative responsiveness of different CREs to cAMP-induced transcriptional activation. Nucleotides flanking or overlapping the CRE may serve as DNA regulatory sites for other DNA binding proteins. The CRE-containing promoters of the rat somatostatin and the human gonadotropin α-subunit genes were highly responsive to cAMP elevating agents, while the CRE-containing rat glucagon and bovine parathyroid hormone promoters were less responsive. All promoters have the consensus TGACGTCA CRE motif, but differ in the nucleotides flanking this sequence. The rat glucagon promoter contains a 5' upstream TCATT motif, which seems to negatively regulate the cAMP response of the promoter. Similarly, mutations of sequences adjacent to the CRE of human glycoprotein hormone α-subunit gene and rat somatostatin gene suppressed cell-specific transcriptional enhancer activities of these CREs (Miller et al., 1993 and references therein). Overlapping CRE and glucocorticoid-response element binding motifs are found in the promoters of e.g. the phosphoenolpyruvate carboxykinase and glycoprotein α-subunit (Miller et al., 1993 and references therein). Hence, binding of proteins to other DNA regulatory sites may prevent CREB from interacting with CRE. Alternatively, the DNA binding proteins may compete for the same co-factors, or they may modulate CREB its transcriptional activity through protein-protein interaction. On the other hand, adjacent binding motifs may be essential for a full cAMP response as was shown for the *phosphoenol-pyruvate carboxykinase* gene (see next section) and the rat tissue plasminogen activator promoter. NF-1 and CRE sites juxtaposed on the rat tissue plasminogen activator promoter are essential for full response in granulosa cells (Ohlsson et al., 1993). The CRE consensus sequence has also been found in the context of other distinct promoter elements, where they apparently confer different transcriptional properties. For example, MHC class II genes contain a CRE in the X-box motif. This element acts as a non-cAMP-inducible pre-B-cell enhancer (reviewed in Foulkes and Sassone-Corsi, 1996). In conclusion, the sequence context of the CRE motif can affect the functionality of the element in mediating cAMP responses.

7. A cAMP Response Unit: Cooperativity Between Multiple Cis-Acting Elements is Required for full cAMP-Responsiveness

Most promoters are composed of several different *cis*-acting elements that can cooperate, thereby allowing a fine-tuned regulation by distinct stimuli. Full cAMP-responsiveness of a promoter may require the action of several transcription factors binding to distinct regulatory elements. In an earlier section the hCGα promoter was described as an example of a promoter that was cAMP responsive due to cooperation with Ets and CREB. Studies with the promoter of the *phosphoenolpyruvate carboxykinase* (PEPCK) gene have illustrated that several promoter elements in addition to the CRE are required for induction of PEPCK gene transcription by the cAMP/PKA signalling pathway. The single CRE motif was insufficient to mediate the cAMP-response, however a composite cAMP responsive unit containing this CRE and additional binding sites rendered the promoter cAMP responsive. In HepG2 cells, two C/EBP binding sites at position –355/-300 (with +1 the start site of transcription), an AP-1 and a C/EBP site at –271/-200, as well as the CRE and TATA box sequences composed the cAMP responsive unit, while in H4IIe cells, a tripartite binding region encompassing the AP1 and C/EBP motifs at –271/-200 plus the CRE were necessary to mediate cAMP-induced activation of the PEPCK promoter (Yeagley et al., 1998 and references therein). The binding of specific transcription factors to these different *cis*-acting elements also determines the degree of cAMP responsiveness of the PEPCK promoter. Maximal cAMP responsiveness requires the constitutive presence of AP-1 and was obtained when CREB is bound to the CRE, while C/EBPα is bound to the three C/EBP sites. An intermediate response is observed when C/EBPα is bound to the CRE and the three C/EBP sites or when 1 or 2 of the C/EBP sites is occupied by C/EBPβ, or when CREB is bound to CRE and C/EBPβ to the three C/EBP sites. A weak cAMP-response is measured when C/EBPβ is bound to CRE, regardless of the C/EBP isoform bound to the C/EBP sites or when the C/EBP sites are incompletely or not occupied. The latter situation is operational in kidney cells where C/EBP protein concentrations are low (reviewed in Roesler, 2000). Thus the specific occupancy of CRE and the additional C/EBP regulatory elements by distinct transcription factors contribute to cell-specific and amplitude-specific activity of the PEPCK promoter in response to increased cAMP levels.

In accordance with the cAMP responsiveness of the PEPCK promoter, a single CRE binding site is insufficient to mediate cAMP-induced activation of the *CYP 19* aromatase and of the α-glycoprotein promoters. The *CYP 19* promoter also required a steroidogenic factor-1 (SF-1) binding sequence, while the α-glycoprotein promoter depended on a tandem CRE repeat and on four other regulatory elements, including a CCAAT box, to render the promoter cAMP responsive (Budworth et al., 1997; Carlone and Richards, 1997). Another example of a promoter where multiple *cis*-acting regulatory elements are required for induction by cAMP is the *CYCB1* promoter. A far upstream enhancer region contains 2 CRE sites and 4 binding sites for the transcription factor SF-1. Mutation of CRE2 completely abrogated cAMP-induction of the promoter in Y-1 cells. Interestingly, mutation of all 4 SF-1 motifs also completely impaired cAMP-responsiveness of the promoter, even if the critical CRE2 motif remained intact. Hence the cAMP-responsiveness depended on intact CRE and SF-1 sites. Moreover, mutation of the AP-1 site in the proximal promoter region contributed to the cAMP response as mutation decreased the cAMP stimulation by almost 60%. The SF-1/CREB interaction was further confirmed by studies in C3H10T1/2 cells that do express CREB, but lack SF-1. Wild-type *CYC1B1* promoter was not activated by cAMP in these cells, whereas ectopic expression of SF-1 restored the cAMP response. HEK293 cells contain CREB, but lack SF-1. The *CYC1B1* promoter was, however, not induced by cAMP in these cells even after co-expression of SF-1. They authors proposed that other co-activators or co-repressors are involved in cell-specific induction of the *CYC1B1* promoter by cAMP. They found that CBP and SRC-1, but not p/CAF potentiated cAMP-induced activation of the *CYC1B1* promoter. These factors are probably not responsible for the cell-specific stimulation of this promoter by cAMP as they are both expressed in HEK293 cells. The identity of these critical co-factors and/or co-repressors remains elusive (Zheng and Jefcoate, 2005).

D. Regulation at Post-Transcriptional Level

Increased intracellular cAMP concentrations have been demonstrated to affect the stability of certain mRNA species such as lactate dehydrogenase A, phosphoenolpyruvate carboxykinase, surfactant protein A, PKA regulatory subunit RIα, GLUT1, GLUT5, and the glucocorticoid receptor mRNA (Hod and Hanson, 1988; Dong et al., 1989; Cornelius et al., 1991; Knutsen et al., 1992; Tian et al., 1998; Gouyon et al., 2003). Regulation of the rennin gene represents an example of dual regulation of gene expression by cAMP. The promoter of this gene contains a CRE motif which is occupied by unphosphorylated ATF1:CREB heterodimers under basal conditions in the human pulmonary carcinoma cell line Calu-6. Increase in cAMP levels resulted in displacement of the ATF-1:CREB complex with phosphoCREB dimer, leading to enhanced transcription. However, expression of this gene is also controlled at the post-transcriptional level. The 3′untranslated region of rennin mRNA contains a 34 nucleotide AU-rich region immediately preceding the poly(A) sequence. This motif destabilizes rennin mRNA and heterogenous mRNA when linked to it. Treatment of Calu-6 cells with forskolin increased rennin mRNA half-life. Forskolin augmented the levels of RNA-binding proteins that can associate with this destabilizing signal, as well as increase the binding of such proteins to this sequence, and this prolonged the half-life of rennin mRNA (reviewed in Morris et al., 2004).

However, cAMP may also be involved in destabilization of mRNA. Human β_2-adrenergic receptor mRNA is destabilized in a cAMP-dependent manner. The promoter of this gene contains a CRE and expression of the gene is initially upregulated in cells stimulated with cAMP elevating agents, while prolonged stimulation leads to destabilization of the mRNA via cAMP-dependent RNA-binding proteins (Collins et al., 1990; Danner et al., 1998). Hence, cAMP provides a mode for autostimulation of human β_2-adrenergic receptor gene expression, as well as receptor desensitization/downregulation through mRNA destabilization. A similar mechanism is operational for the gene encoding pituitary adenylate cyclase activating polypeptide (PACAP). The promoter of this gene contains a CRE (TGACGTCT), approximately 200 bp upstream the transcription start site, which can bind CREB. Transient transfection studies in primary rat cortical neurons showed that forskolin or co-expression of the catalytic subunit of PKA stimulated PACAP promoter activity. However, forskolin seems also to exert an opposite effect on the expression of PACAP since it decreased PACAP transcript levels in these cells (Fukuchi et al., 2004). Forskolin strongly enhanced (100-fold) the transcriptional activity of the PACAP promoter as monitored in transient transfection studies in PC12 cells with a PACAP promoter-luciferase reporter plasmid. However, this strong induction of PACAP promoter activity did not correlate with increased PACAP mRNA levels as only a 1.4-fold increase, which was not statistical significant, in PACAP transcript levels was measured after forskolin treatment (Hashimoto et al., 2000). These findings may also indicate that the PACAP promoter is a cAMP responsive promoter, but that cAMP is involved in destabilization of PACAP mRNA. On the other hand, exposure of human neuroblastoma IMR-32 cells to dibutyryl cAMP increased PACAP transcript levels and protein levels, but the effect on PACAP promoter activity was not investigated (Suzuki et al., 1990). Thus cAMP-mediated regulation of PACAP mRNA stability may be cell-specific. This post-transcriptional regulation accounts another mechanism that explains why cAMP-responsive genes can be expressed in cell- or tissue-specific induction upon cAMP stimulation.

CONCLUSION

Activation of CREB-mediated transcription by the cAMP/PKAsignalling pathway is generally accepted to occur through phosphorylation of CREB at serine-133 by PKA and subsequent recruitment of the coactivators CBP/p300 by phosphoCREB. Although correct, this paradigm is incomplete. Additional mechanisms involving PKA, CREB interacting proteins, CRE-binding and other DNA binding proteins, chromatin structure, and post-transcriptional modifications will all affect the outcome of the expression of CREB target gene in response to activated cAMP/PKA signalling pathway. Moreover, phosphorylation-induced activation of CREB is controlled by various pathways that converge to CREB. Although not the focus of this review, they also contribute to the regulation of CREB-mediated transcription. Furthermore, additional, incomplete understood mechanisms affect the function of CREB. For example, Ubc9, an E2-like enzyme specific for small ubiquitin-like modifier 1 (Sumo-1), functionally interacts with RNA helicase A. Ectopic co-expression of wild-type Ubc9 plus RNA helicase A in 293T cells stimulated transcription of the *luciferase* reporter gene under control of a minimal promoter and 15 CRE motifs. The exact mechanism for the jointly activation of CRE-dependent transcription by Ubc9 and RNA helicase A remains elusive, but it does not seem to require the Sumo-1 conjugation activity because a mutant Ubc9 defective in this catalytic activity in conjunction with RNA helicase A could also stimulate CRE-mediated transcription (Argasinska et al., 2004). Another poorly understood mechanism that influences CREB-mediated transcription was described for doublecortin (DCX) protein. This protein, a microtubule-associated protein that stabilizes microtubules, reduced phosphoCREB immunoreactivity when overexpressed in PC12 cells and decreased CREB-mediated transcription. The mechanism by which DCX modulates CREB's activity remains elusive, but overexpression of DCX did not affect the total levels or phospholevels of MEK1/2, ERK1/2, JNK, p38 and Akt/PKB (Shmueli et al., 2001). Activation of these pathways has been shown to trigger CREB phosphorylation, suggesting that DCX does not prevent CREB phosphorylation by inhibiting these pathways.

Thus several mechanisms govern the fine-tuned and stimulus- and cell-specific regulation of gene expression by the cAMP/PKA/CREB signalling pathway. Realising the versatile role CREB fullfils in numerous crucial biological processes, it is not suprising that a vast array of mechanisms control the activity of CREB. Accordingly, disturbances in normal CREB function have been associated with various dysfunctions in specific cellular proteins and are associated with severe clinical conditions (Shaywitz and Greenberg, 1999; Mayr and Montminy, 2001; Lonze and Ginty, 2002). Complete understanding of the regulation of CREB may help us to design specific and efficient therapies for conditions with perturbed CREB function.

References

The authors apologize to those authors whose work was not referred due to space limitations.

Alberts, AS; Montminy, M; Shenolikar, S; Feramisco, JR. Expression of a peptide inhibitor of protein phosphatase 1 increases phosphorylation and activity of CREB in NIH 3T3 fibroblasts. *Molecular and Cellular Biology*. 1994, 14:4398-4407.

for neuronal responses to ethanol. *Journal of Biological Chemistry*. 2002, 277:18810-18816.

Cornelius, P; Marlowe, M; Call, K; Pekala, PH. Regulation of glucose transport as well as glucose transporter and immediate early gene expression in 3T3-L1 preadipocytes by 8-bromo-cAMP. *Journal of Cell Physiology*. 1991, 146:298-308.

Cowell, IG. E4BP4/NFIL3, a PAR-related bZIP factor with many roles. *Bioessays*. 2002, 24:1023-1029.

Danner, S; Frank, M; Lohse, MJ. Agonist regulation of human beta2-adrenergic receptor mRNA stability occurs via a specific AU-rich element. *Journal of Biological Chemistry*. 1998, 2736:3223-3229.

De Cesare, D; Sassone-Corsi, P. Transcriptional regulation by cyclic AMP-responsive factors. *Progress in Nucleic Acid Research and Molecular Biology*. 2000, 64:343-369.

Delghandi, MP; Johannessen, M; Moens, U. The cAMP signalling pathway activates CREB through PKA, p38 and MSK1 in NIH 3T3 cells. *Cellular Signalling*. 2005, 17:1343-1351.

Deutsch, PJ; Hoeffler, JP; Jameson, JL; Habener, JF. Cyclic AMP and phorbol ester-stimulated transcription mediated by similar DNA elements that bind distinct proteins. *Proceedings of the National Academy of Sciences of the United States of America*. 1988, 85:7922-7926.

Dong, Y; Aronsson, M; Gustafsson, JA; Okret, S. The mechanism of cAMP-induced glucocorticoid receptor expression. Correlation to cellular glucocorticoid response. *Journal of Biological Chemistry*. 1989, 264:13679-13683.

Elliot, MR; Tolnay, M; Tsokos, GC; Kammer, GM. Protein kinase A regulatory subunit type IIβ directly interacts with and suppresses CREB transcriptional activity in activated T cells. *Journal of Immunology*. 2003, 171:3636-3644.

Elliot, MR; Shanks, RA; Khan, IU; Brooks, JW; Burkett, PJ; Nelson, BJ; Kyttaris, V; Juang, YT; Tsokos, GC; Kammer, GM. Down-regulation of IL-2 production in T lymphocytes by phosphorylated protein kinase A-RIIβ. *Jorunal of Immunology*. 2004, 172:7804-7812.

Ellis, MJC; Lindon, AC; Flint, KJ; Jones, NC; Goodbourn, S. Activating transcription factor-1 is a specific antagonist of the cyclic adenosine 3',5'-monophosphate (cAMP) response element-binding protein-1-mediated response to cAMP. *Molecular Endocrinology*. 1995, 9:255-265.

Euskirchen, G; Royce, TE; Bertone, P; Martone, R; Rinn, JL; Nelson, FK; Sayward, F; Luscombe, NM; Miller, P; Gerstein, M; Weissman, S; Snyder, M. CREB binds to multiple loci on human chromosome 22. *Molecular and Cellular Biology*. 2004, 24:3804-3814.

Fass, DM; Butler, JE; Goodman, RH. Deacetylase activity is required for cAMP activation of a subset of CREB target genes. Journal of Biological Chemistry. 2003, 278:43014-43019.

Fimia, GM; De Cesare, D; Sassone-Corsi, P. A family of LIM-only transcriptional coactivators: tissue-specific expression and selective activation of CREB and CREM. *Molecular and Cellular Biology*. 2000, 20:8613-8622.

FitzGerald, PC; Shlyakhtenko, A; Mir, AA; Vinson, C. Clustering of DNA sequences in human promoters. *Genome Research*. 2004, 14:1562-1574.

Foulkes, NS; Laoide, BM; Schlotter, F; Sassone-Corsi, P. Transcriptional antagonist cAMP-responsive element modulator (CREM) down-regulates c-*fos* cAMP-induced expression.

Proceedings of the National Academy of Sciences of the United States of America. 1991, 88:5448-5452.

Foulkes, NS; Sassone-Corsi, P. CREM: a bZip transcription factor central to the cAMP nuclear response. In: Goodbourn S, editor. *Eukaryotic gene transcription.* Oxford, UK: IRL Press, Oxford University; 1996; 59-101.

Freeland, K; Boxer, LM; Latchman, DS. The cyclic AMP response element in the Bcl-2 promoter confers inducibility by hypoxia in neuronal cells. *Brain Research and Molecular Brain Research.* 2001, 92:98-106.

Fukuchi, M; Tabuchi, A; Tsuda, M. Activity-dependent transcriptional activation and mRNA stabilization for cumulative expression of pituitary adenylate cyclase-activating polypeptide mRNA controlled by calcium and cAMP signals in neurons. *Journal of Biological Chemistry.* 2004, 279:47856-47865.

Gagelin, C; Toru-Delbauffe, D; Gavaret, JM; Pierre, M. Effects of cyclic AMP on components of the cell cycle machinery regulating DNA synthesis in cultured astrocytes. *Journal of Neurochemistry.* 1999, 73:1799-1805.

Ghosh, D; Sachdev, S; Hannink, M; Roberts, RM. Coordinate regulation of basal and cAMP-activated expression of hCGα by Ets-2 and CREB. *Molecular Endocrinology.* 2005, 19:1049-1066.

Gouyon, F; Onesto, C; Dalet, V; Pages, G; Leturque, A; Brot-Laroche, E. Fructose modulates GLUT5 mRNA stability in differentiated Caco-2 cells: role of cAMP-signalling pathway and PABP (polyadenylated-binding protein)-interacting protein (Paip) 2. *Biochemical Journal.* 2003, 375:167-174.

Gyles, SH; Burns, CJ; Whitehouse, BJ; Sugden, D; Marsh, PJ; Persaud, SJ; Jones, PM. ERKs regulate cyclicAMP-induced steroid synthesis through transcription of the steroidogenic acute regulatory (*StAR*) gene. *Journal of Biological Chemistry.* 2001, 276:34883-34895.

Hagiwara, M; Brindle, P; Harootunian, A; Armstrong, R; Rivier, J; Vale, W; Tsien, R; Montminy, MR. Coupling of hormonal stimulation and transcription via the cyclic AMP-responsive factor CREB is rate limited by nuclear entry of protein kinase A. *Molecular and Cellular Biology.* 1993, 13:4852-4859.

Hai, T; Hartman, MG. The molecular biology and nomenclature of the activating transcription factor/cAMP responsive element binding family of transcription factors: activating transcription factor proteins and homeostasis. *Gene.* 2001, 273:1-11.

Hashimoto, H; Hagihara, N; Koga, K; Yamamoto, K; Shintani, N; Tomimoto, S; Mori, W; Koyama, Y; Matsuda, T; Baba, A. Synergistic induction of pituitary adenylate cyclase-activating polypeptide (PACAP) gene expression by nerve growth factor and PACAP in PC12 cells. *Journal of Neurochemistry.* 2000, 74:501-507.

Hod, Y; Hanson, RW. Cyclic AMP stabilizes the mRNA for phosphoenolpyruvate carboxykinase (GTP) against degradation. *Journal of Biological Chemistry.* 1988, 263:7747-7752.

Iannello, RC; Gould, JA; Young, JC; Giudice, A; Medcalf, R; Kola, I. Methylation-dependent silencing of the testis-specific Pdha-2 basal promoter occurs through selective targeting of an activating transcription factor/cAMP-responsive element-binding site. *Journal of Biological Chemistry.* 2000, 275:19603-19608.

Iguchi-Ariga, SMM; Schaffner, W. CpG methylation of the cAMP-responsive enahncer/promoter sequence TGACGTCA abolishes specific factor binding as well as transcriptional activation. *Genes and Development.* 1989, 3:612-619.

Impey, S; McCorkle, SR; Cha-Molstad, H; Dwyer, JM; Yochum, GS; Boss, JM; McWeeney, S; Dunn, JJ; Mandel, G; Goodman, RH. Defining the CREB regulon: a genome-wide analysis of transcription factor regulatory regions. *Cell.* 2004, 119:1041-1054.

Iourgenko, V; Zhang, W; Mickanin, C; Daly, I; Jiang, C; Hexham, JM; Orth, AP; Miraglia, L; Meltzer, J; Garza, D; Chirn, GW; McWhinnie, E; Cohen, D; Skelton, J; Terry, R; Yu, Y; Bodian, D; Buxton, FP; Zhu, J; Song, C; Labow, MA. Identification of a family of cAMP response element-binding protein coactivators by genome-scale functional analysis in mammalian cells. *Proceedings of the National Academic of Sciences of the United States of America.* 2003, 100:12147-12152.

Ishii, S; Merlino, GT; Pastan, I. Promoter region of the human Harvey ras proto-oncogene: similarity to the EGF receptor proto-oncogene promoter. *Science.* 1985, 230:1378-1381.

it-Si-Ali, S; Carlisi, D; Ramirez, S; Upegui-Gonzalez, LC; Duquet, A; Robin, P; Rudkin, B; Harel-Bellan, A; Trouche, D. Phosphorylation by p44 MAP Kinase/ERK1 stimulates CBP histone acetyl transferase activity *in vitro. Biochemical and Biophysical Research Communications.* 1999, 262:157-162.

Janknecht, R; Nordheim, A. Regulation of the c-*fos* promoter by the ternary complex factor Sap-1a and its coactivator CBP. *Oncogene.* 1996, 12:1961-1969.

Johannessen, M; Olsen, PA; Sorensen, R; Johansen, B; Seternes, OM; Moens, U. A role of the TATA box and the general co-activator hTAF(II)130/135 in promoter-specific trans-activation by simian virus 40 small t antigen. *Journal of General Virology.* 2003, 84:1887-1897.

Johannessen, M; Delghandi, MP; Moens, U. What turns CREB on? *Cellular Signalling.* 2004, 16:1211-1227.

Jungling, S; Cibelli, G; Czardybon, M; Gerdes, HH; Thiel, G. Differential regulation of chromogranin B and synapsin I gene promoter activity by cAMP and cAMP-dependent protein kinase. *European Journal of Biochemistry.* 1994, 226:925-935.

Kanei-Ishii, C; Ishii, S. Dual enhancer activities of the cyclic-AMP responsive element with cell type and promoter specificity. *Nucleic Acids Research.* 1989, 17:1521-1536.

Karpinski, BA; Morle, GD; Huggenvik, J; Uhler, MD; Leiden, JM. Molecular cloning of human CREB-2: an ATF/CREB transcription factor that can negatively regulate transcription from the cAMP response element. *Proceedings of the National Academic of Sciences of the United States of America.* 1992, 89:4820-4824.

Kemp, DM; George, SE; Kent, TC; Bungay, PJ; Naylor, LH. The effect of ICER on screening methods involving CRE-mediated reporter gene expression. *Journal of Biomolecular Screening.* 2002, 7:141-148.

Kim, SW; Harney, JW; Larsen, PR. Studies of the hormonal regulation of type 2 5'-iodothyronine deiodinase messenger ribonucleic acid in pituitary tumor cells using semiquantitative reverse transcription-polymerase chain reaction. *Endocrinology.* 1998, 139:4895-4905.

Knutsen, HK; Tasken, KA; Eskild, W; Jahnsen, T; Hansson, V. Half-lives of different sized mRNAs for the PKA subunit RI alpha are regulated differently in response to inhibition of transcription and translation. *Biochemical and Biophysical Research Communications.* 1992, 184:454-460.

Kost, TA; Theodorakis, N; Hughes, SH. The nucleotide sequence of the chick cytoplasmic beta-actin gene. *Nucleic Acids Res.* 1983, 11:8287-8301.

Kudo, Y; Boyd, CAR; Sargent, IL; Redman, CWG; Lee, JM; Freeman, TC. An analysis using DNA microarray of the time course of gene expression during syncytialization of a human placental cell line (BeWo). *Placenta*. 2004, 25:479-488.

Kwok, RP; Lundblad, JR; Chrivia, JC; Richards, JP; Bachinger, HP; Brennan, RG; Roberts, SG; Green, MR; Goodman, RH. Nuclear protein CBP is a coactivator for the transcription factor CREB. *Nature*. 1994, 370:223-226.

Ledo, F; Kremer, L; Mellstrom, B; Naranjo, JR. Ca2+-dependent block of CREB-CBP transcription by repressor DREAM. *EMBO Journal*. 2002, 21:4583-4592.

Lee, YN; Park, YG; Choi, YH; Cho, YS; Cho-Chung, YS. CRE-transcription factor decoy oligonucleotide inhibition of MCF-7 breast cancer cells: cross-talk with p53 signaling pathway. *Biochemistry*. 2000, 39:4863-4868.

Lee, H; Rezai-Zadeh, N; Seto, E. Negative regulation of histone deacetylase 8 activity by cyclic AMP-dependent protein kinase A. *Molecular and Cellular Biology*. 2004, 24:765-773.

Legube, G; Trouche, D. Regulating histone acetyltransferases and deacetylases. *EMBO Reports*. 2003, 4:944-947.

Leonard, J; Serup, P; Gonzalez, G; Edlund, T; Montminy, M; The LIM family transcription factor Isl-1 requires cAMP response element binding protein to promote somatostatin expression in pancreatic islet cells. *Proceedings of the National Academic of Sciences of the United States of America*. 1992, 89:6247-6251.

Li, J; Guo, Y; Schroeder, FA; Youngs, RM; Schmidt, TW; Ferris, C; Konradi, C; Akbarian, S. Dopamine D2-like antagonists induce chromatin remodeling in striatal neurons through cyclic AMP-protein kinase A and NMDA receptor signaling. *Journal of Neurochemistry*. 2004, 90:1117-1131.

Lim, J; Yang, C; Hong, SJ; Kim, KS. Regulation of tyrosine hydroxylase gene transcription by the cAMP-signaling pathway: involvement of multiple transcription factors. *Molecular and Cellular Biochemistry*. 2000, 212:51-60.

Liu, YZ; Chrivia, JC; Latchman, DS. Nerve growth factor up-regulates the transcriptional activity of CBP through activation of the p42/p44(MAPK) cascade. *Journal of Biological Chemistry*. 1998, 273:32400-32407.

Liu, YZ; Thomas, NS; Latchman, DS. CBP associates with the p42/p44 MAPK enzymes and is phosphorylated following NGF treatment. *Neuroreport*. 1999, 10:1239-1243.

Lonze, BE; Ginty, DD. Function and regulation of CREB family transcription factors in the nervous system. *Neuron*. 2002, 35:605-623.

Lu, Q; Hutchins, AE; Doyle, CM; Lundblad, JR; Kwok, RPS. Acetylation of cAMP-responsive element-binding protein (CREB) by CREB-binding protein enhances CREB-dependent transcription. *Journal of Biological Chemistry*. 2003, 278:15727-15734.

Mao, D; Warner, EA; Gurwitch, SA; Dowd DR. Differential regulation and transcriptional control of immediate early gene expression in forskolin-treated WEHI7.2 thymoma cells. *Molecular Endocrinology*. 1998, 12:492-503.

Mayr, B; Montminy, M. Transcriptional regulation by the phosphorylation-dependent factor CREB. *Nature Reviews Molecular Cell Biology*. 2001, 2:599-609.

Mathis, DJ; Chambon, P. The SV40 early region TATA box is required for accurate *in vitro* initiation of transcription. *Nature*. 1981, 290:310-315.

McFadyen, MC; Melvin, WT; Murray, GI. Cytochrome P450 enzymes: novel options for cancer therapeutics. *Molecular Cancer Therapeutics*. 2004, 3:363-371.

Meijer, LK; Schesser, K; Wolf-Watz, H; Sassone-Corsi, P; Pettersson, S. The bacterial protein YopJ abrogates multiple signal transduction pathways that converge on the transcription factor CREB. *Cellular Microbiology*. 2000, 2:231-238.

Michael, LF; Asahara, H; Shulman, AI; Kraus, WL; Montminy, M. The phosphorylation status of a cyclic AMP responsive activator is modulated via a chromatin-dependent mechanism. *Molecular and Cellular Biology*. 2000, 20:1596-1603.

Miller, CP; Lin, JC; Habener, JF. Transcription of the rat glucagon gene by the cyclic AMP response element-binding protein CREB is modulated by adjacent CREB-associated proteins. *Molecular and Cellular Biology*. 1993, 13:7080-7090.

Moens, U; Subramaniam, N; Johansen, B; Aarbakke, J. The c-*fos* cAMP-response element: regulation of gene expression by a beta 2-adrenergic agonist, serum and DNA methylation. *Biochimica et Biophysica Acta*. 1993, 1173:63-70.

Monroy, MA; Ruhl, DD; Xu, X; Granner, DK; Yaciuk, P; Chrivia, JC. Regulation of cAMP-responsive element-binding protein-mediated transcription by the SNF2/SWI-related protein, SRCAP. *Journal of Biological Chemistry*. 2001, 276:40721-40726.

Morimura, T; Tsunekawa, K; Kasahara, T; Seki, K; Ogiwara, T; Mori, M; Murakami, M. Expression of Type 2 Iodothyronine Deiodinase in Human Osteoblast is Stimulated by Thyrotropin. *Endocrinology*. 2005, 146:2077-2084.

Morris, BJ; Adams, DJ; Beveridge, DJ; van der Weyden, L; Mangs, H; and Leedman, PJ. 2004. cAMP controls human rennin mRNA stability via specific RNA-binding proteins. *Acta Physiologica Scandinavica*. 2004, 181:369-373.

Morris, RC; Morris, GZ; Zhang, W; Gellerman, M; Beebe, SJ. Differential transcriptional regulation by the α- and γ-catalytic subunit isoforms of cAMP-dependent protein kinase. *Archives of Biochemistry and Biophy*sics. 2002, 403: 219-228.

Motokura, T; Arnold, A. PRAD1/cyclin D1 proto-oncogene: genomic organization, 5' DNA sequence, and sequence of a tumor-specific rearrangement breakpoint. *Genes, Chromosomes and Cancer*. 1993, 7:89-95.

Mulholland, NM; Snyder, SK; Kolla, SS; Smith, CL. Chromatin-dependent regulation of the MMTV promoter by cAMP signalling is mediated through distinct pathways. *Experimental Cell Research*. 2003, 287:361-373.

Musa, NL; Ramakrishnan, M; Li, J; Kartha, S; Liu, P; Pestell, RG; Hershenson, MB. Forskolin inhibits cyclin D1 expression in cultured airway smooth-muscle cells. *American Journal of Respiratory Cell and Molecular Biology*. 1999, 20:352-358.

Newman, JR; Keating, AE. Comprehensive identification of human bZIP interactions with coiled-coil arrays. *Science*. 2003, 300:2097-2101.

Nitsch, D; Stewart, AF; Boshart, M; Mestril, R; Weih, F; Schutz, G. Chromatin structures of the rat tyrosine aminotransferase gene relate to the function of its cis-acting elements. *Molecular and Cellular Biology*. 1990, 10:3334-3342.

Ohlsson, M; Leonardsson, G; Jia, XC; Feng, P; Ny, T. Transcriptional regulation of the rat tissue type plasminogen activator gene: localization of DNA elements and nuclear factors mediating constitutive and cyclic AMP-induced expression. *Molecular and Cellular Biology*. 1993, 13:266-275.

Oikawa, T. ETS transcription factors: possible targets for cancer therapy. *Cancer Science*. 2004, 95:626-633.

Ozkurt, IC; Tetradis, S. Parathyroid hormone-induced E4BP4/NFIL3 down-regulates transcription in osteoblasts. *Journal of Biological Chemistry*. 2003, 278:26803-26809.

Ozkurt, IC; Pirih, FQ; Tetradis S. Parathyroid hormone induces E4bp4 messenger ribonucleic acid expression primarily through cyclic adenosine 3',5'-monophosphate signaling in osteoblasts. *Endocrinology*. 2004, 145:3696-3703.

Paolillo, M; Feliciello, A; Porcellini, A; Garbi, C; Bifulco, M; Schinelli, S; Ventra, C; Stabile, E; Ricciardelli, G; Schettini, G; Avvedimento, EV. The type and the localization of cAMP-dependent protein kinase regulate transmission of cAMP signals to the nucleus in cortical and cerebellar granule cells. *Journal of Biological Chemistry*. 1999, 274:6546-6552.

Park, GH; Choe, J; Choo, HJ; Park, YG; Sohn, J; Kim, MK. Genome-wide expression profiling of 8-chloroadenosine- and 8-chloro-cAMP-treated human neuroblastoma cells using radioactive human cDNA microarray. *Experimental and Molecular Medicine*. 2002, 34:184-193.

Park, YG; Park, S; Lim, SO; Lee, MS; Ryu, CK; Kim, I; Cho-Chung, YS. Reduction in cyclin D1/Cdk4/retinoblastoma protein signaling by CRE-decoy oligonucleotide. *Biochemical and Biophysical Research Communications*. 2001, 281:1213-1219.

Peterson, CL; Laniel, MA. Histones and histone modifications. *Currurent Biology*. 2004, 14:R546-551.

Pickering, CS; Watkins, RH; Dickson, AJ. Rat primary hepatocytes and H4 hepatoma cells display differential sensitivity to cyclic AMP at the level of expression of tyrosine aminotransferase. *Biochemical and Biophysical Research Communications*. 1998, 252:764-769.

Pongubala, JM; Atchison, ML. Activating transcription factor 1 and cyclic AMP response element modulator can modulate the activity of the immunoglobulin kappa 3′enhancer. *Journal of Biological Chemistry*. 1995, 270:10304-10313.

Popovici, RM; Kao, LC; Giudice, LC. Discovery of new inducible genes in *in vitro* decidualized human endometrial stromal cells using microarray technology. *Endocrinology*. 2000, 141:3510-3513.

Ramji, DP; Foka, P. CCAAT/enhancer-binding proteins: structure, function and regulation. *Biochemical Journal*. 2002, 365:561-575.

Reddy, UR; Basu, A; Bannerman, P; Ikegaki, N; Reddy, CD; Pleasure, D. ZPK inhibits PKA induced transcriptional activation by CREB and blocks retinoic acid induced neuronal differentiation. *Oncogene*. 1999, 18:4474-4484.

Reusch, JE; Klemm, DJ. Inhibition of cAMP-response element-binding protein activity decreases protein kinase B/Akt expression in 3T3-L1 adipocytes and induces apoptosis. *Journal of Biological Chemistry*. 2002, 277:1426-1432.

Roesler, WJ. What is a cAMP response unit? *Molecular and Cellular Endocrinology*. 2000, 162:1-7.

Saluja, D; Vassallo, MF; Tanese, N. Distinct subdomains of human $TAF_{II}130$ are required for interactions with glutamine-rich transcriptional activators. *Molecular and Cellular Biology*. 1998, 18:5734-5743.

Schild-Poulter, C; Sassone-Corsi, P; Granger-Schnarr, M; Schnarr, M. Nucleosome assembly on the human c-*fos* promoter interferes with transcription factor binding. *Nucleic Acids Research*. 1996, 24:4751-4758.

Screaton, RA; Conkright, MD; Katoh, Y; Best, JL; Canettieri, G; Jeffries, S; Guzman, E; Niessen, S; Yates, JRI; Takemori, H; Okamoto, M; Montminy, M. The CREB coactivator

TORC2 functions as a calcium- and cAMP-sensitive coincidence detector. *Cell.* 2004, 119:61-74.

Servillo, G; Fazia, MAD; Sassone-Corsi, P. Coupling cAMP signalling to transcription in the liver: pivotal role of CREB and CREB. *Experimental Cell Research.* 2002, 275:143-154.

Seternes, OM; Sørensen, R; Johansen, B; Moens, U. Activation of PKA by dibuyryl cAMP treatment of NIH3T3 cells inhibits proliferation but fails to induce Ser-133 phosphorylation and transcriptional activation of CREB. *Cellular Signaling* 1999, 11:211-219.

Sharrocks, AD. The ETS-domain transcription factor family. *Nature Reviews Molecular Cell Biology.* 2001, 2:827-837.

Shaywitz, AJ; Greenberg, ME. CREB: a stimulus-induced transcription factor activated by a diverse array of extracellular signals. *Annual Review of Biochemistry.* 1999, 68:821-861.

Shmueli, O; Gdalyahu, A; Sorokina, K; Nevo, E; Avivi, A; Reiner, O. DCX in PC12 cells: CREB-mediated transcription and neurite outgrowth. *Human Molecular Genetics.* 2001, 10:1061-1070.

Silver, BJ; Bokar, JA; Virgin, JB; Vallen, EA; Milsted, A; Nilson JH. 1987. Cyclic AMP regulation of the human glycoprotein hormone alpha-subunit gene is mediated by an 18-base-pair element. *Proceedings of the National Academic of Sciences of the United States of America* 1987, 84:2198-2202.

Skålhegg, BS; Taskén, K. Specificity in the cAMP/PKA signalling pathway. Differential expression, regulation, and subcellular localization of subunits of PKA. *Frontiers in Bioscience.* 2000, 5:d678-693.

Srivastava, RK; Lee, YN; Noguchi, K; Park, YG; Ellis, MJ; Jeong, JS; Kim, SN; Cho-Chung, YS. The RIIbeta regulatory subunit of protein kinase A binds to cAMP response element: an alternative cAMP signaling pathway. *Proceedings of the National Academic of Sciences of the United States of America.* 1998, 95:6687-6692.

Suen, TC; Goss, PE. Transcription of BRCA1 is dependent on the formation of a specific protein-DNA complex on the minimal BRCA1 Bi-directional promoter. *Journal of Biological Chemistry.* 1999, 274:31297-31304.

Suzuki, N; Harada, M; Hosoya, M; Fujino, M. Enhanced production of pituitary adenylate-cyclase-activating polypeptide by 1, N6-dibutyryladenosine 3',5'-monophosphate, phorbol 12-myristate 13-acetate and by the polypeptide itself in human neuroblastoma cells, IMR-32. *European Journal of Biochemistry.* 1990, 223:147-153.

Tanese, N; Saluja, D; Vassallo, MF; Chen, JL; Admon, A. 1996. Molecular cloning and analysis of two subunits of the human TFIID complex: $hTAF_{II}130$ and $hTAF_{II}100$. *Proceedings of the National Academic of Sciences of the United States of America.* 1996, 93:13611-13616.

Taskén, K; Aandahl, EM. Localized effects of cAMP mediated by distinct routes of protein kinase A. *Physiological Reviews.* 2004, 84:137-167.

Tenbrock, K; Tsokos, GC. Transcriptional regulation of interleukin 2 in SLE T cells. *International Reviews of Immunology.* 2004, 23:333-345.

Thiel, G; Al Sarraj, J; Vinson, C; Stefano, L; Bach, K. Role of basic region leucine zipper transcription factors cyclic AMP response element binding protein (CREB), CREB2, activating transcription factor 2 and CAAT/enhancer binding protein alpha in cyclic AMP response element-mediated transcription. *Journal of Neurochemistry.* 2005, 92:321-336.

Tian, D; Huang, D; Brown, RC; Jungmann, RA. Protein kinase A stimulates binding of multiple proteins to a U-rich domain in the 3'-untranslated region of lactate dehydrogenase A mRNA that is required for the regulation of mRNA stability. *Journal of Biological Chemistry*. 1988, 273:28454-28460.

Tierney, EP; Tulac, S; Huang, ST; Giudice, LC. Activation of the protein kinase A pathway in human endometrial stromal cells reveals sequential categorical gene regulation. *Physiological Genomics*. 2003, 16:47-66.

Tinti, C; Yang, C; Seo, H; Conti, B; Kim, C; Joh, TH; Kim, KS. Structure/function relationship of the cAMP response element in tyrosine hydroxylase gene transcription. *Journal of Biological Chemistry*. 1997, 272:19158-19164.

Tortora, G; Cho-Chung, YS. Type II regulatory subunit of protein kinase restores cAMP-dependent transcription in a cAMP-unresponsive cell line. *Journal of Biological Chemistry*. 1990, 265:18067-18070.

Ventra, C; Porcellini, A; Feliciello, A; Gallo, A; Paolillo, M; Mele, E; Avvedimento, VE; Schettini, G. The differential response of protein kinase A to cyclic AMP in discrete brain areas correlates with the abundance of regulatory subunit II. *Journal of Neurochemistry*. 1996, 66:1752-1761.

Wadleigh, DJ; Reddy, ST; Kopp, E; Gosh, S; Herschman, HR. 2000. Transcriptional activation of the cyclooxygenase-2 gene in endotoxin-treated RAW 264.7 macrophages. *Journal of Biological Chemistry*. 2000, 275:6259-6266.

Wadzinski, BE; Wheat, WH; Jaspers, S; Peruski, LF Jr; Lickteig, RL; Johnson, GL; Klemm, DJ. Nuclear protein phosphatase 2A dephosphorylates protein kinase A-phosphorylated CREB and regulates CREB transcriptional stimulation. *Molecular and Cellular Biology*. 1993, 13:2822-34.

Wang, Z; Zhang, B; Wang, M; Carr, BI. Persistent ERK phosphorylation negatively regulates cAMP response element-binding protein (CREB) activity via recruitment of CREB-binding protein to pp90RSK. *Journal of Biological Chemistry*. 2003, 278:11138-11144.

Wilson, HL; Roesler, WJ. CCAAT/enhancer binding proteins: do they possess intrinsic cAMP-inducible activity? *Molecular and Cellular Endocrinology*. 2002, 188:15-20.

Won, KA; Xiong, Y; Beach, D; Gilman MZ. Growth-regulated expression of D-type cyclin genes in human diploid fibroblasts. 1992. *Proceedings of the National Academy of Science of the United States of America*. 1992, 89:9910-9914.

Xu, J; De, ZJ; Ni, M; Wan, F; Gu, JR. The ATF/CREB site is the key element for transcription of the human RNA methyltransferase like 1(RNMTL1) gene, a newly discovered 17p13.3 gene. *Cell Research*. 2002, 12:177-197.

Xu, L; Lavinsky, RM; Dasen, JS; Flynn, SE; McInerney, EM; Mullen, TM; Heinzel, T; Szeto, D; Korzus, E; Kurokawa, R; Aggarwal, AK; Rose, DW; Glass, CK; Rosenfeld, MG. Signal-specific co-activator domain requirements for Pit-1 activation. *Nature*. 1998, 395:301-306.

Xu, W; Chen, H; Du, K; Asahara, H; Tini, M; Emerson, BM; Montminy, M; Evans, RM. A transcriptional switch mediated by cofactor methylation. *Science*. 2001, 294:2507-2511.

Yan, XD; Hanson, AJ; Nahreini, P; Koustas, WT; Andreatta, C; Prasad, KN. 2002. Altered expression of genes regulating cell growth, proliferation, and apoptosis during adenosine 3',5'-cyclic monophosphate-induced differentiation of neuroblastoma cells in culture. *In Vitro Cellular and Developmental Biology - Animal*. 2002, 38:529-537.

experiments allowed to further insight in the intracellular signalling modalities used by these RET mutants. Although MEN2A- and MEN2B-associated RET mutants involve similar intracellular signalling pathways they differ in the extent of induction of enzymatic cascades and in the gene expression pattern. Indeed, a recent report on expression profiling on MEN2-associated medullary thyroid carcinomas provide molecular evidence on the clinical differences between the two syndromes. In this review we will discuss several approaches used to understand the molecular mechanisms induced by Ret mutants that dictate the clinical phenotype of these two closely related but distinct neoplastic diseases.

Keywords: MEN2; RET; Tyrosine kinase receptor; GDNF; oncogene

1. INTRODUCTION

Receptor-tyrosine kinases are transmembrane proteins that in response to growth factors stimulate intracellular pathways that control important cellular functions as differentiation and proliferation. In tumors, mutated receptors may become active without growth factor-binding, which results in uncontrolled cell proliferation. This review will highlight the most recent reports concerning Ret as a paradigm of a single receptor-tyrosine kinase that induces different types of human cancer depending on which residue is mutated. For instance, mutations in the extracellular domain of RET, responsible for MEN2A syndrome, lead to constitutive dimerization of two mutated RET molecules, whereas, a mutation within the catalytic domain, that causes the MEN2B syndrome, involves an intramolecular mechanism to convert *RET* into a dominant transforming gene.

The Ret protein is a transmembrane receptor of the protein tyrosine kinase family consisting of three functional regions: the extracellular region including four N-terminal cadherin-like domains (CLD1-4) followed by a single cysteine rich domain (CRD), the transmembrane region, and the intracellular region formed by a bipartite tyrosine kinase domain [1] (Fig. 1A and B). Ret is stimulated by growth factors of the glial cell line–derived neurotrophic factor (GDNF) family ligands (GFLs): GDNF, Neurturin (NRTN), Persephin, and Artemin [2]. The biological actions of the GFLs are mediated through a multi-component receptor complex that contains a common signalling component, the RET receptor-tyrosine kinase, and one of the four members of the GFRα family (GFRα1 to α4). These are glycosyl-phosphatidyl-inositol (GPI)-linked proteins that are clustered in the lipid raft structures of the extracellular cell membrane [reviewed in 3]. Despite a cross-talk between the different ligands-GFRαs pairs, a preferred co-receptor molecule exists for each ligand, GDNF being the preferred high affinity ligand for GFRα1 [2,4]. GFLs support *in vivo* the survival of many types of neuronal cells, both in central and in peripheral nervous system [5]. GDNF has potent trophic effects on dopaminergic nigral neurons, indicating this factor as a promising protective agent in neurodegenerative diseases. In various animal models of Parkinson's disease, GDNF has been shown to prevent the neurotoxin-induced death of dopamine neurons and can promote functional recovery [for a review see 6].

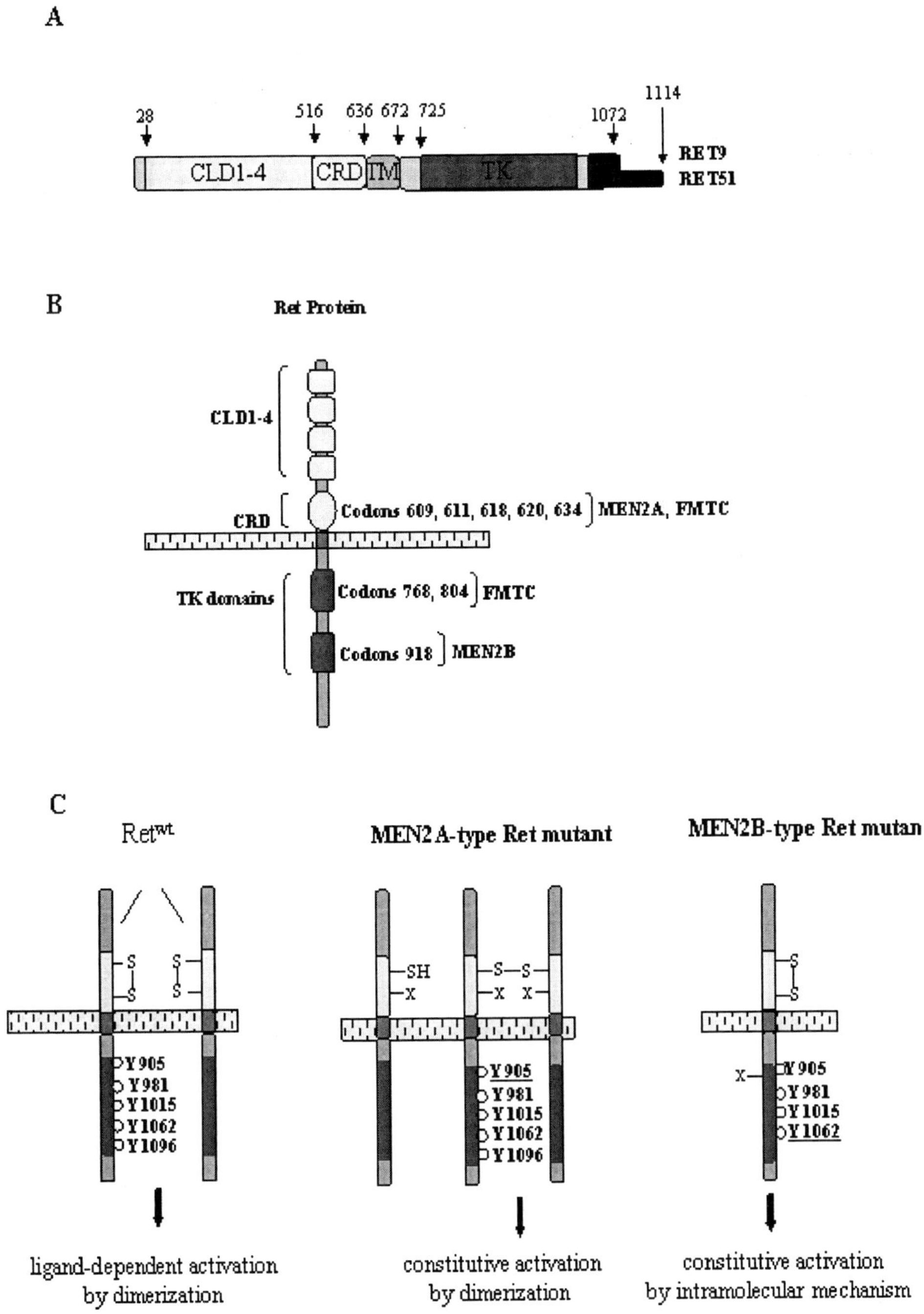

Fig. 1. A. Schematic representation of the domain architecture of Ret. Amino acids numbering in Ret is indicated. CLD, cadherin-like domain; CRD, cysteine-rich domain; TM, transmembrane region; TK, tyrosine kinase domain. The two spliced forms, Ret9 and Ret51, are indicated. B. Most common mutations in MEN2A, MEN2B and FMTC syndromes. C. Mechanism of activation of MEN2-type Ret mutants. Ret autophosphorylation sites are indicated. Amongst these, the tyrosine residues that play a different role in MEN2A- and in MEN2B-type mutants are underlined (see text for details).

RET (REarranged during Transformation), originally identified following a transfection assay of mouse fibroblasts, NIH 3T3, with the DNA of a human T-cell lymphoma, represents an important example of a single gene causing different human neoplastic and developmental diseases. Germline mutations in the RET gene are associated with multiple endocrine neoplasia type 2A (MEN2A; 171400), multiple endocrine neoplasia type 2B (MEN2B; 162300), and familial medullary thyroid carcinoma (FMTC; 155240). Beside the *gain-of-function* mutations in MEN2 another "face" of Ret exists of *loss-of-function* mutations leading to a different developmental syndrome, the Hirschsprung's disease (HSCR; aganglionic megacolon; 142623). Indeed, heterozygous, incompletely penetrant point mutations and deletions of the *RET* proto-oncogene have been described in sporadic and familial cases of HSCR [7-9].

Furthermore, Ret activation is frequent in papillary thyroid carcinoma (PTC), a neoplasm of the follicular cells of thyroid. Ret activation in PTC results from chromosomal inversions or translocations that cause the juxtaposition of heterologous genes with the catalytic intracellular domain of Ret thereby generating the chimeric RET/PTC oncogenes [10].

Because its implication in several diseases, studying Ret has been an attracting model to investigate the mechanisms that determine how distinct mutations in the same gene may lead to distinct disease phenotypes.

2. MEN2 Syndromes

2.1. General Description of the Phenotypes

As summarized in Table I, MEN2A, MEN2B and FMTC are distinct hereditary neoplastic diseases [11,12]. MEN2A, MEN2B are both characterized by the presence of medullary thyroid carcinoma (MTC), and in half of patients, pheochromocytoma (a malignancy of the adrenal medulla). MEN2A also features hyperplasia or adenoma of parathyroid cells (present in approximately 25% of cases). MEN2B is a more severe disease, being associated with skeletal abnormalities, ganglioneuromas of the intestinal tract, mucosal neuromas and characterized by an earlier age of tumor onset [13] but hyperparathyroidism is rare. Skeletal abnormalities include a long, thin body shape, pes cavus pectus excavatum, bifid ribs and slipped femoral epiphyses. Peripheral nerve abnormalities include hyperplasia of extrinsic autonomic nerve fibres in the intestinal wall that causes troubles of intestinal motility, disorganized nerve tissues that results in neuromas which give rise to characteristically full lips and lumps on the tongue, and thickened corneal nerve fibres. FMTC is characterized by the occurrence of only MTC.

2.2. Ret Mutations that Cause the MEN2 Syndromes

As shown in Figure 1B, missense mutations in one of five conserved cysteine residues encoded in exon 10 of the RET gene (codons 609, 611, 618, and 620) and exon 11 (codon 634) have been found in more than 92% of MEN2A (MTC and pheochromocytoma and/or hyperparathyroidism) [13-17]. These mutations lead to constitutive dimerization of the mutated receptor and ligand-independent activation (Fig.1C and see below for details).

The MEN2B mutations M918T in exon 16 (94% cases) and A883F in exon 15 lie in the distal half of the split kinase domain, either side of the highly conserved `activation loop' which is seen in receptor and non-receptor-tyrosine kinases [14,18-21]. The M918T mutation leads to ligand-independent activation of RET by an uncertain mechanism, and also to altered tyrosine phosphorylation of the receptor and altered catalytic substrate specificity [22-24].

Eighty-eight percent of families with FMTC (medullary thyroid carcinoma only) are caused by mutations similar to those seen in MEN2A exons 10 and 11. In addition, there have been seven families reported with mutations (E768D, V804L in exons 13 and 14, respectively) in the intracellular kinase domain [14,15,25,26]. The mechanism of RET activation by these intracellular mutations is uncertain [27].

Table 1. MEN2 syndromes and their clinical features.

Syndrome	Characteristic phenotypic manifestations
MEN2A	Thyroid C cells (MTC) Adrenal medulla (pheochromocytoma) Parathyroid glands (hyperplasia)
FMTC	Thyroid C cells (MTC)
MEN2A or FMTC with Hirschsprung's disease	MEN2A or FMTC with Hirschsprung's disease
MEN2B	Thyroid C cells (MTC) Adrenal medulla (pheochromocytoma) Intestinal and mucosal ganglioneuromatosis Marfanoid habitus

3. RET Implication in Hirschsprung's Disease

HSCR is a multigenic disorder resulting from the arrest of the craniocaudal migration of enteric neurons from the neural crest along gastrointestinal segments of variable length (frequency of one in 5000 live births of which 15–20% of cases are familial). It is characterized by the absence of ganglion cells in the ganglion plexuses of the most caudal gut. In the aganglionic colon, the plexus remnants are replaced by aggregates of glial cells and hypertrophied nerve fibers leading to intestinal obstruction or chronic constipation during childhood. Mutations in RET gene are common in Hirschsprung's patients and estimates of the frequency to vary, ranging in different populations from 5% to 50% [reviewed by 28,29]. RET mutations are more common in the familial form than in sporadic HSCR disease (about 50% of familial and 10–20% of sporadic cases). Heterozygous, incompletely penetrant point mutations, insertion, frameshift ,and deletions of the *RET* proto-oncogene have been described in sporadic and familial cases of HSCR [7-9,30,31].

Heterozygous mutations of GDNF and NRTN have been identified at low frequency in the germline of HSCR patients. Two of the GDNF mutations have been shown to reduce the affinity of GDNF for GFRα1 without reducing the activity of the Ret kinase [32,33]. Whether these mutations contribute to the pathogenesis of HSCR disease seem uncertain. However, genetic evidence support the notion that these mutations are not sufficient to cause HSCR but modifies the disease severity caused by a RET mutation in the pedigree [34].

Implication of defects of Ret signalling in HSCR has been supported by several studies in animal models. Homozygous mice with targeted disruption of either Ret [35], GDNF [36-38] or GFRα1 [39,40] all die soon after birth and share a similar phenotype of kidney agenesis and absence of enteric neurons below the stomach leading to Hirschsprung-type intestinal obstruction [3]. Surprisingly, GFRα1, 2 and 3 mutations have never been found in Hirschsprung's patients [41].

4. Responsiveness of Ret Mutants to the Presence of Ligand

4.1. The 2B-type Mutation Preserves Ret Sensitiveness to Ligand Stimulation

As stated above, the inheritance of specific RET mutations causes distinct disease phenotypes, thus suggesting that specific cell types undergo abnormal proliferation depending on the type of Ret activation (*via* a FMTC/MEN2A or *via* a MEN2B-type mutation).

One suggestive possibility to explain the differences in tissues affected is that cell- or tissue-specific biological constraints, such as the local density of the available ligand, modulate the extent of activity of mutated Ret receptors. It seems plausible to predict that the different mechanisms by which MEN2B and FMTC/MEN2A mutations activate Ret may cause different responsiveness of these mutants to ligand stimulation.

This possibility has been strongly supported by results obtained either by GDNF stimulation of Ret^{M918T} and by epidermal growth factor (EGF) stimulation of an EGF receptor (EGFR)-based chimera, EGFR/Ret^{M918T}. The chimeric receptor, EGFR/Ret^{M918T}, consists of the transmembrane and ligand-binding domains of the receptor fused to the catalytic domain of Ret. These experiments have shown that Ret carrying the metionine-918 to threonine mutation retains the ability to be further stimulated by an extracellular ligand thus indicating that such mutation does not cause full activation of Ret [42,43].

In addition, consistent with the notion that MEN2B causes a *gain-of-function* of Ret, the EGFR/ Ret^{M918T} construct was able to transform NIH 3T3 cells and induce differentiation in the rat pheochromocytoma-derived cells, PC12, even in the absence of EGF. PC12 cells transfected with EGFR/Ret^{M918T} showed a phenotype indistinguishable from PC12 transfected with Ret^{M918T}, thus confirming that the expression in PC12 cells, of a constitutive active Ret version results in a MEN2 phenotype. However, despite the fact that the metionine-918 to threonine mutation constitutively activates the chimera, the biological effects of the EGFR/Ret^{M918T} construct were markedly sensitive to EGF triggering. EGF stimulation caused a marked increase of the transforming ability of the EGFR/Ret^{M918T} construct and modified the phenotype of PC12-EGFR/Ret^{M918T} cells, determining the overgrowth of long neuritic processes and a dramatic phosphorylation of the docking protein FRS2 (fibroblast growth factor receptor substrate-2). These effects were consistent with the stimulation of tyrosine phosphorylation of EGFR/Ret^{M918T} caused by EGF [42].

Accordingly, GDNF stimulation of Ret^{M918T} in NIH 3T3 cells resulted in an increased autophosphorylation of the receptor and tyrosine phosphorylation of SHC (Src-homology-2 containing transforming protein) thus indicating that the MEN2B disease phenotype could be influenced by the tissue distribution of Ret ligands. A further confirmation of these data came from the observation that in tumors with a Ret^{M918T} mutation the levels of both *GDNF* and *RET* genes where highly enhanced. Indeed, it seems plausible that ligand availability can further stimulate the receptor kinase activity and thus autocrine or paracrine expression of Ret ligand will give a growth advantage in tumors caused by the Ret^{M918T}.

These reports show that the MEN2B-type mutation does not abrogate ligand responsiveness of Ret thus suggesting that high local density of available ligand can modulate the Ret^{M918T} receptor kinase activity. Even though it still remains to be elucidated whether the stimulation of the mutated Ret enhances the activity of the receptor without changing the substrate specificity or, more likely, uncovers docking sites for new substrates. In the latter case ligand responsiveness may have important implications in the human diseases associated with MEN2B-type mutations. Indeed, some of the differences in the disease phenotype between MEN2A and MEN2B syndromes could depend on the tissue distribution of the Ret ligand coupled to the different susceptibility of MEN2A-type and MEN2B-type alleles to ligand stimulation.

4.2. What About the 2A-type Mutations?

At difference of the MEN2B-like mutations that are located in the substrate recognition pocket of the receptor, the activating mutations of extracellular cysteines involves the formation of intermolecular disulfide bonds between two Ret monomers (Fig.1C). Therefore, it is conceivable that mutations of the cysteine 634 of Ret are sufficient to induce complete activation of the Ret receptor [43], and it seems unlikely that it can be further stimulated by binding of the ligand.

High expression levels of *GDNF* and *RET* genes were also observed in the MEN2A pheochromocytomas carrying a C634 mutation [44]. Although it is conceivable that enhanced expression of mutated *RET* alleles would confer a growth advantage to MEN2A associated tumors, the presence of high levels of GDNF in these tumors was quite unexpected and difficult to reconcile with the ligand-independent activation of the Ret^{C634} mutants. On the other hand, mutations of the cysteine 634 are believed to be sufficient to induce complete activation of the Ret receptor, and it seems unlikely that it can be further stimulated by binding of the ligand. However, these findings are consistent with the recently described *in vitro* effects of GDNF [45]. Indeed, primary cultures from human MEN2-associated pheochromocytomas with a Ret^{C634} mutant protein still respond to GDNF, which likely stimulates the *wild-type* Ret encoded by the nonmutated allele [45]. Furthermore, we recently demonstrated that a protein consisting of the entire ectodomain of the Ret mutant carrying the cysteine 634 to tyrosine substitution ($EC\text{-}Ret^{C634Y}$) competes with the *wild-type* Ret receptor for ligand binding thus supporting the notion of a possible sensitivity of the Ret^{C634} mutant to GDNF stimulation [46]. However, whether in these C634-carrying MEN2A tumors the acute stimulation of the *wild-type* Ret can participate to define the final phenotype remains to be determined.

4.3. The Paradox of the HSCR/2A Cosegregation

Less is known about the sensitivity to ligand stimulation of Ret mutants carrying mutations in cysteines other than C634 (cys 609, 611, 618, 620). Families have been reported that developed both MEN-2A/FMTC and HSCR diseases. In these families the inherited Ret allele is mutated in Cys 609, 618 or 620 [47]. We and other have shown that these mutations have a dual effect: 1) to cause formation of active covalent homodimers (similarly to those caused by the C634 mutations); 2) to reduce the net amounts of Ret molecules on cell surface probably altering the folding of Ret, thus partially interfering with its maturation or intracellular trafficking [48,49]. Therefore the nature of Cys 609, 618 or 620 appears as unique among all HSCR mutations. Indeed, HSCR mutations are dispersed all along the Ret gene and impair its activity by different biochemical mechanisms (see above, 3), whereas Cys 609 618 or 620 mutations are the only mutations that activate the Ret kinase activity by causing dimer formation. The distinctive effects of Cys 609, 618 or 620 mutations may thus provide a plausible explanation for this apparent paradox that relies on the different need of target cells, or tissues, for active Ret molecules [49].

Indeed, two implicit conditions for this model should be considered. The first is to assume that the levels of available functional Ret in enteric neuroblasts are critical and cells do not survive under a threshold level of Ret signalling. The second implicates that the amounts of active Ret dimers resulting from mutations of cysteines 609, 618 or 620 that are expressed at the plasma membrane even low is still sufficient to trigger the uncontrolled proliferation of thyroid C-cell, leading to the development of MTCs, but is not sufficient to balance the decrease of Ret expressed at the cell surface and would transmit a signal below the threshold required to elicit survival or differentiation in the precursors of enteric neuron, thereby leading to HSCR. If true, this model has the advantage to be consistent with the indication that RET haploinsufficiency may result in HSCR and in good agreement with the possibility that in MEN2 each type of endocrine cells affected has different sensitivities to Ret mutations [17] (see section 9).

4.4. Does the Ligand Density make the Difference?

A corollary to this model is that during development of enteric ganglia the amounts of available Ret ligand is not the limiting factor for survival, whereas in adrenal chromaffin cells and in C-cells of thyroid even low levels of Ret on the cell surface (as it is the case for mutants in Cys 609, 618 or 620) are sufficient to disturb the growth control and drive cells toward uncontrolled proliferation. An indirect confirmation to this hypothesis is given by the observation that in around 50 % of sporadic pheochromocytomas, not associated with any Ret mutation, the *wild-type* Ret allele is overexpressed and strongly active likely because the sustained overexpression (> 20 times) of GDNF [44]. In this context it is important to note that, in contrast to C634, Ret mutants with C620 are still sensitive to GDNF stimulation *in vitro* [48]. A prediction of this model would be that the ligand amount in thyroid and chromaffin cells is rather limiting and in MEN2A patients with the C620 mutation a more severe disease should correlate with higher expression of GFLs.

5. THE MEN2A- *VERSUS* MEN2B-TYPE RET KINASE MUTATIONS

5.1. Preferential Substrate Specificity

The 2B-like mutations in the kinase domain cause Ret activation by a mechanism that doesn't involve formation of stable dimers but rather *via* an intramolecular mechanism. Therefore, it seems plausible that the occurrence of qualitative changes in Ret^{M918T} kinase activity could be considered as the most promising candidates responsible for the specific neoplastic phenotype caused by this Ret mutant. Indeed, the presence of a methionine in position 918 is highly conserved in receptor-tyrosine kinases, whereas in cytosolic non-receptor-tyrosine kinases a threonine is found at the equivalent position [27,50]. In addition the comparison with other tyrosine kinases revealed that the M918T mutation converts the substrate-binding pocket of Ret to resemble those of non-receptor Src-related tyrosine kinases [19,20]. This prediction was confirmed by a number of studies showing that this mutation results in increased activity toward the optimal peptide substrates of src and abl [23,51]. The first indications that the clinical difference between the MEN2B and the less aggressive MEN2A or FMTC diseases may at least partially rely on the substrate specificities of the Ret mutants were obtained in the laboratory of P.P Di Fiore and in that of JA Dixon [22,24]. Indeed, by using a two-dimensional gel electrophoresis of phosphotyrosine-containing proteins, Santoro *et al.* have tested this possibility in NIH 3T3 cells transformed with either a C634Y or M918T Ret mutant in which they showed that, although the pattern of intracellular phosphorylated proteins is similar, the extent of phosphorylation of several proteins differs between the two transformed cell lines, thus suggesting that the substrate specificity of these mutants is different and can thus affect phosphorylation of intracellular Ret effectors [22].

5.2. Tyrosine(s) Autophosphorylation of the Receptor

Conversely, Liu et al. addressed the same issue by utilising a peptide mapping approach coupled to site directed mutagenesis of the Ret^{C634S} and Ret^{M918T} active mutants. They first identified the tyrosine residues of Ret that are phosphorylated following oncogenic activation. Interestingly, they showed that the pattern of tyrosine residues that are phosphorylated in the M918T mutant does not exactly superpose with that of MEN2A-type mutants. Indeed, they demonstrated that among the six autophosphorylation sites found in the C634S Ret receptor, the M918T mutant lacks phosphorylation at the Y1096 (pY*X*N*X*), that leads to less efficient GRB2 (Growth factor receptor bound protein 2) binding to the MEN2B mutant as compared to the *wild-type* Ret. Even tough, the biological effects for which there is a need of the phosphorylation to Y1096 are not obvious, as indicated by the observation that deleting the Y1096 didn't affected the transforming efficiency of neither C634S nor M918T Ret mutants [52]. Furthermore, they found that the M918T mutation creates a new phosphorylation site [24].

Conversely, Takahashi and colleagues have shown that different tyrosine residues in the kinase domain are essential for the transforming activity of Ret^{C634R} or Ret^{M918T} mutants. Indeed, substitution of phenylalanine for tyrosine 905 (Y905F) completely abolished the transforming activity of Ret-MEN2A (C634R) but not Ret-MEN2B (M918T) [53]. This tyrosine is known to be conserved in all tyrosine kinases and play crucial roles in their catalytic and/or biological activities [54].

By using phosphorylation-specific antibodies directed against Y1015 and Y1062 of Ret [55] demonstrated that although both residues are constitutively phosphorylated in MEN2A- and in MEN2B-type Ret mutants, Y1062 autophosphorylation is significantly increased by the MEN2B mutation with respect to the MEN2A-type mutation. In agreement with this finding, the Ret^{M918T} kinase phosphorylated Y1062 more efficiently than did a MEN2A-type kinase in an *in vitro* kinase assay. They suggest that the increased phosphorylation of Y1062 depends on the altered specificity of the MEN2B-type Ret kinase. Indeed, the Ret sequence that flanks Y1062 (Leu-Tyr-Gly-Arg-Ile) corresponds to the optimal substrate (Ile/Leu/Val-Tyr-Gly-X-hydrophobic) of non-receptor tyrosine kinase (to which the Ret^{M918T} functionally belongs; see Ref. 23) and thus, Y1062 has the structural requisites for efficient phosphorylation by a Ret-MEN2B kinase. On the other hand the experiments described above were all performed in the absence of ligand. Therefore, since the activity of Ret^{M918T} is highly sensitive to ligand stimulation (see section 4.1) an obvious prediction of these experiments would be that ligand binding to Ret^{M918T} may induce the phosphorylation on tyrosine residues other that the Y1062 (Y1096?). If true, the increase of tyrosine phosphorylation induced by GDNF would be specific for few residues and not merely reflect a general increase in receptor activity. Therefore, the M918T mutation alters Ret autophosphorylation sites as well as the pattern of intracellular proteins undergoing tyrosine phosphorylation with respect to wild-type ligand-stimulated Ret and to MEN2A-type Ret mutants. At least three intrinsic dramatic changes in tyrosine phosphorylation have been identified that may distinguish the downstream signalling pathways triggered by the Ret^{M918T} mutant: 1) a change in substrate specificity of Ret caused by the presence of the M918 to T mutation in the activation loop; 2) in the Ret^{M918T} mutant, phosphorylation of tyrosine 1096 is not present and conversely that of tyrosine 1062 is increased; 3) phosphorylation of tyrosine 905 is not necessary for the transforming activity of Ret^{M918T}. Whether these changes in autophosphorylation may alter the signalling specificities of Ret^{M918T} remains to be determined.

Concerning the *gain-of-function* M918T mutation in Ret, several studies have shown that the same M to T substitution at the homologous residues in the activation loop of the kinase domain, is able to activate not only Ret but also other receptor-tyrosine kinases such as Met and Ron [56,57]. Interestingly, the introduction of this mutation in the TrkA tyrosine kinase domain resulted in the MEN2B-like M688T TrkA mutant that displayed a completely abrogated TrkA receptor activity [58]. These results indicate that although this mutation involves conserved amino acids in highly homologous regions, it exerts distinct effects in different receptors, resulting alternatively in a *gain-of-function* or in a *loss–of-function* mutation.

6. Regulation of Intracellular Signaling

6.1. Same Cascades are Utilized by both Types of Ret Mutants

The intracellular signalling pathways triggered by MEN2-associated Ret mutants have been largely investigated mainly by transfection of established cells with constitutively active Ret mutants. Most of the studies have been performed using two main cell systems: the mouse fibroblast NIH 3T3 and the rat pheochromocytoma derived PC12 cell lines. The NIH 3T3 cells are very sensitive to oncogene-induced transformation and have been largely used to determine the enzymatic cascades needed for Ret induced transformation. The PC12 cells display several characteristics of the pheochromocytoma tumor from which this cell line has been originated, thus allowing to study the effects of RET mutants in a cell environment that closely resembles the tumor type that is caused in MEN2 patients by these same active mutants. Furthermore, these cell systems either do not express (the NIH 3T3 cells), or express at very low levels (the PC12 cells), the endogenous Ret, thus representing the most appropriate recipient of Ret mutants. Human neuroblastoma-derived cell lines that express the *wild-type RET* allele and the human MTC-derived cell line (TT cell line) that express the Ret^{C634W} mutant have been also frequently used to investigate the intracellular cascades initiated by Ret wt and Ret^{C634} oncoproteins, respectively.

Comparing the biological effects induced by overexpressing Ret mutants in different systems have shown that Ret triggers growth, differentiation or survival depending on the cell type or developmental stage (see section 9).

RET is alternatively spliced to produce two protein variants (designated Ret9 and Ret51) that differ in the carboxy-terminal sequence (the functional properties of these isoforms are discussed in section 7.1).

In the intracellular domain of Ret9 there are 16 tyrosines and 18 tyrosines in that of Ret51, ten of which are in the kinase domain, one in the kinase insert, two in the juxtamembrane domain and three (or five in the long isoform) in the carboxy terminus. Activation of Ret kinase, either owing to the presence of activating mutations or ligand binding, results in autophosphorylation of intracellular tyrosine residues. To date, nine of Ret intracellular tyrosines have been shown to be phosphorylated upon Ret activation [24] and may serve as docking sites for adaptor proteins and substrate molecules containing SH2 domains or PTB (phosphotyrosine binding) domains. Amongst these, Y905, Y981, Y1015, Y1062, and Y1096 have been shown to provide the binding sites for the SH2 or PTB domains of Grb7-10, c-Src, PLC-γ, SHC/Enigma/FRS2, and Grb2, respectively.

Compelling reports from several laboratories unveiled a complex network of signalling molecules have been described for Ret, that include the activation of RAS/ERK, phosphatidylinositol 3-kinase (PI3-K), AKT, p38MAPK, Jun-associated N-terminal kinase (JNK) and ERK5 pathways. *In vitro* the RAS/ERK and the PI3-K/AKT pathways have been confirmed to be major pathways that control the biological effects of Ret, including cell transformation, even though the upstream events that regulate its activation are still under investigation and involve binding of multiple possible adaptor proteins such as SHC, SNT/FRS2, Insulin receptor substrate, IRS-1, Downstream of kinase protein Dok1, Dok4-5, Grb2-associated binder (Gab) protein Gab1 and Gab2 [for recent reviews 59-61].

6.2. Phosphorylation of Tyrosine 1062 Creates a Promiscuous Binding Site

The tyrosine 1062 appears to play a key role in determining downstream signalling by Ret. Indeed by substituting this tyrosine with phenylalanine, several laboratories have reported that phosphorylation of this tyrosine is needed for the transforming activity of all Ret mutants (either those associated with the MEN2 syndromes and the truncated PTC oncogenes) as well as for the function of the wild type Ret in cell differentiation [52]. Furthermore, phosphorylation of this tyrosine residue of Ret has been shown to uncover a binding site for multiple partners, including SHC, Rai (N-SHC), IRS-1, Dokl, Dok4-5, FRS2 and Enigma (in the case of Enigma binding is independent on phosphorylation), thus leading to propagation of critical intracellular signals including the activation of the Ras, JNK, PI3-K and ERK5 pathways. Furthermore, it has been reported that a few mutations identified in Hirschsprung's disease could affect signalling *via* Y1062 in Ret [62,63] thus confirming a key role played by this tyrosine for Ret function.

However, the extent of the contribution of Y1062 phosphorylation to downstream signalling seems to be rather cell type specific than an absolute need for Ret function. Recently, it has been described a model of knock-in mice in which Y1062 was replaced with phenylalanine [64]. In this report Takahashi and coworkers, even confirming the importance of signalling *via* Y1062, observed considerable differences in phenotype between Y1062F knock-in and *Ret*-deficient mice. For instance, null-mutant *RET* mice die at birth and lack the enteric nervous system in the entire bowel accompanied by renal agenesis or hypoplastic kidneys [reviewed 5]. Homozygous knock-in mice born normally, dye by day 27 after birth and show growth retardation. Furthermore, although severely impaired, kidneys developed in all of the homozygous Y1062F mice and the development of the enteric nervous system has a less severe phenotype. Comparing the milder phenotype of homozygous mutant mice to those of Ret-deficient mice, the authors suggest that the signal *via* tyrosine 1062 plays an important role in histogenesis of the enteric nervous system and nephrogenesis but, in addition, signalling via other tyrosine residues also plays a role in organogenesis [64].

6.3. Which are the Biological Constraints that Determine the Winner Partner for Tyrosine 1062?

In analogy with other tyrosine phosphorylated proteins, differences in spatial localization and concentration of the active Ret molecules likely controls the timing and the specificity of the proteins recruited to the receptor and thus the kind and the intensity of the intracellular signalling [65]. Compartmentalization of signal transduction at the cell surface allows a single class of receptors to transmit different signals from alternate locations in the cell. Lipid rafts are membrane microdomains constituted of sphyngolypids and cholesterol that are responsible of the compartmentalization in the plasma membrane of different signalling molecules. This event promotes the identification and interaction of proteins localized in these microdomains preventing them recognition with other proteins excluded from the rafts [66]. Indeed, the active Ret receptor can be isolated in multiprotein complexes together with several substrates and/or effectors, the identity of which likely dictate the different biological effects triggered by the receptor. These include the membrane bound multidoking proteins, as

Gab2, Gab1, FRS2, IRS-1, members of the p62Dok and SHC protein families, and intracellular effectors or enzymes as the regulatory subunit of the PI3-kinase, p85 and the SH2 domain-containing tyrosine phopshatases, Shp2 and Shp1. Binding to Ret of any of these proteins require phosphorylation of Y1062, but FRS2, Dok4-5, IRS-1, SHC and Rai bind directly to pY1062, thus indicating that these adaptor proteins compete for binding to the same site on Ret, whereas recruitment of Gab1-2 is mediated by other proteins. In turn, Shp1, Shp2, and the regulatory subunit of PI3-K, p85, are recruited to the complex via binding to phosphotyrosines on either of the multidocking proteins. The result is that several distinct Ret-bound protein complexes can be isolated that are simultaneously present in the same cell. Whether any of these complexes are enriched in specific cell compartments is still unknown, however, recent reports from several laboratories are suggestive that at least the raft-containing microdomains may play an important role in partitioning Ret-bound protein complexes. Indeed, following ligand stimulation, the Ret^{wt} is specifically recruited to membrane rafts where it recruits distinct effector molecules [67]. Active Ret mutants (of MEN2A- and MEN2B-type) are as well partitioned between the soluble and raft-containing fractions where they form multiprotein complexes with distinct effectors that differentially associate depending on whether the activated receptor is in raft or non-raft compartments. Interestingly the specific enrichment in rafts of Shp1-bound mutant Ret proteins (both Ret^{C634Y} and Ret^{M918T}) raises the question of the biological meaning of these interactions. Indeed, given the inhibitory action of Shp1 on Ret signalling [68], it seems plausible that this phosphatase may contribute to generate a negative feedback mechanism to attenuate downstream signalling. If this is the case, the enrichment of Shp1-bound Ret molecules in rafts might be part of this negative mechanism. Since Ret neither directly binds to Shp1 nor it is a good substrate for this phosphatase, identifying in raft the Shp1 substrates in the Ret complex will be thus a future major challenge [69].

Even though partitioning in raft of signalling molecules and receptors has been already described, the meaning still remains elusive. Indeed, nerve growth factor (NGF) and EGF receptors, are constitutively associated with rafts, but while NGF treatment has no effects on TrkA localization [70], EGF stimulation induces rapid exit of EGFR from raft [71], thus indicating that partitioning of receptor-tyrosine kinases in raft-containing membranes is a receptor specific mechanism. In the case of Ret mutants, the relative amounts in rafts of Shp1-bound Ret^{C634Y} were clearly higher than those of Shp1-bound Ret^{M918T}. A possible explanation for the preferential enrichment of Shp1-bound Ret^{C634Y} active proteins in rafts could rely on the homodimer conformation of this mutant that likely stabilize its permanence within the rafts.

In order to fully understand which are the molecular constraints that decide which complex will be recruited to Y1062, a further degree of complexity should be considered that comes from the fact that Y1062 is located just upstream the boundary between the short (Ret9) and the long isoform (Ret51). As a consequence the pY1062 in the short isoform provides both a ShcSH2 and a ShcPTB high affinity binding sites, while in the long isoform it provides a Shc PTB high affinity but a Shc SH2 low affinity binding site [72]. Thus the relative abundance of the two isoforms in the cell contributes to dictate which substrates will be recruited to Y1062 and thus the function of Ret in the cell (see section 7.1 and [73]). Indeed, compelling evidence indicate that the balance existing within the cell between the two isoforms contribute to dictate the biological effects of Ret (for more details see section 7.1).

Therefore, it seems plausible to imagine a general scenario in which different protein partners of Ret compete for binding to Y1062. The effective interactions that will take place are determined by the balance between the local concentration of the proteins involved and the affinity for Tyr-1062 either in the Ret9 or in the Ret51 context. However, deciphering how the spatial and time density of these proteins is coordinated within a specific cell type is still poorly understood.

6.4. Do Intracellular Signals by Ret Mutants Really Superpose?

As discussed above the fact that a different pattern of phosphorylated tyrosine residues may distinguish the two kinds of Ret mutants suggests that different signalling pathways may be initiated by each of the mutants. Indeed, the different mutated Ret proteins differ in their preferential substrate specificity, depending on the presence of either methionine or threonine at codon 918. However, an understanding of whether and how the different Ret mutations (either associated with MEN2A or MEN2B syndromes) activate distinct enzyme cascades thus leading to diverse clinical neoplastic phenotypes remain largely elusive.

However, the most confident and consistent example of a pathway preferentially activated by Ret^{M918T} is the activation of the PI3-K pathway. PI3-K are crucial transducers that control multiple biologic responses, including protection from programmed cell death, cellular proliferation and motility. Both Ret mutants induce activation of the PI3-K/AKT pathway, that has been shown to be required for the transforming activity and promotion of cell survival [74,75]. However, the extent of activation of this pathway is dictated in vitro by which is the Ret mutant used. Indeed, as reported by Murakami et al [76] in NIH 3T3 stable transfectants, expression of the M918T mutant induces the PI3-K activity at higher levels than that of C634R. Furthermore, the PI3-K dependent phosphorylation of PKB/AKT was as well enhanced in M918T transfectants.

In good agreement with these results we have recently shown that the expression of Ret^{M918T} rescued PC12 cells from apoptosis more efficiently as compared to Ret^{C634Y}, and further delayed the genomic DNA fragmentation [77]. In addition, the expression of Ret^{M918T}, and barely of Ret^{C634Y}, rapidly induces activation of the AKT/Bad pathway, and, at least if triggered by Ret^{M918T}, activation of this pathway requires continuous activation of the non-transmembrane tyrosine phosphatase, Shp2, thus suggesting that enhanced PI3-K signalling may be involved in the development of the aggressive phenotype of MEN2B.

A different tyrosine phosphatase, the leukocyte common antigen-related (LAR) has been also recently implicated in signalling by Ret mutants. By expressing into NIH 3T3 cells the LAR phosphatase together with Ret^{C634Y} or Ret^{M918T} the group of Takahashi demonstrated that the constitutive tyrosine autophosphorylation of Ret^{C634Y} but not of Ret^{M918T} was impaired. Furthermore, high levels of LAR expression significantly reduced proliferation of Ret-MEN2A expressing cells *in vitro* as well as tumor growth *in vivo*, clearly emphasizing potentially important roles for LAR in the regulation of neoplastic cell proliferation and of tumorigenicity Ret^{C634Y}- but not of Ret^{M918T}-transfected cells [78].

As stated in the preceding section, 6.1, several multi-tyrosine phosphorylated scaffolding proteins have been shown to mediate Ret signalling. In several cases, including FRS2 and Gab2, the extent to which the Ret mutants bind to these proteins seems to be similar independently of which mutant is considered [77,79]. Nevertheless, as reported by Murakami

H et al [80], p62Dokl and the complex formation consisting of Dokl, RasGAP and Nck were clearly induced at higher levels in the MEN2B transfectant than in the MEN2A transfectant. Binding of Nck to Dokl probably leads to a negative regulation for the Ras/ERK signalling and positively regulate the JNK [80].

The implication of different members of Dok family as linkers for Ret intracellular signalling may suggest a path to address the issue of which are the specific pathways generated by the structural changes caused by the M918T mutation. In fact, despite that the M918T mutation alters Ret autophosphorylation, including the extent of phosphorylation on Y1062, the ability to recruit the Gab1-2 , as well as FRS2, seems not to be affected. Indeed, in contrast to Gabl and Gab2 that are intracellular adaptor molecules expressed ubiquitously [81], Dokl and Dok2 are predominantly expressed in tissues of hematopoietic origin, Dok4 and Dok5 are highly expressed in non-hematopoietic tissues, and Dok5 is specifically expressed in the nervous system. Furthermore, the Dokl-3 act as negative regulators of tyrosine kinase-activated signalling pathways whereas Dok4 and Dok5 positively regulate receptors-tyrosine kinases signalling pathways. Therefore it seems plausible that signalling pathways mediated by Dok proteins may more closely mirror the specific Ret mutation, whereas those mediated by Gabs seems to be less specific and not affected by the specific Ret mutation. Even small differences between Ret mutants in binding affinity for members of the Dok family, as observed *in vitro*, may be translated *in vivo* in more dramatic effects on downstream signalling and thus differentially specify the tissue affected in the corresponding MEN2 syndromes. This possibility is further supported by the observation that Dok members display a strong binding specificity for some receptors-tyrosine kinases since do not interact directly to several receptors, such as Met, Kit, Fms, Ros, TrkA, and ErbB-2 [82].

Indeed, despite the manifest biochemical differences between the two kinds of Ret mutants (described in section 5) the identification of a specific enzyme whose activation may discriminate between the signalling initiated by each mutant has failed. Therefore, the evidence by these *in vitro* studies indicate that the relative extent and persistence to which are activated the same pathways, rather than the involvement of specific effectors, makes the difference between the intracellular signalling initiated by the two types of Ret mutants. Whether these are as well sufficient to determine the clinical differences (in severity and tissues affected) observed between the MEN2A and MEN2B syndromes remains to be assessed. On the other hand, this result should not be surprising because the intracellular signalling pathways initiated by even much far related receptor-tyrosine kinases largely superpose in a given cell type. In same case it rather differs in the extent or persistence of signaling than in the sage of specific enzyme cascades. A paradigmatic example of such difference is represented by comparing in PC12 cells signalling initiated by EGF or NGF. Both ligands induce stimulation of the downstream ERKs activity but the kinetic of stimulation by EGF is rapid and rapidly decrease, whereas upon NGF stimulation, ERK activity remains persistent for hours. Persistence of ERK activity is necessary to dictate the biological effects induced by these receptors that are dramatically different, in that NGF, but not EGF induces PC12 cells to differentiate toward a neuronal-like phenotype.

7. Alternative Splicing as a Regulatory Mechanism for RET Function

7.1. Ret9 and Ret51 acts as Distinct Receptors

A number of *RET* splicing variants are produced by alternative splicing of the *RET* pre-messenger RNA [83-85]. A group of 3' variants is obtained by combination of alternative exons 20 and 21 and retention of intron IVS19-20, with four alternative polyadenylation sites [83,84]. These transcripts code for two different Ret proteins, Ret9 (1072 aa) and Ret51 (1114 aa), which differ in the region coding for the carboxy-terminal domain (Fig. 1A). A third 3' variant, Ret43, has been described although, in both normal and tumor samples, the RET43 transcripts represent only minor species [86,87]. The short (Ret9) and the long (Ret51) isoforms present distinct biochemical and biological properties and might play different roles in development and in tumorigenesis. In particular, these isoforms differ for the presence of tyrosine residues, and *in vitro* transforming potential [88-90]. Recent and well-designed experiments by Tsui-Pierchala [73] provide evidence that Ret9 and Ret51 acts as distinct receptors in the same cell. Indeed, Ret9 and Ret51 differ in the relative kinetics of autophosphorylation after GDNF stimulation, phosphorylation of Y905 and Y1062 being more sustained in Ret9 over time than in Ret51, while phosphorylation of Y1015 remained higher in Ret51 as compared to that in Ret9. This indicated that protein recruited to these phosphotyrosine residues have different binding kinetics if recruited by Ret9 or Ret51. Furthermore, in dissociated sympathetic neurons the long and the short isoforms are unable to form homodimers upon activation, even though they share identical ligand-binding extracellular domains. However, the ability to form homodimers seems to be cell type specific. Indeed, Ret9 and Ret51 associate each other if expressed in CHP126 cells, consistent with data from other neuroblastoma cells [91]. The mechanism that accounts for the inability of Ret9 and Ret51 to associate in sympathetic neurons is unclear.

Furthermore, several reports have provided convincing *in vitro* evidence that Ret51 and Ret9 carrying MEN2 mutations, mainly in the cytoplasmic region, display different biological effects in pheochromocytoma PC12 cells. Moreover, the transforming efficiency of the two isoforms measured in NIH 3T3 is different, being higher for the Ret51 than for the Ret9 isoform [53,89,90].

The biochemical differences between the two Ret isoforms are supported in vivo by the relative rates of expression during human kidney development and expression in tumors. By using a semi-quantitative reverse- transcriptase polymerase chain reaction (RT-PCR) the group of Lois Mulligan observed that Ret51 is expressed in human kidney at low levels in the earliest gestational ages and the relative expression levels dramatically increased by nine weeks of gestation. Thus suggesting that the long Ret51 isoform is not required in kidney for early inductive events, but rather may contribute to later events responsible of kidney morphogenesis [86]. However, by performing a microarray analysis in stable cell cultures the same group was unable to further strengthen these results. By using kidney-derived cell lines, Myers and Mulligan [92] investigated gene expression associated with either RET9 or RET51. Surprisingly, they detected only four genes up-regulated and three genes down-regulated in RET9 cells as compared with RET51.

Furthermore, we investigated whether the differences in the transforming potential observed in vitro between the Ret9 and Ret51 isoforms may be the mirror of the *in vivo* ability to elicit in MEN2 tumors different biological effects as compared to normal tissues or sporadic pheochromocytomas [87]. By quantitative RNases protection assays, we measured the absolute abundance of the 3' splice variants in pheochromocytomas and in normal tissues. In pheochromocytomas of familial origin (MEN2A and MEN2B) the relative amounts of the long isoform are consistently higher than in the majority of sporadic tumors and normal tissues. These results suggest that the presence of higher amounts of the long isoform would confer a selective growth advantage for MEN2 tumors and are consistent with the evidence that the corresponding Ret51 protein is more transforming than the Ret9 isoform [87].

Still, the most direct evidence of the different functions played by the two Ret isoforms comes from transgenic mice models. By targeted mutagenesis transgenic mice have been generated that express either the short or the long isoform [93]. The effects of depriving the Ret9 isoform resulted in loss of enteric ganglia from the colon and kidney hypodisplasia, whereas mice lacking the Ret51 allele were viable and apparently normal, thus indicating that the function of the two Ret isoforms *in vivo* is qualitatively different [94-95].

7.2. Ret51 but not Ret9 is Transactivated by the NGF Receptor, TrkA

It has been recently reported that in developing sympathetic neurons, the Ret receptor can be activated in a GFL-independent manner [94]. Intriguingly, a cross-talk between the Ret and TrkA signalling pathways has been already described that leads to an increase in Ret expression and phosphorylation as a consequence of TrkA activation by NGF and in complete absence of GDNF stimulation. This mechanism of GDNF-independent Ret activation is exclusive for Ret51 isoform since no change in Ret9 phosphorylation was observed in sympathetic neurons maintained in vitro up to 21 days in the presence of NGF, whereas the tyrosine phosphorylation of Ret51 increased markedly with age in culture. The fact that Ret9 and Ret51 share identical extracellular domains and, in sympathetic neurons, were stimulated equally well by GFLs, supports the finding that the transactivation of Ret is independent of the extracellular domain and, therefore, independent of GDNF ligands. This raises the possibility that neurotrophic factor receptors such as TrkA or Ret may have physiologic functions independent of their natural ligands. Different mechanisms could be evoked to explain this interplay among these neuronal growth factor receptors. A plausible explanation is that NGF may act on Ret phosphorylation *via* a downregulation of a protein tyrosine phosphatase. Alternatively, or in addition, it could act by increasing the activity of a tyrosine kinase that in turn phosphorylates Ret. Furthermore, NGF may induce the expression of an adaptator molecule able to induce Ret dimerization independently of GDNF and GFRα. In all the cases, the final consequence is an increase of basal levels of Ret phosphorylation in a NGF-dependent manner. If it is crucial to understand the physiologic function of this phenomenon, it is undoubting that great advantages could arise from the existence of interplay between receptor-tyrosine kinases in the nervous system. First, a single growth factor can activate different pathways providing intracellular signal amplification. In addition, activation of multiple receptor-tyrosine kinases by a single growth factor may prolong the activity of a signal pathway, thus prolonging a functional outcome. This is the case reported for the functional collaboration of Ret and TrkA signal pathways which are required for differentiation of some malignant neuroblastoma cells where the presence of GDNF stimulation enhances TrkA expression and acts to strengthen the NGF differentiation signal [96]. If other examples of Ret transactivation and/or interplay among Ret and other receptor-tyrosine kinases exist, remains fundamental to demonstrate.

7.3. 5′ splicing Variants of Ret

A distinct group of three transcript variants has been described in the region coding for the Ret extracellular domain, the 5' splicing variants RET2/4, -2/5 and -2/6. These 5' variants may code for Ret proteins which differ in their extracellular ligand-binding domains and display putative biological differences during kidney development [85]. However, by quantitative RT-PCR, we have recently shown that the abundance and the distribution of the 5' variants in a panel of normal and tumor tissues as rare transcripts never exceeding 1 % of the full length *RET* transcripts [97].

8. Gene Expression Profiling for MEN2A and MEN2B Syndromes

Recent studies using gene expression profiling have been instrumental in defining subtypes of tumors within previously described categories that are based on clinical or histological characteristics [98,99]. In some instances, tumor subtypes defined by expression fingerprints are also associated with specific genetic alterations, thus implying that these alterations underlie the changes in the expression profile and in tumor behavior. With the aim to explain the different phenotype and early malignancy in MEN2B compared with MEN2A syndrome, an expression microarray analysis has recently been performed on tumors from patients with MEN2A and MEN2B characterized by specific RET mutations [100]. Importantly, the expression of some genes has been found altered only in MTCs from MEN2B patients. For examples, genes associated with matrix remodeling and involved in the process of epithelial to mesenchymal transition were upregulated in MEN2B MTCs. These genes have been associated with an increased metastatic potential in a variety of others tumors, suggesting that they may contribute to the early malignant potential of MEN2B specific tumors. Furthermore, chondromodulin-1 (CHM1), a known regulator of cartilage and bone growth, was expressed at high levels specifically in MEN2B MTCs and its high expression was associated with the presence of skeletal abnormalities in MEN2B patients.

Anyway, the design of the experimental approach, the cell type used, or temporal events may be critical for the identification of specific genes eventually associated with MEN2B tumors. In fact, a differential display analysis of gene expression performed on NIH 3T3 cells transfected with RET mutants associated with MEN2A and MEN2B [101,102], identified some genes with differential expression by MEN2A-RET and MEN2B-RET, but these genes did not overlap with those identified by the above group [100]. In particular, a role for stanniocalcin-1 (STC1) as a possible regulator of skeletal growth has been suggested in the development of the MEN2B phenotype since it was specifically induced in fibroblasts expressing Ret^{M918T}. New insights into the molecular basis of the MEN2B phenotype is needed to confirm the possibility that the aberrant high level expression of STC1, CHM1 or other genes may contribute to some, if not all, associated skeletal abnormalities and early malignancy.

9. TRANSGENIC MICE EXPRESSING THE MEN2A- AND MEN2B-TYPE RET MUTANT

The *in vivo* biological effects of expressing the MEN2-associated Ret mutants have been directly investigated by generating transgenic mice models. Transgenic mice expressing the human RET9 [103] as well as the RET51 [104] isoform carrying the MEN2A mutation C634R, under the control of the C-cell specific calcitonin promoter, developed C-cell tumor biologically and morphologically similar to human MTC. The CT gene-related peptide (CGRP)/CT promoter drives the expression both in thyroid C-cells and in chromaffin cells of adrenal medulla [105]. Despite the expression of the transgene in adrenal glands, these mice did not exhibit pheochromocytomas nor histological abnormalities of adrenal medulla. This observation is surprising since pheochromocytoma occurs in 50% of *MEN2A* gene carriers. One explanation that might account for this lack of concordance between the mouse model and the MEN2 situation is that the Ret^{C634R} protein is expressed at a level insufficient to exert an effect in adrenal gland.

Furthermore, to study the mechanism of tissue-specific tumor development by MEN2A-type Ret mutants, transgenic mice have been generated by introducing the mutated human RET (C634R) gene fused to Moloney murine leukemia virus long terminal repeat [106]. The animals developed thyroid C-cell hyperplasia and medullary carcinoma, however, no tumor development was observed in several tissues despite high levels of expression of the RET transgene. In addition, development of mammary or parotid gland adenocarcinoma was observed in one-half of the transgenic mice. Results indicate that tumor formation was rather tissue specific and strictly dependent on Ret dimerization. Indeed, Ret dimerization and its complex formation with SHC and Grb2 adaptor proteins were detected in tumor tissues samples. In contrast, in those tissues that didn't developed tumors, despite a comparable transgene expression both Ret^{C634R} dimerization and activation remained undetectable. Because these tissues, but not tumors, expressed GFRα at high levels, authors suggested a possible role of the co-receptor molecule in hampering the Ret^{C634R} dimerization, and thus leading to tissue-specific tumor development *in vivo*.

In this respect, the authors suggest a possibility that may partly explain the differences of clinical phenotypes between MEN2A and MEN2B. In fact, MEN2A does not develop the mucosal neuroma and ganglioneuromatosis of the enteric nervous system that are the characteristic features of MEN2B. These neuronal cells affected in MEN2B are known to express GFRα1 at high levels [107]. Because the MEN2B mutations activate RET in a monomeric form, this fact implied that expression of GFRα1 may not impair activation of the MEN2B-type Ret mutant protein in neuronal cells. These findings, indicate that the expression pattern of GFRαs could be one of factors that determine the clinical features of MEN2A and MEN2B, in addition to possible differences of intracellular signalling activated by MEN2A- and MEN2B-type proteins (discussed above 6.4).

To produce an animal model to study the dominant activity of the MEN2B-type mutation in tumorigenesis and tumor progression, gene targeting techniques in conjunction with Cre/loxP site-specific recombination have been used to introduce the typical MEN2B-type codon alteration (M918T) into the murine RET gene. Mice heterozygous for the mutant allele, displayed C-cell hyperplasia, pheochromocytoma and sympathoadrenal malformation although they were less severely affected than humans with MEN2B [108]. This model is the first MEN2 mouse model to develop pheochromocytoma, while MTC was not observed. This is surprising given the fact that in the human MEN2B syndrome, MTC has an earlier onset and greater penetrance than pheocromocytoma. One possible explanation of the authors for this finding is that the lack of tumor formation in the thyroids of RET-MEN2B mice may be due to the relatively short life-span of mice versus humans. It seems rather improbable that older mice will develop MTC in addition to pheocromocytoma. Alternatively it remains possible that breeding the mutation onto different genetic backgrounds might shorten the latency of tumor development (for examples, alleles of modifier genes may act to inhibit the effect of the M918T mutation or decrease the frequency of additional somatic mutation in the thyroid).

Furthermore, mice homozygous for RET carrying the MEN2B mutation did not display any developmental abnormalities indicating that, while causing the MEN2B phenotype, the M918T mutation does not interfere with the normal development of the kidney and enteric nervous system. These results seems to exclude that in addition to the *gain-of-function* nature of the MEN2B mutation, it might create concomitant *loss-of-function* effects by failing to phosphorylate some of Ret normal substrates.

In contrast to the above mouse model, it has been found that overexpression of a human MEN2B transgene with the M918T mutation under the dopamine β-hydroxilase promoter, in addition to cause ganglioneuromas, induces developmental renal malformations analogous to those reported with loss-of -function mutation in the *RET* gene [109,110]. These observations indicate that activating mutations of the *RET* gene can produce a phenotype usually associated with *RET* deficiency. A possible explanation of the apparent discrepancy among the above studies could be found in the existence of important interspecies differences between humans and mice in the function of Ret oncogene. Accordingly, it has been recently reported that the introduction into mice of a YAC (yeast artificial chromosome)-based transgene containing the human RET oncogene with the mutation associated with MEN2B, resulted in generation of transgenic mice that did not develop any of the neoplasms associated with MEN2B in humans [111]. Moreover, introduction of the transgene into mice lacking RET resulted in only partial correction of murine knockout phenotype consisting of renal aplasia and partial intestinal aganglionosis. These results are in contrast to the case in which a normal murine RET allele is present; such mice exhibit complete correction of the knockout phenotype.

10. Ret Mutants as Target for Therapy

Targeting receptor tyrosine kinases has become a major challenge for therapeutic intervention. In this respect, Ret constitutes a model system of choice in that the MEN2A-type transforming mutations located in the extracellular domain simplify the issue of intracellular accessibility for a targeting molecule, whereas the MEN2B-type mutations are intracellular and induce dimerization- and ligand-independent signalling. In the recent few years, several molecules that inhibit Ret activity have been developed, these include small synthetic chemical inhibitors of tyrosine kinases [112-117], anti-Ret monoclonal antibodies [118,119], protein-based Ret mimics [46,120] and RNA-based specific inhibitors of Ret that belong to a new class of molecules, called aptamers [121] (Table 2). Unlike ribozymes and antisense oligonucleotides, aptamers function by folding into a specific globular structure that dictates high-affinity binding to the target. Single stranded RNA-based aptamers can be selected for their ability to bind directly and tightly to specific proteins. These molecules exhibit many characteristics desired for non-invasive *in vivo* tumour imaging and therapy. Their low molecular weight (8-15 kDa) is one order of magnitude lower than that of antibodies, hence aptamers are expected to have rapid tissue and tumour penetration, as well as fast blood clearance. In contrast to antibodies, aptamers are non-immunogenic and can display favorable target-to-noise ratios at early time points. Finally, in some cases aptamers have the potential to inhibit the biological function of the target resulting in useful reagents for target validation in a variety of disease models [122]. We adopted a whole cell SELEX (Systematic Evolution of Ligands by EXponential enrichment) strategy to target Ret^{C634Y} in a complex environment that is expected to expose a native protein to the selection procedure, thus best mimicking *in vivo* conditions. We obtained RNA-based aptamers that not only recognize at high affinity and specificity the extracellular domain of Ret, but also block Ret downstream signaling and subsequent molecular and cellular events [121].

Table 2. Ret activity inhibiting compounds.

Name	Description	Target	Ref.
Imatinib, Glivec, STI571	2-Phenylaminopyrimidine	Brc-Abl,c-Kit, PDGFR, Ret	[112,113]
CEP-751 CEP-701	Indolocarbazole alkaloid	FLT3, VEGFR2, Trk, Ret	[114]
ZD6474	Quinazoline	VEGFR2, Ret	[115]
PP1	Pyrazolo-pyrimidine	Src kinases, Ret	[116,117]
Anti-Ret antibodies	MAbs	Hu-Ret	[118,119]
RET$^{\Delta TK}$	Dominant-negative peptide inhibitor	Hu-Ret	[120]
EC-Ret^{C634Y}	Competitive peptide inhibitor	Hu-Ret	[46]
D4	RNA-based aptamer	Hu-Ret	[121]

11. Conclusion

The direct correspondence between the RET oncogene genotypes and the MEN2 neoplastic syndromes each caused by distinct Ret mutants provides us with a quite unique model to study the molecular mechanisms that dictate the clinical phenotype of these diseases. As shown in this review, despite the huge amount of information and the different approaches utilized a comprehensive understanding of the mode of action of the different Ret mutants remains largely elusive. Nevertheless, the ability to elicit a specific cellular response, meant as the pattern of gene expression, reveals to be a promising approach to understand the events initiated by the active Ret mutants and leading to distinct diseases. Indeed, the use of DNA array for gene expression profiling provided evidence that it is possible to identify small set of genes specifically regulated by each Ret mutant. However, the extent of each set is probably largely underestimated thus making hard to extrapolate from the identity of the genes involved an exhaustive indication of the events leading to the different phenotypes in MEN2A *versus* MEN2B diseases. On the other hand, the differences in the transducing signalling pathways triggered by each of the two kinds of Ret mutants seems to be rather tuned by quantitative differences of the same cascades (either in time and/or space compartmentalization) than by the usage of specific alternative pathways, thus making critical the choice of the biological model system utilized. Furthermore, the more direct approach of introducing the M918T mutation in a transgenic mouse model didn't caused a plain MEN2B phenotype probably because of some specie specificity that caused the *in vivo* models to deceive the initial hopes. Even though transgenic mice will be essential to understand the cell-type specific constraints that determine the tissues affected in the two different MEN2 diseases.

A fine understanding of the molecular mechanisms involved will certainly take great advantage of a computational description of the network interplay of signalling molecules, and will reveal crucial to develop innovative differential diagnostic therapeutic instruments for these two closely related diseases.

Acknowledgements

This work was supported by the European Molecular Imaging Laboratory (EMIL) network. We wish to thank M. Chiariello and R. Frunzio for critical reading of the manuscript and members of our laboratory M. Incoronato, CL. Esposito, I. Amelio for scientific advice and fruitful discussions.

References

[1] Anders J, Kjar S, Ibanez CF. Molecular modeling of the extracellular domain of the RET receptor tyrosine kinase reveals multiple cadherin-like domains and a calcium-binding site. *J Biol Chem*. 2001, 276, 35808-35817.

[2] Baloh RH, Enomoto H, Johnson EM Jr, Milbrandt J. The GDNF family ligands and receptors - implications for neural development. *Curr Opin Neurobiol* 2000,10, 103-110.

[3] Airaksinen MS, Titievsky A, Saarma M. GDNF family neurotrophic factor signaling: four masters, one servant? *Mol Cell Neurosci* 1999, 13, 313-325.

[4] Creedon DJ, Tansey MG, Baloh RH, Osborne PA, Lampe PA, Fahrner TJ, Heuckeroth RO, Milbrandt J, Johnson EM Jr. Neurturin shares receptors and signal transduction pathways with glial cell line-derived neurotrophic factor in sympathetic neurons. *Proc Natl Acad Sci U S A* 1997, 94, 7018-7023.

[5] Airaksinen M. S. and Saarma M. The GDNF family: signalling, biological functions and therapeutic value. *Nat Rev Neurosci* 3, 2002, 383-394.

[6] Kordower, J.H. In vivo gene delivery of glial cell line--derived neurotrophic factor for Parkinson's disease. *Ann Neurol.* 2003, 53, 120-132

[7] Edery P, Lyonnet S, Mulligan LM, Pelet A, Dow E, Abel L, Holder S, Nihoul-Fekete C, Ponder BA. Munnich A Mutations of the RET proto-oncogene in Hirschsprung's disease. *Nature.* 1994,367, 378-380.

[8] Romeo G, Ronchetto P, Luo Y, Barone V, Seri M, Ceccherini I, Pasini B, Bocciardi R, Lerone M, Kaariainen H, et al. Point mutations affecting the tyrosine kinase domain of the RET proto-oncogene in Hirschsprung's disease. *Nature*, 1994, 367, 377-378.

[9] Angrist M, Bolk S, Thiel B, Puffenberger EG, Hofstra RM, Buys CH, Cass DT, Chakravarti A. Mutation analysis of the RET receptor tyrosine kinase in Hirschsprung disease. *Hum Mol Genet.* 1995, 4, 821-830.

[10] Alberti L, Carniti C, Miranda C, Roccato E, Pierotti MA. RET and NTRK1 proto-oncogenes in human diseases. *J Cell Physiol.* 2003, 195, 168-186

[11] Farndon JR, Leight GS, Dilley WG, Baylin SB, Smallridge RC, Harrison TS, Wells SA Jr. Familial medullary thyroid carcinoma without associated endocrinopathies: a distinct clinical entity. *Br J Surg.* 1986, 73, 278-281.

[12] Schimke RN. Genetic aspects of multiple endocrine neoplasia. *Annu Rev Med.* 1984;35, 25-31.

[13] Ponder BA & Smith D. The MEN II syndromes and the role of the ret proto-oncogene. *Adv Cancer Res.* 1996, 70, 179-222.

[14] Eng C, Clayton D, Schuffenecker I, Lenoir G, Cote G, Gagel RF, van Amstel HK, Lips CJ, Nishisho I, Takai SI, Marsh DJ, Robinson BG, Frank-Raue K, Raue F, Xue F, Noll WW, Romei C, Pacini F, Fink M, Niederle B, Zedenius J, Nordenskjold M, Komminoth P, Hendy GN, Mulligan LM, et al. The relationship between specific RET proto-oncogene mutations and disease phenotype in multiple endocrine neoplasia type 2. International RET mutation consortium analysis. *JAMA* 1996, 276, 1575-1579.

[15] Eng C, & Mulligan LM. Mutations of the RET proto-oncogene in the multiple endocrine neoplasia type 2 syndromes, related sporadic tumours, and hirschsprung disease. *Hum Mutat.* 1997, 9, 97-109.

[16] Mulligan LM, Kwok JB, Healey CS, Elsdon MJ, Eng C, Gardner E, Love DR, Mole SE, Moore JK, Papi L, *et al.* Germ-line mutations of the RET proto-oncogene in multiple endocrine neoplasia type 2A. *Nature*, 1993, 363, 458-460.

[17] Mulligan LM, Eng C, Healey CS, Clayton D, Kwok JB, Gardner E, Ponder MA, Frilling A, Jackson CE, Lehnert H, *et al.* Specific mutations of the RET proto-oncogene are related to disease phenotype in MEN 2A and FMTC. *Nat Genet.*, 1994, 6, 70-74.

[18] Eng C, Smith DP, Mulligan LM, Nagai MA, Healey CS, Ponder MA, Gardner E, Scheumann GF, Jackson CE, Tunnacliffe A, *et al.* Point mutation within the tyrosine kinase domain of the RET proto-oncogene in multiple endocrine neoplasia type 2B and related sporadic tumours. *Hum Mol Genet.*, 1994, 3, 237-241.

[19] Hofstra RM, Landsvater RM, Ceccherini I, Stulp RP, Stelwagen T, Luo Y, Pasini B, Hoppener JW, van Amstel HK, Romeo G, *et al.* A mutation in the RET proto-oncogene associated with multiple endocrine neoplasia type 2B and sporadic medullary thyroid carcinoma *Nature*, 1994, 367, 375-376.

[20] Carlson KM, Dou S, Chi D, Scavarda N, Toshima K, Jackson CE, Wells SA Jr, Goodfellow PJ, Donis-Keller H. Single missense mutation in the tyrosine kinase catalytic domain of the RET protooncogene is associated with multiple endocrine neoplasia type 2B. *Proc Natl Acad Sci U S A.*, 1994, 91, 1579-1583.
[21] Hanks SK, Quinn AM, Hunter T. The protein kinase family: conserved features and deduced phylogeny of the catalytic domains. *Science*, 1988, 241, 42-52.
[22] Santoro M, Carlomagno F, Romano A, Bottaro DP, Dathan NA, Grieco M, Fusco A, Vecchio G, Matoskova B, Kraus MH, *et al.* Activation of RET as a dominant transforming gene by germline mutations of MEN2A and MEN2B. *Science*, 1995, 267, 381-383.
[23] Songyang Z, Carraway KL 3rd, Eck MJ, Harrison SC, Feldman RA, Mohammadi M, Schlessinger J, Hubbard SR, Smith DP, Eng C, *et al.* Catalytic specificity of protein-tyrosine kinases is critical for selective signalling. *Nature*, 1995, 373, 536-539.
[24] Liu X, Vega QC, Decker RA, Pandey A, Worby CA, Dixon JE. Oncogenic RET receptors display different autophosphorylation sites and substrate binding specificities. *J Biol Chem.* 1996, 271, 5309-5312.
[25] Eng C, Smith DP, Mulligan LM, Healey CS, Zvelebil MJ, Stonehouse TJ, Ponder MA, Jackson CE, Waterfield MD, Ponder BA. A novel point mutation in the tyrosine kinase domain of the RET proto-oncogene in sporadic medullary thyroid carcinoma and in a family with FMTC *Oncogene*, 1995, 10, 509-513.
[26] Bolino A, Schuffenecker I, Luo Y, Seri M, Silengo M, Tocco T, Chabrier G, Houdent C, Murat A, Schlumberger M, et al. RET mutations in exons 13 and 14 of FMTC patients. *Oncogene*, 1995, 10, 2415-2419.
[27] Pasini A, Geneste O, Legrand P, Schlumberger M, Rossel M, Fournier L, Rudkin BB, Schuffenecker I, Lenoir GM, Billaud M. Oncogenic activation of RET by two distinct FMTC mutations affecting the tyrosine kinase domain. *Oncogene*, 1997, 15, 393-402.
[28] Newgreen D, Young HM. Enteric nervous system: development and developmental disturbances--part 1. *Pediatr Dev Pathol.*, 2002, 5, 224-247.
[29] Newgreen D, Young HM. Enteric nervous system: development and developmental disturbances--part 2. *Pediatr Dev Pathol.*, 2002, 5, 329-349.
[30] Chakravarti, A._Endothelin receptor-mediated signaling in hirschsprung disease._*Hum. Molec. Genet.* 5, 1996, 303-307.
[31] Parisi, M.A. and Kapur, R.P. Genetics of Hirschsprung disease. *Curr. Opin. Pediatr.,* 2000, 12, 610–617.
[32] Borghini S, Bocciardi R, Bonardi G, Matera I, Santamaria G, Ravazzolo R, Ceccherini I. Hirschsprung associated GDNF mutations do not prevent RET activation. *Eur J Hum Genet.,* 2002, 10, 183-187.
[33] Eketjall S, Ibanez CF. Functional characterization of mutations in the GDNF gene of patients with Hirschsprung disease. *Hum Mol Genet.* 2002, 11, 325-329.
[34] Doray B, Salomon R, Amiel J, Pelet A, Touraine R, Billaud M, Attie T, Bachy B, Munnich A, Lyonnet S. Mutation of the RET ligand, neurturin, supports multigenic inheritance in Hirschsprung disease. *Hum Mol Genet.* 1998, 7, 1449-1452.
[35] Schuchardt A, D'Agati V, Larsson-Blomberg L, Costantini F, Pachnis V. Defects in the kidney and enteric nervous system of mice lacking the tyrosine kinase receptor Ret. *Nature*, 1994, 367, 380-383.

[36] Moore MW, Klein RD, Farinas I, Sauer H, Armanini M, Phillips H, Reichardt LF, Ryan AM, Carver-Moore K, Rosenthal A. Renal and neuronal abnormalities in mice lacking GDNF. *Nature*, 1996, 382, 76-79.

[37] Pichel JG, Shen L, Sheng HZ, Granholm AC, Drago J, Grinberg A, Lee EJ, Huang SP, Saarma M, Hoffer BJ, Sariola H, Westphal H. Defects in enteric innervation and kidney development in mice lacking GDNF. *Nature*, 1996, 382, 73-76.

[38] Sanchez MP, Silos-Santiago I, Frisen J, He B, Lira SA, Barbacid M. Renal agenesis and the absence of enteric neurons in mice lacking GDNF. *Nature*, 1996, 382, 70-73.

[39] Cacalano G, Farinas I, Wang LC, Hagler K, Forgie A, Moore M, Armanini M, Phillips H, Ryan AM, Reichardt LF, Hynes M, Davies A, Rosenthal A. GFRalpha1 is an essential receptor component for GDNF in the developing nervous system and kidney. *Neuron*, 1998, 21, 53-62.

[40] Enomoto H, Araki T, Jackman A, Heuckeroth RO, Snider WD, Johnson EM Jr, Milbrandt J. GFR alpha1-deficient mice have deficits in the enteric nervous system and kidneys. *Neuron* 1998, 21, 317-324.

[41] Borrego S, Fernandez RM, Dziema H, Niess A, Lopez-Alonso M, Antinolo G, Eng C. Investigation of germline GFRA4 mutations and evaluation of the involvement of GFRA1, GFRA2, GFRA3, and GFRA4 sequence variants in Hirschsprung disease. *J Med Genet.*, 2003 40, e18.

[42] Rizzo C, Califano D, Colucci-D'Amato GL, De Vita G, D'Alessio A, Dathan NA, Fusco A, Monaco C, Santelli G, Vecchio G, Santoro M, de Franciscis V. Ligand stimulation of a Ret chimeric receptor carrying the activating mutation responsible for the multiple endocrine neoplasia type 2B. *J Biol Chem.*, 1996, 271, 29497-29501.

[43] Carlomagno F, Melillo RM, Visconti R, Salvatore G, De Vita G, Lupoli G, Yu Y, Jing S, Vecchio G, Fusco A, Santoro M. Glial cell line-derived neurotrophic factor differentially stimulates ret mutants associated with the multiple endocrine neoplasia type 2 syndromes and Hirschsprung's disease. *Endocrinology*, 1998, 139, 3613-3619.

[44] Le Hir H, Colucci-D'Amato LG, Charlet-Berguerand N, Plouin PF, Bertagna X, de Franciscis V, Thermes C. High levels of tyrosine phosphorylated proto-ret in sporadic phenochromocytomas. *Cancer Res.* 2000, 60, 1365-1370.

[45] Powers, J. F., Tsokas, P., and Tischler, A. S. The ret-Activating Ligand GDNF Is Differentiative and Not Mitogenic for Normal and Neoplastic Human Chromaffin Cells In Vitro. *Endocr. Pathol.*, 1998, 9, 325–331.

[46] Cerchia L, Libri D, Carlomagno MS, de Franciscis V. The soluble ectodomain of RetC634Y inhibits both the wild-type and the constitutively active Ret. *Biochem J.* 2003, 372, 897-903.

[47] Takahashi M, Iwashita T, Santoro M, Lyonnet S, Lenoir GM, Billaud M. Co-segregation of MEN2 and Hirschsprung's disease: the same mutation of RET with both gain and loss-of-function? *Hum Mutat.*, 1999,13, 331–336.

[48] Carlomagno F, Salvatore G, Cirafici AM, De Vita G, Melillo RM, de Franciscis V, Billaud M, Fusco A, Santoro M. The different RET-activating capability of mutations of cysteine 620 or cysteine 634 correlates with the multiple endocrine neoplasia type 2 disease phenotype. *Cancer Res.* 1997, 57, 391-395.

[49] Ito S, Iwashita T, Asai N, Murakami H, Iwata Y, Sobue G, Takahashi M. Biological properties of Ret with cysteine mutations correlate with multiple endocrine neoplasia

type 2A, familial medullary thyroid carcinoma, and Hirschsprung's disease phenotype. *Cancer Res.,* 1997, 57, 2870-2872.

[50] Marengere LE, Songyang Z, Gish GD, Schaller MD, Parsons JT, Stern MJ, Cantley LC, Pawson T. SH2 domain specificity and activity modified by a single residue. *Nature (Lond.),* 1994, 369, 502–505.

[51] Pandit SD, Donis-Keller H, Iwamoto T, Tomich JM, Pike LJ. The multiple endocrine neoplasia type 2B point mutation alters long-term regulation and enhances the transforming capacity of the epidermal growth factor receptor. *J Biol Chem.*, 1996, 271, 5850-5858.

[52] Asai N, Murakami H, Iwashita T, Takahashi M. A mutation at tyrosine 1062 in MEN2A-Ret and MEN2B-Ret impairs their transforming activity and association with shc adaptor proteins. *J Biol Chem.* 1996, 271, 17644-17649.

[53] Iwashita T, Asai N, Murakami H, Matsuyama M, Takahashi M. Identification of tyrosine residues that are essential for transforming activity of the ret proto-oncogene with MEN2A or MEN2B mutation. *Oncogene.*, 1996, 12, 481-487.

[54] Iwashita T, Kato M, Murakami H, Asai N, Ishiguro Y, Ito S, Iwata Y, Kawai K, Asai M, Kurokawa K, Kajita H, Takahashi M. Biological and biochemical properties of Ret with kinase domain mutations identified in multiple endocrine neoplasia type 2B and familial medullary thyroid carcinoma. *Oncogene.*, 1999, 18, 3919-3922.

[55] Salvatore D, Melillo RM, Monaco C, Visconti R, Fenzi G, Vecchio G, Fusco A, Santoro M. Increased in vivo phosphorylation of ret tyrosine 1062 is a potential pathogenetic mechanism of multiple endocrine neoplasia type 2B. *Cancer Res.*, 2001, 61, 1426-1431.

[56] Schmidt L, Duh FM, Chen F, Kishida T, Glenn G, Choyke P, Scherer SW, Zhuang Z, Lubensky I, Dean M, Allikmets R, Chidambaram A, Bergerheim UR, Feltis JT, Casadevall C, Zamarron A, Bernues M, Richard S, Lips CJ, Walther MM, Tsui LC, Geil L, Orcutt ML, Stackhouse T, Zbar B, et al. Germline and somatic mutations in the tyrosine kinase domain of the MET proto-oncogene in papillary renal carcinomas. *Nat Genet.*, 1997, 16, 68-73.

[57] Santoro MM, Penengo L, Minetto M, Orecchia S, Cilli M, Gaudino G. Point mutations in the tyrosine kinase domain release the oncogenic and metastatic potential of the Ron receptor. *Oncogene.* 1998, 17, 741-749.

[58] Miranda C, Zanotti G, Pagliardini S, Ponzetto C, Pierotti MA, Greco A. Gain of function mutations of RTK conserved residues display differential effects on NTRK1 kinase activity. *Oncogene.* 2002, 21, 8334-8339.

[59] Jhiang SM. The RET proto-oncogene in human cancers. *Oncogene*, 2000, 19, 5590-5597

[60] Takahashi M. The GDNF/RET signaling pathway and human diseases. *Cytokine Growth Factor Rev.* 2001, 12, 361-373.

[61] Manie S, Santoro M, Fusco A, Billaud M. The RET receptor: function in development and dysfunction in congenital malformation. *Trends Genet.*, 2001, 17, 580-589.

[62] Geneste O, Bidaud C, De Vita G, Hofstra RM, Tartare-Deckert S, Buys CH, Lenoir GM, Santoro M, Billaud M. Two distinct mutations of the RET receptor causing Hirschsprung's disease impair the binding of signalling effectors to a multifunctional docking site. *Hum Mol Genet.* 1999, 8, 1989-1999.

[63] Ishiguro Y, Iwashita T, Murakami H, Asai N, Iida K, Goto H, Hayakawa T, Takahashi M. The role of amino acids surrounding tyrosine 1062 in ret in specific binding of the shc phosphotyrosine-binding domain. *Endocrinology*. 1999, 140, 3992-3998.

[64] Jijiwa M, Fukuda T, Kawai K, Nakamura A, Kurokawa K, Murakumo Y, Ichihara M, Takahashi M. A targeting mutation of tyrosine 1062 in Ret causes a marked decrease of enteric neurons and renal hypoplasia. *Mol Cell Biol*. 2004, 24, 8026-8036.

[65] Pawson T. Specificity in signal transduction: from phosphotyrosine-SH2 domain interactions to complex cellular systems. *Cell* 2004, 116, 191-203.

[66] Simons K, & Toomre D. Lipid rafts and signal transduction. *Nat Rev Mol Cell Biol.* 2000, 1, 31-39.

[67] Paratcha G, Ledda F, Baars L, Coulpier M, Besset V, Anders J, Scott R, Ibanez CF. Released GFRalpha1 potentiates downstream signaling, neuronal survival, and differentiation via a novel mechanism of recruitment of c-Ret to lipid rafts. *Neuron.*, 2001, 29, 171-184.

[68] Hennige AM, Lammers R, Hoppner W, Arlt D, Strack V, Teichmann R, Machicao F, Ullrich A, Haring HU, Kellerer M. Inhibition of Ret oncogene activity by the protein tyrosine phosphatase SHP1. *Endocrinology*. 2001, 142, 4441-4447.

[69] Incoronato M, D'Alessio A, Paladino S, Zurzolo C, Carlomagno MS, Cerchia L, de Franciscis V. Cell Signal. 2004 Jul;16(7):847-56.

[70] Huang CS, Zhou J, Feng AK, Lynch CC, Klumperman J, DeArmond SJ, Mobley WC. Nerve growth factor signaling in caveolae-like domains at the plasma membrane. *J Biol Chem*. 1999, 274, 36707-36714.

[71] Mineo C, Gill GN, Anderson RG. Regulated migration of epidermal growth factor receptor from caveolae. *J. Biol. Chem.* 1999, 274, 30636–30643.

[72] Lorenzo MJ, Gish GD, Houghton C, Stonehouse TJ, Pawson T, Ponder BA, Smith DP. RET alternate splicing influences the interaction of activated RET with the SH2 and PTB domains of Shc, and the SH2 domain of Grb2. *Oncogene.*, 1997, 14, 763-771.

[73] Tsui-Pierchala BA, Ahrens RC, Crowder RJ, Milbrandt J, Johnson EM Jr. The long and short isoforms of Ret function as independent signaling complexes. *J Biol Chem.*, 2002, 277, 34618-34625.

[74] Segouffin-Cariou C, Billaud M. Transforming ability of MEN2A-RET requires activation of the phosphatidylinositol 3-kinase/AKT signaling pathway. *J Biol Chem.* 2000 275, 3568-3576.

[75] De Vita G, Melillo RM, Carlomagno F, Visconti R, Castellone MD, Bellacosa A, Billaud M, Fusco A, Tsichlis PN, Santoro M. Tyrosine 1062 of RET-MEN2A mediates activation of Akt (protein kinase B) and mitogen-activated protein kinase pathways leading to PC12 cell survival. *Cancer Res.* 2000, 60, 3727-3731.

[76] Murakami H, Iwashita T, Asai N, Shimono Y, Iwata Y, Kawai K, Takahashi M. Enhanced phosphatidylinositol 3-kinase activity and high phosphorylation state of its downstream signalling molecules mediated by ret with the MEN 2B mutation. *Biochem Biophys Res Commun.* 1999, 262, 68-75.

[77] D'Alessio A, Califano D, Incoronato M, Santelli G, Florio T, Schettini G, Carlomagno MS, Cerchia L, de Franciscis V. The tyrosine phosphatase Shp-2 mediates intracellular signaling initiated by Ret mutants. *Endocrinology.*, 2003, 144, 4298-4305.

[78] Qiao S, Iwashita T, Furukawa T, Yamamoto M, Sobue G, Takahashi M. Differential effects of leukocyte common antigen-related protein on biochemical and biological

activities of RET-MEN2A and RET-MEN2B mutant proteins. *J Biol Chem.,* 2001, 276, 9460-9467.

[79] Kurokawa K, Iwashita T, Murakami H, Hayashi H, Kawai K, Takahashi M. Identification of SNT/FRS2 docking site on RET receptor tyrosine kinase and its role for signal transduction. *Oncogene.*, 2001, 20, 1929-1938.

[80] Murakami H, Yamamura Y, Shimono Y, Kawai K, Kurokawa K, Takahashi M. Role of Dok1 in cell signaling mediated by RET tyrosine kinase*J Biol Chem.*, 2002, 277, 32781-32790.

[81] Gu H, & Neel BG. The "Gab" in signal transduction. *Trends Cell Biol.*,, 2003 13, 122-130.

[82] Grimm J, Sachs M, Britsch S, Di Cesare S, Schwarz-Romond T, Alitalo K, Birchmeier W J. Novel p62dok family members, dok-4 and dok-5, are substrates of the c-Ret receptor tyrosine kinase and mediate neuronal differentiation. *Cell Biol.,* 2001, 154, 345-354.

[83] Tahira T, Ishizaka Y, Itoh F, Sugimura T, Nagao M. Characterization of ret proto-oncogene mRNAs encoding two isoforms of the protein product in a human neuroblastoma cell line. *Oncogene.* 1990, 5, 97-102.

[84] Myers SM, Eng C, Ponder BA, Mulligan LM. Characterization of RET proto-oncogene 3' splicing variants and polyadenylation sites: a novel C-terminus for RET. *Oncogene*, 1995, 11, 2039-2045.

[85] Lorenzo MJ, Eng C, Mulligan LM, Stonehouse TJ, Healey CS, Ponder BA, Smith DP Multiple mRNA isoforms of the human RET proto-oncogene generated by alternate splicing. *Oncogene*, 1995, 10, 1377-1383.

[86] Ivanchuk SM, Myers SM, Mulligan LM. Expression of RET 3' splicing variants during human kidney development. *Oncogene*. 1998, 16, 991-996.

[87] Le Hir H, Charlet-Berguerand N, Gimenez-Roqueplo A, Mannelli M, Plouin P, de Franciscis V, Thermes C. Relative expression of the RET9 and RET51 isoforms in human pheochromocytomas. *Oncology*., 2000, 58, 311-318.

[88] Asai N, Iwashita T, Matsuyama M, Takahashi M. Mechanism of activation of the ret proto-oncogene by multiple endocrine neoplasia 2A mutations. *Mol Cell Biol.,* 1995, 15, 1613-1619.

[89] Rossel M, Pasini A, Chappuis S, Geneste O, Fournier L, Schuffenecker I, Takahashi M, van Grunsven LA, Urdiales JL, Rudkin BB, Lenoir GM, Billaud M. Distinct biological properties of two RET isoforms activated by MEN 2A and MEN 2B mutations. *Oncogene.*, 1997, 14, 265-275.

[90] Pasini A, Geneste O, Legrand P, Schlumberger M, Rossel M, Fournier L, Rudkin BB, Schuffenecker I, Lenoir GM, Billaud M. Oncogenic activation of RET by two distinct FMTC mutations affecting the tyrosine kinase domain. *Oncogene*. 1997, 15, 393-402.

[91] Nozaki C, Asai N, Murakami H, Iwashita T, Iwata Y, Horibe K, Klein RD, Rosenthal A, Takahashi M. Calcium-dependent Ret activation by GDNF and neurturin. *Oncogene.*, 1998, 16, 293-299.

[92] Myers SM, Mulligan LM. The RET receptor is linked to stress response pathways. *Cancer Res.* 2004, 64, 4453-4463.

[93] de Graaff E, Srinivas S, Kilkenny C, D'Agati V, Mankoo BS, Costantini F, Pachnis V. Differential activities of the RET tyrosine kinase receptor isoforms during mammalian embryogenesis. *Genes Dev.* 2001, 15, 2433-2444

[94] Lee DC, Chan KW, Chan SY. RET receptor tyrosine kinase isoforms in kidney function and disease. *Oncogene*, 2002, 21, 5582-5592.

[95] Tsui-Pierchala BA, Milbrandt J, Johnson EM Jr. NGF utilizes c-Ret via a novel GFL-independent, inter-RTK signaling mechanism to maintain the trophic status of mature sympathetic neurons. *Neuron*, 2002, 33, 261-273.

[96] Peterson S, Bogenmann E. The RET and TRKA pathways collaborate to regulate neuroblastoma differentiation. *Oncogene*, 2004, 23, 213-225.

[97] Le Hir H, Charlet-Berguerand N, de Franciscis V, Thermes C. 5'-End RET splicing: absence of variants in normal tissues and intron retention in pheochromocytomas. *Oncology*. 2002;63, 84-91.

[98] Hedenfalk I, Duggan D, Chen Y, Radmacher M, Bittner M, Simon R, Meltzer P, Gusterson B, Esteller M, Kallioniemi OP, Wilfond B, Borg A, Trent J. Gene-expression profiles in hereditary breast cancer. *N Engl J Med.,* 2001, 344, 539-548.

[99] Yeoh EJ, Ross ME, Shurtleff SA, Williams WK, Patel D, Mahfouz R, Behm FG, Raimondi SC, Relling MV, Patel A, Cheng C, Campana D, Wilkins D, Zhou X, Li J, Liu H, Pui CH, Evans WE, Naeve C, Wong L, Downing JR. Classification, subtype discovery, and prediction of outcome in pediatric acute lymphoblastic leukemia by gene expression profiling. *Cancer Cell,* 2002, 1, 133-143

[100] Jain S, Watson MA, DeBenedetti MK, Hiraki Y, Moley JF, Milbrandt J. Expression profiles provide insights into early malignant potential and skeletal abnormalities in multiple endocrine neoplasia type 2B syndrome tumors. *Cancer Res.*, 2004, 64, 3907-3913.

[101] Watanabe T, Ichihara M, Hashimoto M, Shimono K, Shimoyama Y, Nagasaka T, Murakumo Y, Murakami H, Sugiura H, Iwata H, Ishiguro N, Takahashi M. Characterization of gene expression induced by RET with MEN2A or MEN2B mutation. *Am J Pathol.*, 2002, 161, 249-256.

[102] Kurokawa K, Kawai K, Hashimoto M, Ito Y, Takahashi M. Cell signalling and gene expression mediated by RET tyrosine kinase. *J Intern Med.,* 2003, 253, 627-633.

[103] Michiels FM, Chappuis S, Caillou B, Pasini A, Talbot M, Monier R, Lenoir GM, Feunteun J, Billaud M. Development of medullary thyroid carcinoma in transgenic mice expressing the RET protooncogene altered by a multiple endocrine neoplasia type 2A mutation. *Proc Natl Acad Sci U S A.*, 1997, 94, 3330-3335.

[104] Reynolds L, Jones K, Winton DJ, Cranston A, Houghton C, Howard L, Ponder BA, Smith DP. C-cell and thyroid epithelial tumours and altered follicular development in transgenic mice expressing the long isoform of MEN 2A RET. *Oncogene*, 2001, 20, 3986-3994.

[105] Baetscher M, Schmidt E, Shimizu A, Leder P, Fishman MC. SV40 T antigen transforms calcitonin cells of the thyroid but not CGRP-containing neurons in transgenic mice. *Oncogene*, 1991, 6, 1133-1138.

[106] Kawai K, Iwashita T, Murakami H, Hiraiwa N, Yoshiki A, Kusakabe M, Ono K, Iida K, Nakayama A, Takahashi M. Tissue-specific carcinogenesis in transgenic mice expressing the RET proto-oncogene with a multiple endocrine neoplasia type 2A mutation. *Cancer Res.,* 2000, 60, 5254-5260.

[107] Treanor JJ, Goodman L, de Sauvage F, Stone DM, Poulsen KT, Beck CD, Gray C, Armanini MP, Pollock RA, Hefti F, Phillips HS, Goddard A, Moore MW, Buj-Bello A,

Davies AM, Asai N, Takahashi M, Vandlen R, Henderson CE, Rosenthal A. Characterization of a multicomponent receptor for GDNF. *Nature*, 1996, 382, 80-83.

[108] Smith-Hicks CL, Sizer KC, Powers JF, Tischler AS, Costantini F. C-cell hyperplasia, pheochromocytoma and sympathoadrenal malformation in a mouse model of multiple endocrine neoplasia type 2B. *EMBO J.*, 2000, 19, 612-622.

[109] Sweetser DA, Froelick GJ, Matsumoto AM, Kafer KE, Marck B, Palmiter RD, Kapur RP Oncogene. 1999 Jan 28;18(4):877-86.

[110] Gestblom C, Sweetser DA, Doggett B, Kapur RP. Sympathoadrenal hyperplasia causes renal malformations in Ret(MEN2B)-transgenic mice. *Am J Pathol.*, 1999, 155, 2167-2179.

[111] Skinner MA, Kalyanaraman S, Safford SD, Heuckeroth RO, Tourtellotte W, Goyeau D, Goodfellow P, Milbrandt JD, Freemerman A. A human yeast artificial chromosome containing the multiple endocrine neoplasia type 2B Ret mutation does not induce medullary thyroid carcinoma but does support the growth of kidneys and partially rescues enteric nervous system development in Ret-deficient mice. *Am J Pathol.*, 2005, 166, 265-274.

[112] Skinner MA, Safford SD, Freemerman AJ. RET tyrosine kinase and medullary thyroid cells are unaffected by clinical doses of STI571. *Anticancer Res.*, 2003, 23, 3601-3606.

[113] Cohen MS, Hussain HB, Moley JF. Inhibition of medullary thyroid carcinoma cell proliferation and RET phosphorylation by tyrosine kinase inhibitors. *Surgery*, 2002, 132, 960-966.

[114] Strock CJ, Park JI, Rosen M, Dionne C, Ruggeri B, Jones-Bolin S, Denmeade SR, Ball DW, Nelkin BD. CEP-701 and CEP-751 inhibit constitutively activated RET tyrosine kinase activity and block medullary thyroid carcinoma cell growth. *Cancer Res.,* 2003, 63, 5559-5563.

[115] Carlomagno F, Vitagliano D, Guida T, Ciardiello F, Tortora G, Vecchio G, Ryan AJ, Fontanini G, Fusco A, Santoro M. ZD6474, an orally available inhibitor of KDR tyrosine kinase activity, efficiently blocks oncogenic RET kinases. *Cancer Res.*, 2002, 62, 7284-7290.

[116] Carlomagno F, Vitagliano D, Guida T, Napolitano M, Vecchio G, Fusco A, Gazit A, Levitzki A, Santoro M. The kinase inhibitor PP1 blocks tumorigenesis induced by RET oncogenes. *Cancer Res.*, 2002, 62, 1077-1082.

[117] Carniti C, Perego C, Mondellini P, Pierotti MA, Bongarzone I. PP1 inhibitor induces degradation of RETMEN2A and RETMEN2B oncoproteins through proteosomal targeting. *Cancer Res.*, 2003, 63, 2234-2243.

[118] Yano L, Shimura M, Taniguchi M, Hayashi Y, Suzuki T, Hatake K, Takaku F, Ishizaka Y. Improved gene transfer to neuroblastoma cells by a monoclonal antibody targeting RET, a receptor tyrosine kinase. *Hum Gene Ther.*, 2000, 11, 995-1004.

[119] Salvatore G, Nagata S, Billaud M, Santoro M, Vecchio G, Pastan I. Generation and characterization of novel monoclonal antibodies to the Ret receptor tyrosine kinase. Biochem *Biophys Res Commun.*, 2002, 294, 813-817.

[120] Drosten M, Hilken G, Bockmann M, Rodicker F, Mise N, Cranston AN, Dahmen U, Ponder BA, Putzer BM Role of MEN2A-derived RET in maintenance and proliferation of medullary thyroid carcinoma. *J Natl Cancer Inst.,* 2004, 96, 1231-1239.

[121] Cerchia L, Duconge F, Pestourie C, Boulay J, Aissouni Y, Gombert K, Tavitian B, de Franciscis V, Libri D. Neutralizing Aptamers from Whole-Cell SELEX Inhibit the RET Receptor Tyrosine Kinase. *PLoS Biol.*, 2005, 3, e123

[122] Cerchia L, Hamm J, Libri D, Tavitian B, de Franciscis V. Nucleic acid aptamers in cancer medicine. *FEBS Lett.*, 2002, 528, 12-16.

In: Trends in Cellular Signaling
Editor: Dave E. Caplin, pp. 111-141

ISBN 1-59454-633-9

Chapter IV

SIGNALING TO HYPOXIA-INDUCIBLE FACTOR-1 ACTIVATION AND ITS ROLE IN THE PATHOGENESIS OF DISEASES

Mayuko Osada-Oka, Satoshi Akiba and Takashi Sato*
Department of Pathological Biochemistry, Kyoto Pharmaceutical University Misasagi, Yamashina-ku, Kyoto 607-8414, Japan.

ABSTRACT

Hypoxia-inducible factor-1 (HIF-1) is a transcription factor responsible for hypoxia and up-regulates more than 100 target genes to adapt to oxygen deprivation. However, even under non-hypoxic conditions HIF-1 is activated by a number of stimuli and accelerates gene expression to affect cellular function. A variety of signal transduction pathways are involved in hypoxia- or extracellular stimuli-induced HIF-1 activation in a stimulus and cell type-specific manner. Furthermore, HIF-1 is implicated in the pathogenesis of some diseases.

ABBREVIATION

HIF-1: hypoxia-inducible factor-1
VEGF: vascular endothelial growth factor
PDGF: platelet-derived growth factor
pVHL: von Hippel-Lindau tumor-seppressor protein
MAPK: mitogen-activated protein kinase
PI3K: phosphatidylinositol 3-kinase

* Correspondence concerning this article should be addressed to Takashi Sato, Department of Pathological Biochemistry, Kyoto Pharmaceutical University, Misasagi, Yamashina-ku, Kyoto 607-8414, Japan. Tel.: +81-75-595-4655; Fax.: +81-75-595-4759; E-mail: sato@mb.kyoto-phu.ac.jp.

PKC: protein kinase C
HRE: hypoxia-responsive element
ARNT: aryl hydrocarbon receptor
N-TAD: N-terminal transactivation domain
C-TAD: C-terminal transactivation domain
CREB: cAMP responsive element binding protein
CBP: CREB binding protein
ODD domain: oxygen-dependent degradation domain
PHD: HIF-1α prolyl hydroxylase
Mdm2: mouse double minute 2
NLS: nuclear localization signal
Hsp90: heat shock protein 90
ERK: extracellular signal-regulated kinase
NF-κB: nuclear factor-κB
ROS: reactive oxygen specis
MAFP: methyl arachidonyl fluorophosphonate
GLUT1: glucose transporter 1
GSK: glycogen synthase kinase
HSF-1: heat shock factor-1

1. Introduction

It is well known that under oxygen deprivation, hypoxia-inducible factor-1 (HIF-1), a transcription factor for adaptation to hypoxia, is stabilized and up-regulates a number of oxygen-sensitive genes involved in propagating energy metabolism, ventilation of blood flow, and erythropoiesis, such as erythropoietin and vascular endothelial growth factor (VEGF). HIF-1 is a heterodimeric protein consisting of two subunits: the regulatory HIF-1α and the constitutively expressed HIF-1β. HIF-1α is rapidly degraded under normoxic conditions by E3 ubiquitin ligase complex containing von Hippel-Lindau tumor-suppressor protein (pVHL), whereas under hypoxia, HIF-1α is not degraded and is stabilized to up-regulate gene expression. On the other hand, HIF-1α gene expression and the protein levels are increased under non-hypoxic conditions by a variety of stimuli including growth factors, cytokines, hormones, thrombin, angiotensin II, and high concentrations of glucose.

The expression and activation of HIF-1α under both hypoxia and normoxia are known to be regulated by a variety of kinases including mitogen-activated protein kinase (MAPK), phosphatidylinositol 3-kinase (PI3K), and protein kinase C (PKC), and by other factors including NADPH-cytochrome P-450 reductase (NPR), intracellular Ca^{2+}, and phospholipases. However, the involvement of these signaling pathways to HIF-1 activation differs with the stimuli and cell types used [1].

A number of observations have implicated HIF-1 in the pathogenesis of diseases. In local regions of damage due to disease, including inflammation, tumor, atherosclerosis, and renal disease, oxygen deprivation occurs, which results in the stabilization of HIF-1, thus leading to pathogenetic gene expression.

On the basis of these observations together with our findings obtained recently, we will focus in this review on the signal transduction pathway required for HIF-1 activation followed by gene expression under hypoxic and normoxic conditions, and the implication of these signaling pathways in diseases (Fig. 9).

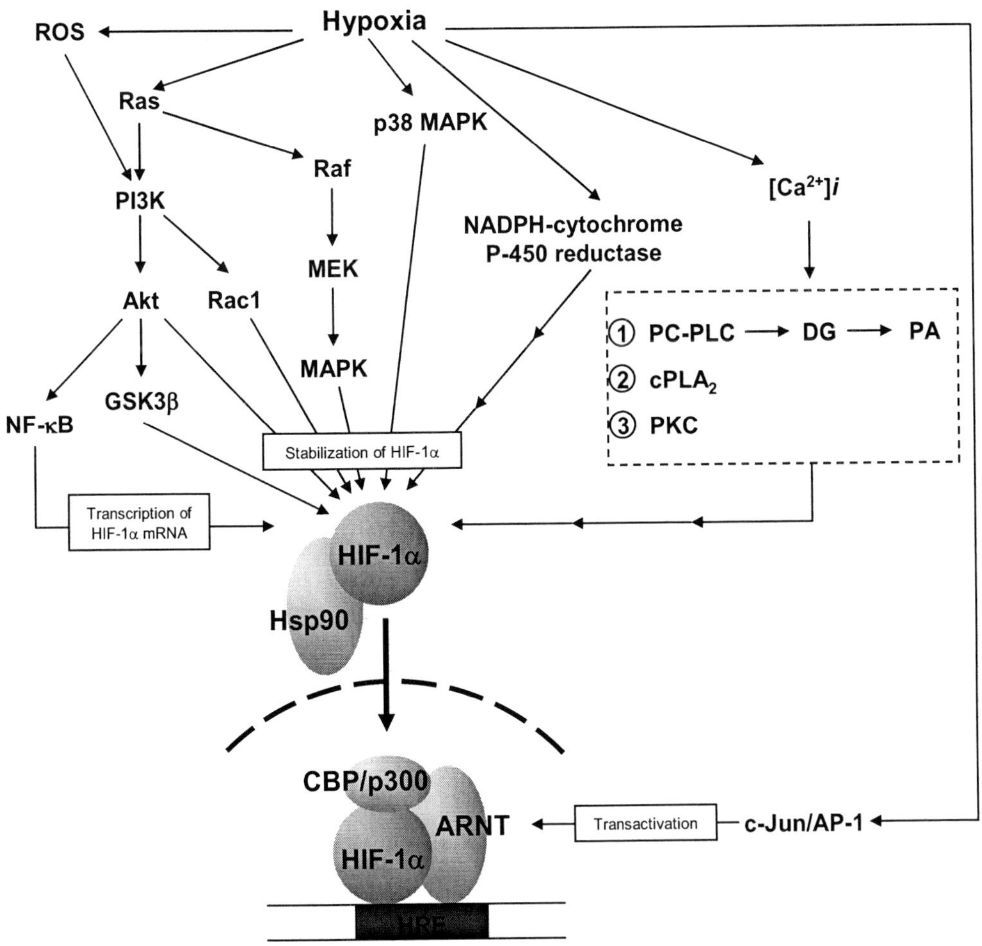

Fig. 9. Schematic representation of the putative signaling pathway lading to HIF-1 activation under hypoxic conditions.

2. Expression and Transcriptional Activation of HIF-1

2-1. Response of HIF-1 to Oxygen in Mammalian Cells

Oxygen is essential to all aerobic organisms because it serves as the terminal electron acceptor in mitochondrial oxidative phosphorylation and because several enzymatic processes require molecular oxygen as a substrate. Even slight reductions in normal oxygen concentrations (hypoxia) can cause the induction of specific genes involved in mammalian oxygen homeostasis such as erythropoiesis in individuals who are anemic or at high attitude, neovascularization in ischemic myocardium, and glycolysis in cells cultured at a reduced O_2 tension (Table). Hypoxia-inducible gene products that participate in these responses include erythropoietin [2], VEGF [3], and glycolytic enzymes [4,5]. A hypoxia-inducible enhancer was defined in the 3'-flanking sequence of the *EPO* gene, and then a hypoxia-induced DNA-binding protein HIF-1, binding specifically to site 1 of the hypoxia-inducible enhancer was identified [2]. Furthermore, HIF-1-binding sites are present in several hypoxia-inducible genes, and there is a consensus sequence (5'-RCGTG-3') for binding HIF-1 termed the hypoxia-responsible element (HRE). In addition to the *epo* gene, HIF-1 binds to cis-acting binding sites located within the 5'-flanking region of the *glut1* gene [6,7]. Moreover, HIF-1 has a role in the induction of VEGF and its receptor Flt-1 (VEGFR-1), which stimulate angiogenesis [8-10]. HIF-1 also stimulates another angiogenic mediator, endothelin-1 [11], platelet-derived growth factor (PDGF) [12], and basic fibroblast growth factor [13].

Recently, the existence of a common hypoxia signal transduction pathway and a central role for HIF-1 in the transcriptional regulation of hypoxia-responsive genes, were reported. HIF-1 is an αβ heterodimer, the subunits of which belong to the rapidly growing PAS (Per-ARNT-AHR-Sim) family of basic helix-loop-helix (bHLH) transcription factors [14]. The β subunit is identical to the 91, 93 and 94 kDa aryl hydrocarbon nuclear translocator (ARNT) that also serves as a heterodimeric partner of the aryl hydrocarbon receptor. ARNT which is constitutively expressed at the protein level in the nucleus is not significantly affected by oxygen, and acts as a common subunit of multiple bHLH-PAS family proteins [15]. In contrast, it appears that the 120 kDa α subunit is remarkably unstable in cells exposed to oxygen, whereas hypoxia induces a striking increase in the abundance of HIF-1α protein. These observations suggest that the accumulation of HIF-1α is a prerequisite to the activation of HIF-1 and lead to the hypothesis that the signal transduction under hypoxic conditions depends primarily on hypoxia-induced stabilization of HIF-1α, which is otherwise rapidly degraded in oxygenated cells. HIF-1α has two independent minimal transcriptional activation domains; one designated the N-terminal transactivation domain (N-TAD) (AA531-575), and the other the C-terminal transactivation domain (C-TAD) (AA 786-826). These two activation domains behaved differently in response to hypoxia. While N-TAD became stabilized under hypoxic conditions, C-TAD was modified by Ref-1 to interact with cAMP responsive element binding protein (CREB) binding protein (CBP)/p300 [16]. It was shown that unstable HIF-1α is controlled by an oxygen-dependent degradation (ODD) domain (AA390-603) within HIF-1α and degradation is catabolized through the ubiquitin-proteasome pathway in a manner that depends on an E3 ubiquitin ligase complex containing pVHL [17]. The *VHL* gene is characterized by a predisposition to develop tumors of the kidney, central nervous system, retina, pancreas, and adrenal gland, and, further, inactivation of this gene is responsible for most sporadic renal cell carcinomas of the clear cell type. Indeed, pVHL is composed of two domains, α and β; the α domain is required for binding elongin C, which bridges pVHL to elongin B, Cul2 and Rbx1, and the β domain is a substrate-docking interface that is required for binding to a region of HIF-1α located in the ODD domain. Thus, this interaction promotes the rapid degradation of HIF-1α.

Although how cells sense changes in ambient oxygen and how HIF-1α expression stabilizes under hypoxic conditions were a central question in biology, the mechanism of the hypoxic induction of HIF-1α has recently been identified. Under normoxic conditions, specific HIF-1α prolyl hydroxylase (PHD) hydroxylates two proline resides (Pro402 and Pro564) in the ODD domain of HIF-1α, and then pVHL E3 ubiquitin ligase complex associates with the hydroxylated proline residues. Moreover, Leu-574 of HIF-1α is essential for the binding of pVHL to the C-terminal ODD (AA 556-572), because it brings a loss of pVHL binding and gain of protein stability in spite of the presence of Pro-564 [18]. Furthermore, in recent studies the asparaginyl hydroxylase FIH-1 (factor inhibiting HIF) has been found to hydroxylate Asp803 of C-TAD in HIF-1α [19]. Thus, there are at least two major steps involved in the hypoxic induction of HIF-1α protein. The strong constitutive activities of the proteins with both proline and asparagine substitutions confirm that the relevant prolyl and asparaginyl hydroxylases are attractive targets for the therapeutic regulation of HIF-1α. In addition, HIF-1α contains two nuclear localization signals (NLSs), which is localized to the N-terminal (AA 17-33) and the C-terminal parts (718-721). The C-terminal NLS (NLS-C) motif of HIF-1α plays a crucial role in mediating hypoxia-inducible nuclear import of the protein, whereas the N-terminal NLS (NLS-N) motif may be less important (see the HIF-1α structure in Fig. 1).

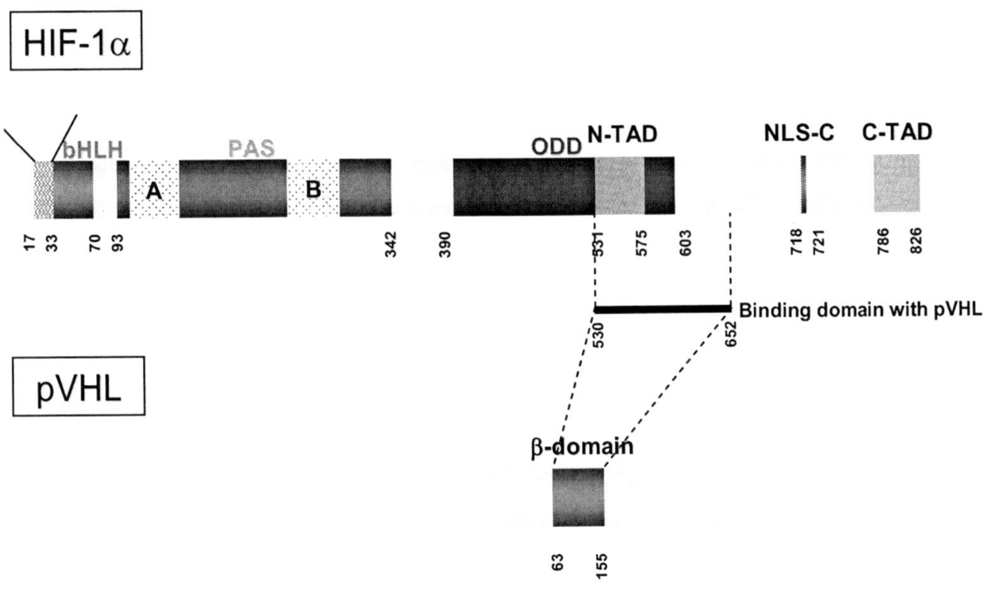

Fig. 1. Structure of HIF-1α, which is basic-helix-loop-helix-PAS protein. It is shown the number of amino acid under box, and A and B, internal homology units within the PAS domain.

2-2. Transcription Factors of HIF Family

Recently, newer members of the alpha family of hypoxia-responsive proteins have been identified. Endothelial PAS protein 1 (EPAS-1) or HIF-2α was originally described in endothelial cells and found to stimulate transcription from the hypoxia-responsive VEGF promoter in a very wide range of cell types [20]. HIF-2α shares 48% sequence homology with HIF-1α and also dimerizes with ARNT. HIF-2α as well as HIF-1α is subject to pVHL-dependent destruction by proteasome [17], and the sequence 517-682 of HIF-2α is indeed sufficient for interaction with pVHL [21].Although the prolyl residues for enzyme-mediated oxygen-dependent hydroxylation differ between HIF-1α and HIF-2α, the domain of HIF-2α required for its activation during hypoxia is an internal domain that extends from amino acids 450-571 and shares homology with the ODD domain of HIF-1α [22,23]. Expression of HIF-1α and HIF-2α and their effects on survival under hypoxia were studied in six human breast cancer cell lines, and it was found that HIF-2α was absent or weekly expressed in the more aggressive cell line compared with the expression of HIF-1α at various levels [24].

Several alternatively spliced variants of HIF-1α mRNA have been found in human cells. Two splicing events were described with alternative splicing between exons1 and 2 and also between exons13 and 14 [25]. The former variant had an additional 3 bp at the junction of exons 1 and 2 without a frame shift. The latter had lost exon 14, resulting in a frame shift and an immediate termination codon, and had most of the essential domains of HIF-1α conserved. Therefore it showed oxygen-tension-dependent regulation and transactivation by the VEGF promoter. Moreover, it was identified as a new variant lacking exons 11 and 12, which produces a frame shift and introduces a shorter form of HIF-1α [26]. This deduced protein contains a bHLH/PAS domain and the NLS-N motif, but is deprived of a part of the ODD domain, both the TADs and NLS-C motif. Thus, the expressed HIF-1α isoform exhibited neither transactivation activity nor hypoxia-induced nuclear translocation, and was demonstrated to work as a dominant-negative isoform to inhibit endogenous HIF-1α.

Another homologous polypeptide, a HIF-3α sequence was cloned from a mouse EST cDNA clone [27]. The N-terminal domain of this protein shares 57% and 53% identity in amino acid sequence with HIF-1α and HIF-2α. A proline residue at position 490 in human HIF-3α has also been reported to be a target for oxygen-regulated prolyl hydroxylation and VHL-mediated ubiquitilation *in vitro* and *in vivo* [28]. An alternative splice variants of HIF-3α gene (hHIF-3α1-6) including IPAS (hHIF-3α2), which lacks the transactivation domain, can act as a dominant inhibitor of the HIF response and seems to be important in suppressing the hypoxic response in certain tissues, such as the cornea [29].

2-3. Degradation and Stabilization of HIF-1α

No mammalian O_2-sensing factor has been conclusively shown to respond directly to O_2 *in vivo* or *in vitro*. Much attention has focused on the HIF-1α protein because unstable HIF-1α was controlled by an ODD domain within HIF-1α. Recently, biochemical studies revealed mammalian 'oxygen sensors', that is, HIF-prolyl-hydroxylases (PHD) which require O_2, iron and oxoglutarate as cofactors for hydroxylation of the proline residue at 564 in HIF-1α [30-32].

PHDs were initially identified as homologous proteins from a conserved HIF-pVHL-prolyl hydroxylase pathway in *Caenorhabditis elegans* Egl-9 [30]. In mammalian cells, there are three isoforms, PHD1, PHD2, and PHD3 [33], and their enzymes belong to the 2-oxoglutarate-dependent dioxygenase superfamily. While PHD1 mRNA is expressed at low levels, PDH2 mRNA is abundantly expressed in all tissues with highest relative expression levels in testis. Relative expression levels of PHD3 mRNA are low to moderate in most tissues but relatively high in the heart and the placenta. It seems likely that there are more specialized and tissue-specific functions between in HIF-2α and HIF-3α in PHDs. It was also shown that PHDs have distinct patterns of subcellular localization [34]. PHD1 is exclusively present in the nucleus, whereas PHD2 is mainly located in the cytoplasm and PHD3 is homogeneously distributed in cytoplasm and nucleus [35]. Moreover, PHD2 mRNA and PHD3 mRNA is hypoxia-inducible, but the expression of PHD1 mRNA is independent of oxygen [35]. It is proposed that HIF-1α can be hydroxylated in the cytoplasm as well as in the cell nucleus, and can be degraded pVHL-dependently [36].

Several reports have suggested pVHL-independent HIF-1α degradation. Although p53 levels are regulated in large part by Mdm2 (mouse double minute 2) ubiquitin E3 ligase, p53 also interacts with HIF-1α and this interaction is mediated by Mdm2 [37,38]. The tumor suppressor, wild-type p53, is expressed at low levels in most cells, but the p53 protein is stabilized in response to various forms of stress such as hypoxia as well as DNA damage and inappropriate oncogene signaling [39,40]. Thus, HIF-1α is degraded by the interaction of p53 under hypoxic conditions. Although hypoxia induces two major forms of HIF-1α: both phosphorylated and dephosphorylated HIF-1α, the dephosphorylated HIF-1α is the major form that binds to p53 and the phosphorylated form preferentially binds to ARNT [41].

Recent studies have also demonstrated that the molecular chaperone heat shock protein 90 (Hsp90) was associated with degradation of HIF-1α independent of the pVHL pathway [42,43], although a definitive biological role for this interaction remains elusive. A study using renal cell carcinomas which lack a functional pVHL, found that HIF-1α protein was highly expressed under normoxia in untreated cells but was down-regulated in a dose-dependent manner in response to a Hsp90 inhibitor, geldanamycin [44]. Geldanamycin is a naturally occurring anasamycin antibiotic that possesses antitumor properties by virtue of its ability to associate and interfere with Hsp90. Since geldanamycin mediates dissociation of the PAS domain B of HIF-1α from Hsp90, HIF-1α is degraded via a pVHL-independent pathway. When renal cell carcinomas were treated with geldanamycin in the presence of the specific proteasome inhibitor, HIF-1α was markedly ubiquitinated in the cells prior to its degradation. Interestingly, even if two proline sites (Pro-402 and Pro564) in HIF-1α were mutated, ubiquitination was detected in the mutated HIF-1α species as well as the wild-type. Geldanamycin selectively suppressed the emergence of dephosphorylated HIF-1α, resulting in inhibition of p53 induction [41]. This raises the possibility that the stability of dephosphorylated HIF-1α may be regulated by Hsp90. Moreover, we also demonstrated that apigenin, a flavonoid, interfered with the interaction of Hsp90 with HIF-1α, and induced to a degree, HIF-1α expression under hypoxic conditions [45]. Interestingly, we observed similar findings even in renal cell carcinomas, which express HIF-2α but not HIF-1α (unpublished data). This means that Hsp90 is important for the stabilization of not only HIF-1α but also HIF-2α.

2-4. Gene Induction Mimicking Hypoxia

It is known that cobalt, nickel, and manganese salts induce the stabilization and transactivation of HIF-1α under normoxic conditions. The biochemical mechanism by which cobalt stabilizes HIF-α was recently elucidated [30,46]. As the group of PHDs has an iron binding-center and iron is critical for its enzymatic activity, the iron can be replaced by cobalt at this site, resulting in inactivation of the hydroxylase activity [30]. Moreover, it was demonstrated that cobalt also stabilizes HIF-1α proteins by binding directly to the ODD and inhibits the interaction between hydroxylated HIF-1α and pVHL [46]. In addition, it was shown that cobalt inhibits the HIF-pVHL interaction even after the hydroxylation of a proline residue (Pro-564). Like HIF-1α, HIF-2α proteins are also stabilized by cobalt [46], because cobalt inhibited the hydroxylation of the proline residue (Pro-531) within the ODD domain of HIF-2α. Iron chelators, such as desferrioxamine, are also capable of mimicking hypoxia. It is proposed that iron chelators can remove iron from the iron-binding center of the enzyme, and thus the iron chelator desferrioxamine most likely inhibits the hydroxylation of HIF-1α by chelating the iron required for activity of the PHDs.

2-5. Role of HIF-1α in Development

Embryonic stem cells lacking HIF-1α are unable to up-regulate several HIF target genes under hypoxic conditions, with the most severe effect seen on several enzymes in the glucolytic pathway [47]. In addition, embryos of HIF-1α-deficient mice arrested in development by E9.0 and died at E10.5 with cardiac, vascular, branchial arch, and neural tube defects and extensive cell death, especially in the branchial and cephalic regions. HIF-1α protein expression in wild-type embryos increased between E8.5 and E9.5, the time at which malformations became apparent in embryos of the HIF-1α-deficient mice. Therefore, the function of HIF-1α is biologically essential and important in the regulation of both development and cellular proliferation.

3. Factors that Participate in the Hypoxic Signaling Cascade

3-1. MAPK Family

Although the precise mechanism underlying signaling cascades leading to hypoxia-induced HIF-1α stabilization is still controversy, several reports have demonstrated that members of the MAPK family, extracellular signal-regulated kinase (ERK) and p38 MAPK, play an important role in the accumulation of HIF-1α protein in response to hypoxia (for a review see Ref. [48]). An *In vitro* study with fibroblasts provided evidence that ERK1/2 (p42/p44 MAPK), but not p38 MAPK or c-Jun N-terminal kinase (JNK), directly phosphorylates HIF-1α to induce an apparent shift of molecular mass in the electrophoretic mobility of HIF-1α [49]. Actually, it is reported in cell systems that HIF-1α is phosphorylated via an ERK-dependent pathway, and also that activation of the ERK pathway is essential for HIF-1-dependent transcriptional activity using a specific ERK pathway inhibitor PD98059 [49,50] or cells expressing an ERK1 dominant negative mutant [51]. In addition to ERK, hypoxia also induced p38 MAPK activation, which is involved in the acceleration of hypoxia-mediated adhesion of vascular smooth muscle cells to fibronectin or collagen via activation of β_1-integrin [52]. Furthermore, hypoxia induced expression of MAPK phosphatase-1, and suppression of the expression using the siRNA technique stimulated HIF-1α phosphorylation as well as transcriptional activity of HIF-1 in human hepatoma HepG2 cells [53]. This may also imply the importance of the MAPK pathway in hypoxia-induced responses. However, the physiological meaning of the phosphorylation of HIF-1α remains obscure, because it was reported that inhibition of HIF-1α phosphorylation does not have any effect on its stabilization under hypoxia and on its DNA binding activity [51,54].

3-2. PI3K/Akt

There is evidence that the PI3K/Akt pathway, but not ERK or p38 MAPK pathway, is involved in the signaling cascade required for hypoxia-dependent HIF-1 stabilization. It has been shown in Ras-transformed NIH3T3cells that exposure to hypoxia causes Ras-dependent activation of PI3K without activation of ERK or JNK/stress-activated protein kinase, and the transcriptional activity of HIF-1 is suppressed by an inhibitor of PI3K or a mutant of the p85 subunit of PI3K, suggesting that the PI3K/Akt pathway is involved in signaling for HIF-1-mediated gene expression [55]. A similar finding was reported with H-*ras*-transformed fibroblasts in which PI3K participated in the accumulation of HIF-1α proteins leading to an up-regulation of HIF-1-mediated *glut1* expression [56].

Interestingly, a mechanism whereby nuclear factor-κB (NF-κB) is implicated in PI3K/Akt pathway-dependent HIF-1α stabilization under hypoxic conditions was reported [57]. In human hepatoma Hep3B cells, a selective inhibitor of PI3K or catalytically inactive or dominant negative Akt abolished NF-κB- and HIF-1-mediated *trans*activation, and, further, inhibition of NF-κB using a selective inhibitor or dominant negative IκB resulted in inhibition of hypoxia-induced HIF-1α protein accumulation and HIF-1-mediated gene expression. These observations suggest that NF-κB participates downstream of PI3K/Akt in the hypoxia signal transduction pathway.

Rac 1, a small GTPase, is shown to be activated by hypoxia leading to HIF-1α protein accumulation, and the Rac 1 activation is inhibited by a specific inhibitor of PI3K, suggesting that Rac 1 is one of the signaling molecules acting downstream of PI3K/Akt activation [58]. Hsp90 is known to interact with HIF-1 to stabilize it by a mechanism independent of the prevention of pVHL-dependent ubiquitination [42,43]. In this respect, it was revealed that in pVHL-deficient renal cell carcinoma, the PI3K/Akt pathway is implicated in the stabilization of HIF-1α [59]. In the present study, suppression of the PI3K/Akt pathway by a specific inhibitor or by the transfection of dominant negative mutants of PI3K or Akt resulted in a decrease in HIF-1 protein with a decrease in Hsp90, suggesting that the PI3K/Akt signaling pathway up-regulates hypoxia-induced HIF-1α stabilization via stimulation of Hsp90 expression.

In addition, reactive oxygen species (ROS), which is generated by hypoxia at mitochondrial complex III, is reported to be a trigger for hypoxia-mediated accumulation of HIF-1α protein [60]. In this study, a specific inhibitor of PI3K attenuated HIF-1α protein accumulation and also H_2O_2-mediated accumulation, suggesting that PI3K acts downstream of ROS and is a signaling molecule required for ROS-mediated HIF-1α stabilization.

Taken together, these findings suggest that PI3K/Akt play a pivotal role in the signal transduction to stabilize HIF-1α under hypoxic conditions.

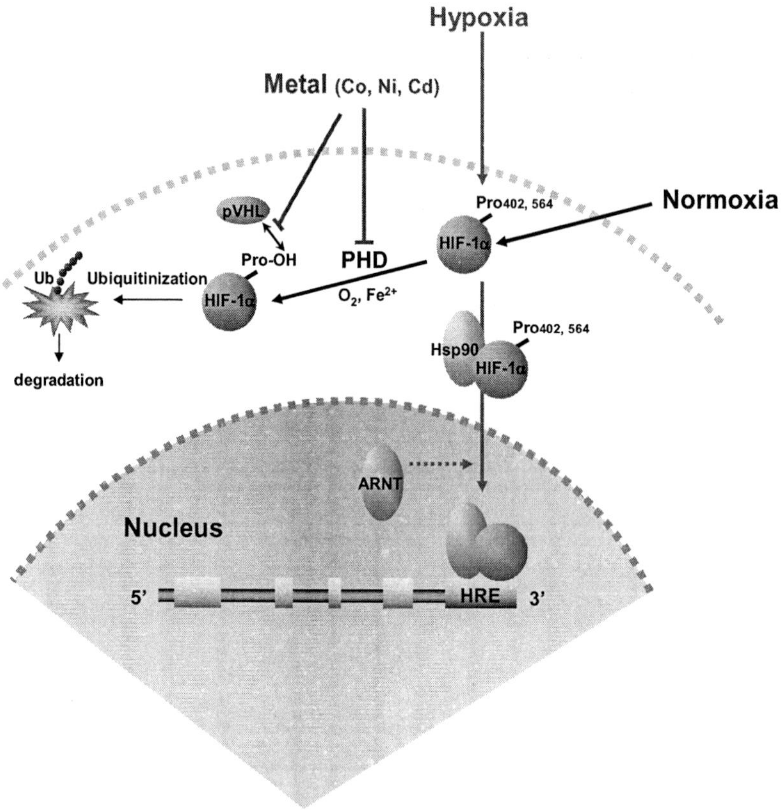

Fig. 2. Stabilization of HIF-1α under hypoxic conditions or by various metals, and degradation of HIF-1α under normoxic conditions.

3-3. NADPH-Cytochrome P-450 Reductase

Osada-Oka, one of the authors of this review, and co-authors reported recently that NADPH-cytochrome P-450 reductase regulates hypoxia-induced gene expression through HIF-1α activation [61]. As shown in Fig. 3, when Hep3B cells were exposed to hypoxia, erythropoietin mRNA expression (Fig. 3 A) and HIF-1α protein accumulation (Fig. 3 B) were increased and the increases were inhibited by diphenyleneiodonium chloride, an inhibitor of NADPH-dependent enzyme. Furthermore, in NADPH-cytochrome P-450 reductase - deficient cells prepared by transfection of the antisense cDNA, hypoxia-induced erythropoietin mRNA expression was abolished (Fig. 3 C). Conversely, overexpression of NADPH-cytochrome P-450 reductase in the cells enhanced HIF-1 transcriptional activity in a luciferase reporter gene assay (Fig. 3 D). This is the first report to demonstrate that NADPH-cytochrome P-450 reductase is an oxygen-sensitive trigger leading to HIF-1α activation followed by target gene expression under hypoxic conditions.

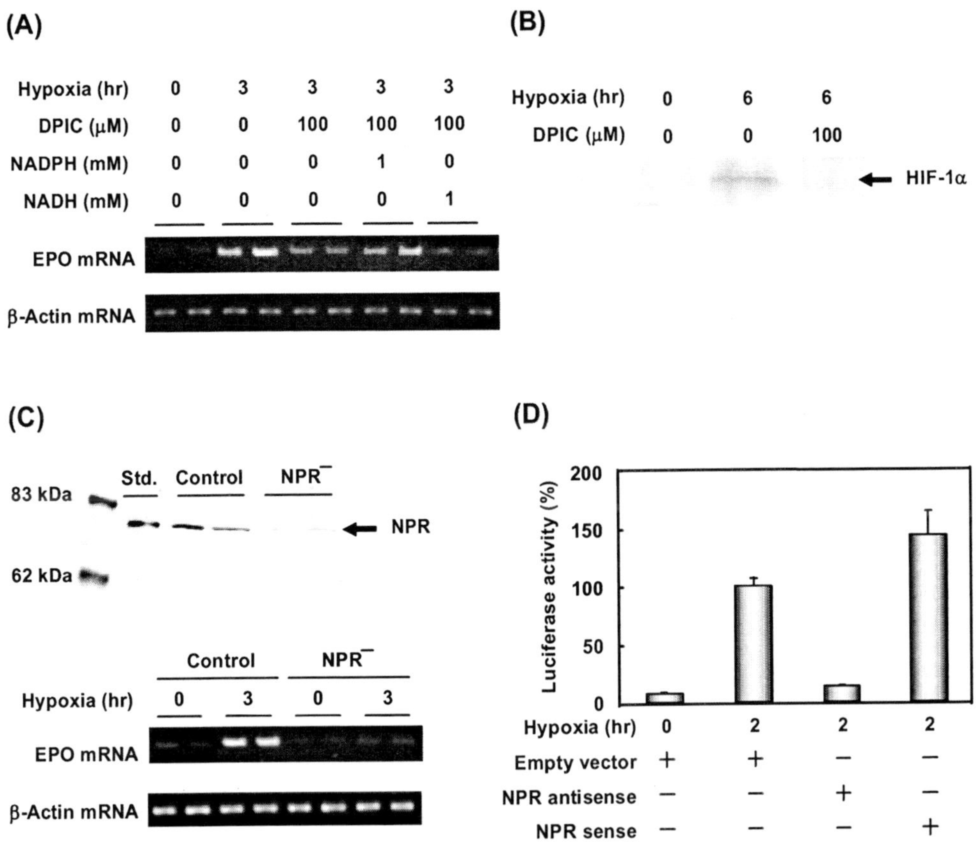

Fig. 3. Involvement of NADPH-P450 reductase (NPR) in the induction of erythropoietin (EPO) under hypoxia. (A) Hep3B cells were exposed to hypoxia (1% O_2, 5% CO_2) for 3 hours in the presence of DPIC (100 μM), DPIC plus NADPH (1 mM), or DPIC plus NADH (1 mM). The expression of EPO mRNA was analyzed by RT-PCR. (B) Hep3B cells were exposed to hypoxia (1% O_2, 5% CO_2) for 6 hours in the presence of DPIC (100 μM). Whole cell lysates were subjected to immunoblotting with the anti-HIF-1α antibody. (C) Hep3B cells were lipofected with pcDNA-NPR antisense (NPR$^-$) or an empty vector (Control). Whole cell lysates were subjected to immunoblotting with the anti-NPR antibody. Purified human NPR was used as a standard (Std.). The cells were exposed to hypoxia (1% O_2, 5% CO_2) for 3 hours. The expression of EPO mRNA was analyzed by RT-PCR. (D) Hep3B cells were co-transfected with a reporter plasmid containing the HIF-1-binding domain of the human EPO gene (HRE) and a control plasmid together with pcDNA-NPR antisense, pcDNA-NPR sense, or an empty vector. The cells were exposed to hypoxia (1% O_2, 5% CO_2) for 2 hours. The activity of luciferase in the cell lysates was measured. The ratio (reporter/control luciferase activity) obtained from the cell lysates under hypoxia was set to 100%. Each value represents the mean ± S.D. of six separate experiments.

3-4. PKC

PKC is one of the key substances regulating cellular responses in the signal transduction pathway. In hypoxic signaling, several isoforms of PKCs are reported to be involved. In cardiac myocytes, hypoxia stimulated activation of PKCα and PKCε, as assessed by translocation from the cytosol to particulate fraction [62]. In Hep3B cells, hypoxia-induced erythropoietin production was shown to be prevented by a PKC inhibitor, long-term treatment with phorbol ester or PKCα antisense oligonucleotides, suggesting a role for PKCα in the signaling pathway [63]. In experiments in which hypoxia up-regulated angiopoietin-2 as well as VEGF expression in mouse mesangial cells, a PKC inhibitor, but not MAPK inhibitor, prevented the mRNA expression, also indicating the participation of PKC in hypoxic signal transduction required for gene expression [64].

3-5. Intracellular Ca^{2+}

Several investigators have reported that intracellular Ca^{2+} increased under hypoxic conditions is involved in hypoxic signaling required for HIF-1-dependent gene expression. Treatment with a Ca^{2+} chelater BAPTA (1,2-bis(2-aminophenoxy)ethane-N,N,N',N'-tetraacetic acid-acetoxymethyl ester) of human fetal kidney cell line 239 cells, inhibited hypoxia-induced VEGF mRNA expression [65]. The findings that the Ca^{2+}-ionophore A23187 induced HIF-1α mRNA expression in HepG2 cells [66], and that the ionophore ionomycin stimulated HIF-1 transcriptional activity [67], indicate that increased Ca^{2+} levels participate in hypoxia-mediated HIF-1 transactivation. In this respect, we demonstrated that A23187 induces the accumulation of HIF-1α and thrombospondin-1, a factor stimulating cell proliferation and growth (see 4-1 Atherosclerosis), in coronary artery smooth muscle cells (Fig. 6, A and B).

Although the precise mechanism by which elevated Ca^{2+} elicits HIF-1α activation is still uncertain, it is suggested that PKC [62,68,69] or ERK [66,67] acts downstream from the elevated Ca^{2+}, thus an increase in intracellular Ca^{2+} results in an up-regulation of PKC or ERK activation followed by HIF-1-dependent gene expression under hypoxia. On the other hand, since an increase in intracellular Ca^{2+} is known to lead to activation of cytosolic phospholipase A_2, we examined whether or not cytosolic phospholipase A_2 acts downstream of Ca^{2+} signaling, resulting in the transcriptional activation of HIF-1. However, the results in Fig. 4 showed that A23187-induced thrombospondin -1 mRNA expression is not affected by a cytosolic phospholipase A_2 inhibitor, methyl arachidonyl fluorophosphonate (MAFP), indicating that cytosolic phospholipase A_2 activation does not contribute to the signal transduction pathway downstream of an increase in Ca^{2+}.

3-6. Phospholipase C/ Diacylglycerol and Phospholipase A_2

It has been shown that hypoxia stimulates an increase in diacylglycerol formation in cardiac myocytes [62] or in Hep3B cells [70]. Exposure of cardiac myocytes to hypoxia is shown to result in the activation of PKCα and PKCε as mentioned above, which was inhibited by an inhibitor of phospholipase C, [62], suggesting a role for diacylglycerol in hypoxic responses. Recently, another role for phospholipase C/ diacylglycerol independent of PKC activation in hypoxia-induced HIF-1α activation was demonstrated in Hela cells and others [71,72]. Treatment of cells with a specific inhibitor of diacylglycerol kinase or phosphatidylcholine- phospholipase C results in the prevention of HIF-1α accumulation and HIF-1 transcriptional activation upon hypoxic stimulation. Therefore, the phospholipase C/

diacylglycerol / phosphatidic acid pathway is suggested to play an important role in the signaling required for HIF-1α activation.

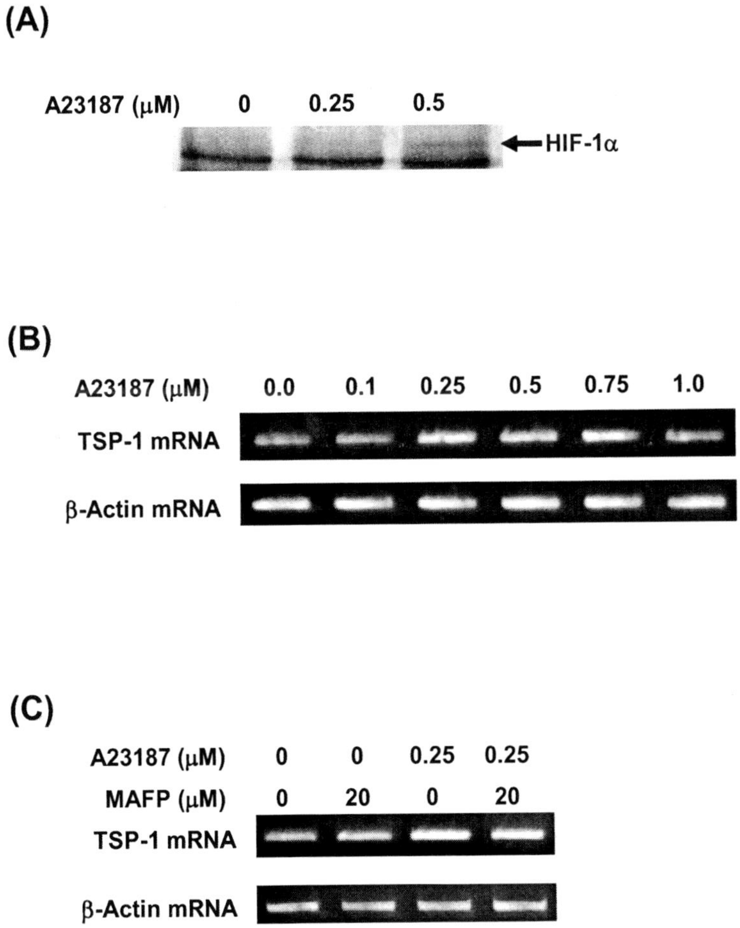

Fig. 4. Calcium ionophore induces the expression of thrombospondin-1 (TSP-1) mRNA through the up-regulation of HIF-1α in coronary artery smooth muscle cells. (A) Coronary artery smooth muscle cells were stimulated with or without A23187 (0.25 or 0.5 μM), a calcium ionophore, for 6 hours under normoxic conditions. Whole cell lysates were subjected to immunoblotting with the anti-HIF-1α antibody. (B) Coronary artery smooth muscle cells were stimulated with or without A23187 (0.1, 0.25, 0.5, 0.75, or 1.0 μM) for 2 hours under normoxic conditions. The expression of TSP-1 mRNA was analyzed by RT-PCR.

We observed the involvement of cytosolic phospholipase A_2 in hypoxic responses. As shown in Fig. 5, when human mesangial cells were exposed to hypoxia, HIF-1α protein accumulation and glucose transporter1 (GLUT1) mRNA expression were increased. Furthermore, GLUT1 gene expression was prevented apparently by a cytosolic phospholipase A_2 inhibitor, MAFP, but HIF-1α accumulation was not. These data indicate, therefore, that cytosolic phospholipase A_2 or a cytosolic phospholipase A_2-derived metabolite has a role in the signal transduction pathway downstream of HIF-1α, although whether or not cytosolic phospholipase A_2 activation occurs in response to hypoxia remains to be elucidated. It has been demonstrated that hypoxia stimulates cyclooxygenase-2 expression via NF-κB, not via HIF-1α, in vascular endothelial cells [73,74] or in pulmonary artery smooth muscle cells [75]. Considering these findings together with our data we suggest that arachidonic acid metabolites associated with the cytosolic phospholipase A_2-cyclooxygenase-2 pathway might be involved in the hypoxia-mediated signal transduction pathway responsible for HIF-1-dependent gene expression.

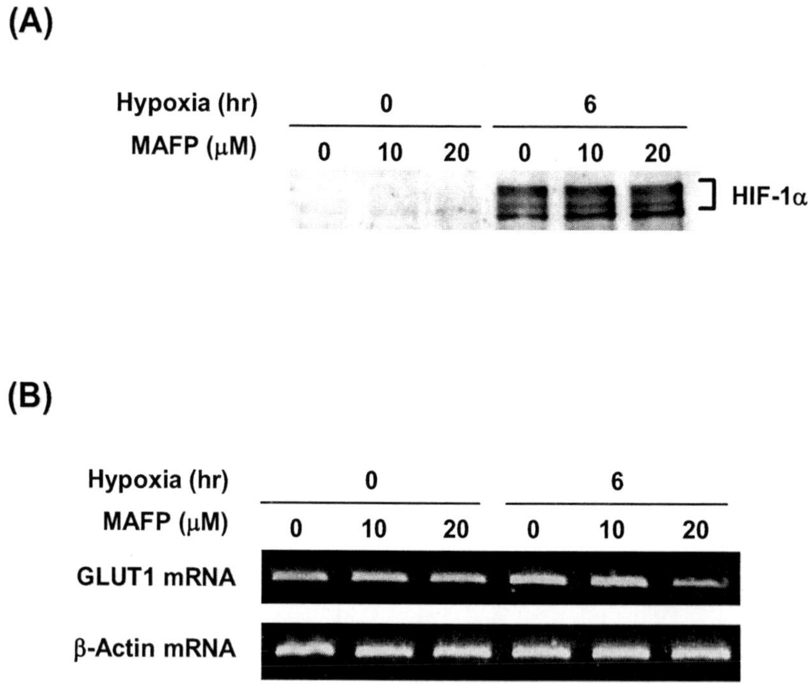

Fig. 5. Effects of $cPLA_2$ inhibitor (MAFP) on the expression of HIF-1α protein and GLUT1 mRNA under hypoxic conditions in human mesangial cells. Human mesangial cells were exposed to hypoxia (1% O_2, 5% CO_2) for 6 hours in the presence or absence of MAFP (10 or 20 μM). (A) Whole cell lysates were subjected to immunoblotting with the anti-HIF-1α antibody. (B) The expression of GLUT1 mRNA was analyzed by RT-PCR.

3-7. Other Factors

The importance of the reduced form of glutathione to the activation of HIF-1α was reported in fetal distal lung epithelium [76]. In the present study, an increase in reduced glutathione by *N*-acetylcysteine, a glutathione precursor, stimulated HIF-1α accumulation and DNA binding, whereas lowering the ratio of reduced/oxidized glutathione that is derived from increased oxidized glutathione abolished HIF-1α activation. These results suggest that glutathione is a key regulator for hypoxia-stimulated signal transduction that leads to the stabilization of HIF-1α.

Glycogen synthase kinase (GSK) 3β that regulates glycogen synthase activity and also the activation of transcription factors such as AP-1, is reported to act as a negative regulator for HIF-1α accumulation [77,78]. Exposure of HepG2 cells to a short period of hypoxia led to the stabilization of HIF-1α which is dependent on the PI3K/Akt pathway, whereas prolonged periods of hypoxia (more than 16 hours), led to a reduction in HIF-1α protein. Prolonged hypoxia attenuated activities of Akt as well as GSK3β, and the specific inhibition of GSK3β restored HIF-1α accumulation. Based on these observations, it is suggested that exposure of cells to a short period of hypoxia results in the activation of Akt followed by phosphorylation of GSK3β that is inactive, whereas exposure to prolonged hypoxia results in inactivation of Akt with the concomitant activation of GSK3b, which in turn induces degradation of HIF-1α. Therefore, GSK3β is implicated in the signaling pathway responsible for the degradation of HIF-1α.

c-Jun, a transcription factor and constituent of AP-1, is known to be phosphorylated in response to several stimuli such as growth factors and cytokines, and to express transcriptional activity in association with other transcription factors such as c-Fos. Recently, the regulatory role in the transcriptional activity of HIF-1 under low oxygen conditions was examined using a dominant negative mutant of c-Jun [79] or *HIF-1α* nullizygous mouse embryonic fibroblasts [80]. The results suggested that c-Jun or c-Jun/AP-1 cooperates with HIF-1 to express its transcriptional activity. Therefore, c-Jun/AP-1 is a factor potentiating the transactivation of HIF-1. Similarly, it is reported that CBP has an enhancing effect on HIF-1-dependent VEGF expression through interaction with HIF-1 in liver and other tissues of diabetic rats [81].

4. Factors that Participate in Signaling Required for Agonist-Dependent HIF-1α Activation During Normoxia

As described above, under normoxic conditions HIF-1α protein is rapidly degraded by the ubiquitin-proteasome pathway. However, numerous findings show that even under normoxia extracellular stimuli increase the accumulation of HIF-1α as well as HIF-1 transcriptional activation, which leads to hypoxia-mediated HIF-1-dependent gene expression such as VEGF. Several factors involved in the agonist-induced signaling pathway that leads to the accumulation of HIF-1α have been characterized.

4-1. MAPK

The G protein-coupled receptor encoded by the Kaposi's sarcoma-associated herpes virus is shown to be constitutively activated and to enhance VEGF expression via HIF-1α activation [82]. In this study, HIF-1α was phosphorylated and the inhibitor of p38MAPK or MAPK kinase inhibited HIF-1α transactivating activity and VEGF expression. In a study of diabetic retinopathy advanced glycation end product (AGE) that increases in patients with diabetes mellitus, HIF-1α protein accumulation and HIF-1 DNA binding activity increased leading to VEGF expression in retinal epithelial cells [83]. In addition, the AGE-induced HIF-1α accumulation was inhibited by an inhibitor of ERK activation but not by an inhibitor of PI3K. These findings suggest that ERK, p38 MAPK or both are factors stimulating agonist-induced signaling to activate HIF-1α. Although the mechanism underlying the kinase-induced stabilization of HIF-1α remains obscure, the molecular mechanism by which p42/44 MAPK, but not PI3K, enhances the stabilization of HIF-1α was proposed in a study using hepatitis B virus X protein that is implicated in hepatocarcinogenesis [84]. The authors suggested that the p42/44 MAPK signaling pathway does not reduce the prolyl hydroxylation of HIF-1α that leads to pVHL binding to and the resultant degradation of HIF-1α, but enhances the association of HIF-1α with CBP, resulting in a potentiation of HIF-1α transcriptional activity.

4-2. PI3K/Akt

Like hypoxic stimulation, the PI3K/Akt signaling pathway has been implicated in HIF-1α expression and subsequent transcriptional activation even under normoxic conditions. It has been reported that overexpression of HER2, a receptor identified in human breast cancer, or stimulation of MCF-7 breast cancer cells with heregulin, the ligand for HER2, results in the expression of HIF-1α protein, which requires activation of the PI3K/Akt signaling pathway [85]. A different signaling pathway is reported to be involved in VEGF expression in *ras*-transformed cells and to be cell-type specific [86]. Up-regulation of VEGF in *ras*-transformed epithelial cells was dependent on the PI3K signaling pathway, while that in *ras*-transformed fibroblasts was dependent on the ERK pathway. On the other hand, both PI3K and ERK pathways are reported to be involved in HIF-1α expression [87]. In human colon carcinoma cells, stimulation with insulin-like growth factor-1 (IGF-1) potentiated HIF-1α protein synthesis and subsequent VEGF mRNA expression, and these responses were prevented by inhibitors of PI3K or ERK signaling pathways.

4-3. ROS

Several lines of evidence have revealed that ROS is implicated in HIF-1α protein accumulation in cooperation with other signaling molecules such as PI3K in response to extracellular stimuli. Exposure of human vascular smooth muscle cells to thrombin or PDGF enhanced HIF-1α accumulation and HIF-1 transcriptional activity leading to the expression of VEGF and plasminogen activator inhibitor-1 (PAI-1), and further, thrombin-induced responses were inhibited by $p22^{phox}$ antisense oligonucleotides or inhibitors of p38 MAPK and PI3K [88]. The authors suggest that ROS generated by stimulation up-regulates p38 MAPK and PI3K signaling pathways, resulting in enhancement of the induction of HIF-1α protein synthesis and HIF-1-dependent target genes. Interestingly, it was reported that in rat vascular smooth muscle cells, stimulation with angiotensin II increases HIF-1α mRNA expression via a mechanism dependent on PKC activation but not on ROS generation, whereas angiotensin II-induced HIF-1α protein accumulation is dependent on the generation of ROS and subsequent activation of the PI3K signaling pathway [89,90]. Similar observations were reported indicating that stimulation of human macrophages with oxidized low-density lipoproteins (oxLDL) elicits HIF-1α protein expression in a ROS-dependent manner as assessed by the inhibition of the expression by several antioxidants, but oxLDL-stimulated HIF-1α mRNA expression is independent of the generation of ROS [91]. These findings suggest that ROS is involved in the mechanism responsible for the translation of agonist-induced HIF-1α protein synthesis and not for the gene expression of HIF-1α.

4-4. Unidentified Factors

There is much evidence that external stimuli induce HIF-1α accumulation and HIF-1-dependent gene expression, although the signaling pathways that are involved in such responses have not been identified. In this regard, it is reported that insulin increased mRNA levels of several glycolytic enzymes, glucose transporter and VEGF, and that HIF-1α/ARNT transcription complex is implicated in the expression of mRNAs in human HepG2 cells and other cell systems [92]. Interleukin-1 and tumor necrosis factor-α are reported to elicit an increase in HIF-1α mRNA expression, HIF-1 DNA binding, and HIF-1-driven reporter gene expression [93,94]. These results suggest that HIF-1 might participate in inflammation.

Under hypoxi conditions, VEGF is a potent mediator responsible for angiogenesis for tumor cell growth. However, it is reported that even during normoxia stimulant-induced VEGF expression plays a role in tumor angiogenesis. In ovarian carcinoma cells, endothelin-1 increased HIF-1α accumulation and HIF-1 transcriptional activity as well as VEGF mRNA and protein levels [95]. This suggests that endothelin-1-induced HIF-1α activation mediates tumor cell growth.

5. Involvement of Hypoxia in Diseases

5-1. Atherosclerosis

In atherosclerotic regions it is demonstrated that areas of decreased oxygen concentration exist in the artery of experimental atherosclerotic animals, and therefore it is possible that hypoxia in the arterial wall might be implicated in the atherogenesis of atherosclerosis. Although it seems difficult to characterize the actual hypoxic region, interesting findings were reported using an immunofluorescence technique with a hypoxia marker, theophyline derivative: hypoxic zones existed at depth in the atherosclerotic plaque in experimental rabbit atherosclerosis [96]. The authors suggest that the thickened arterial wall impairs oxygen diffusion and also foam cells highly consume oxygen. Along this line, considerable *in vitro* effects of hypoxia were revealed. 25-hydroxycholesterol, known to exist in oxLDL, induced the secretion of interluekin-8 in human monocyte-derived macrophages, and hypoxia enhanced the effect [97]. Low oxygen concentrations were reported to stimulate the uptake of LDL by arterial smooth muscle cells and accumulation of cholesterol ester in the cells [98]. Furthermore, it was demonstrated that stimulation of human macrophages with oxLDL elicits HIF-1α protein expression [91].

On the other hand, thrombospondin-1, a platelet-derived matricellular protein, is revealed to be associated with atherosclerotic lesions [99], occluded grafts [100], and vascular smooth muscle cell hyperplasia [101]. These findings together with the observation of hypoxia in atherosclerotic regions, led us to investigate the involvement of thrombospondin-1 in atherosclerosis and also the effect of hypoxia on thrombospondin-1 induction in human coronary artery smooth muscle cells. As shown in Fig. 6, hypoxia increased thrombospondin-1 mRNA and protein levels time-dependently (Fig. 6 A, B). Moreover, in HIF-1α knock-down cells prepared by lipofection of a plasmid expressing Hif-1α-antisense (Fig. 6 C), the induction of thrombospondin-1 mRNA expression under hypoxic conditions was suppressed (Fig. 6 D). We suggest that thrombospondin-1 is a factor atherogenesis and is regulated by hypoxia in atherosclerosis. Thus, these observations suggest that hypoxia that exists in atherosclerotic regions might be responsible for the initiation and development of atherosclerosis.

5-2. Renal Disease

In renal fibrosis, excess collagen production or extracellular matrix (ECM) accumulation contributes to progression of the disease. The accumulation of ECM is caused by increased production and decreased turnover, and hypoxia is known to participate in the accumulation. In this regard, it was demonstrated that in human renal fibroblasts, hypoxia increases the expression of tissue inhibitor of metalloproteinase (TIMP), which decreases ECM turnover, via stimulation of HIF-1 transcriptional activity and that the PKC- and tyrosine kinase-mediated signaling pathway up-regulates TIMP mRNA expression [102]. A similar observation was made in rat proximal tubular cells indicating that hypoxia induces increases in smooth muscle actin and collagen I that lead to the progression of renal fibrosis, although the signaling pathway involved has nor been established [103]. Connective tissue growth factor is also a factor in the development of renal fibrosis, and hypoxic induction of the gene expression of this factor was revealed to be HIF-1α-dependent using tubular epithelial cells from HIF-1α knockout mice [104]. Osteopontin is a matrix protein be implicated in the development of renal fibrosis and glomerulonephritis, and expressed under hypoxic conditions via the PKC or p38 MAPK signaling pathway in rat mesangial cells [69].

As a transcription factor responsible for tissue ischemia, heat shock factor-1 (HSF-1) in addition to HIF-1 is activated by stress, and a role for these two transcription factors was reported in renal ischemia. *In vivo* experiments in rats subjected to renal ischemia or *in vitro* experiments with renal epithelial cells demonstrated a rapid activation of the two factors, and suggested that they might act to promote or to repair injury from renal ischemia either together or separately [105]. These results indicate the importance not only of HIF-1 but also of HSF-1 in renal ischemia such as acute renal failure.

Although a number of factors are implicated in the progression of diabetic nephropathy, VEGF which is known to be involved in the pathogenesis of diabetic retinopathy is suggested to be a factor promoting diabetic nephropathy. In addition, it was demonstrated that a high concentration of glucose up-regulates VEGF mRNA and protein expression and promotes cell proliferation in rat mesangial cells [106]. However, whether or not HIF-1α is involved in the glucose-induced VEGF expression remains unknown. Therefore, we examined this possibility using human mesangial cells. The results shown in Fig. 7 revealed that a high concentration (30 mM) of glucose induced a time-dependent increase in VEGF mRNA expression (Fig. 5 A), and also that it stabilized the HIF-1α protein (Fig. 7 B). We suggest that high glucose-induced HIF-1α activation and subsequent VEGF expression are implicated in the pathogenesis and development of diabetic nephropathy.

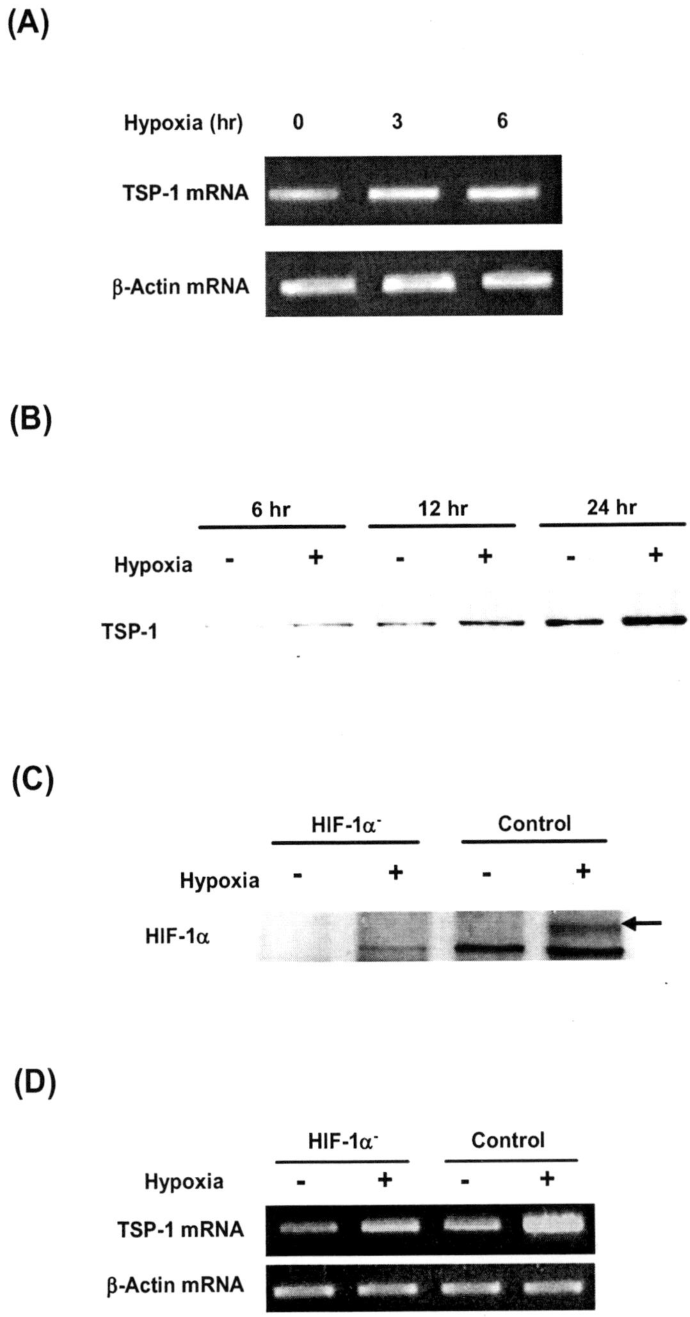

Fig. 6 The expression of thrombospondin-1 (TSP-1) mRNA and protein in HIF-1α-deficient coronary artery smooth muscle cells under hypoxic conditions. (A, B) Coronary artery smooth muscle cells were exposed to hypoxia (1% O_2, 5% CO_2) for 3 or 6 hours. (A) The expression of TSP-1 mRNA (A) and protein in the culture medium (B) was analyzed by RT-PCR and immunoblotting, respectively. (C, D) Coronary artery smooth muscle cells lipofected with pcDNA-HIF-1α antisense (HIF-1α⁻) or an empty vector (Control) were exposed to hypoxia (1% O_2, 5% CO_2) for 6 hours. The expression of HIF-1α protein (C) and TSP-1 mRNA (D) was analyzed by immunoblotting and RT-PCR, respectively.

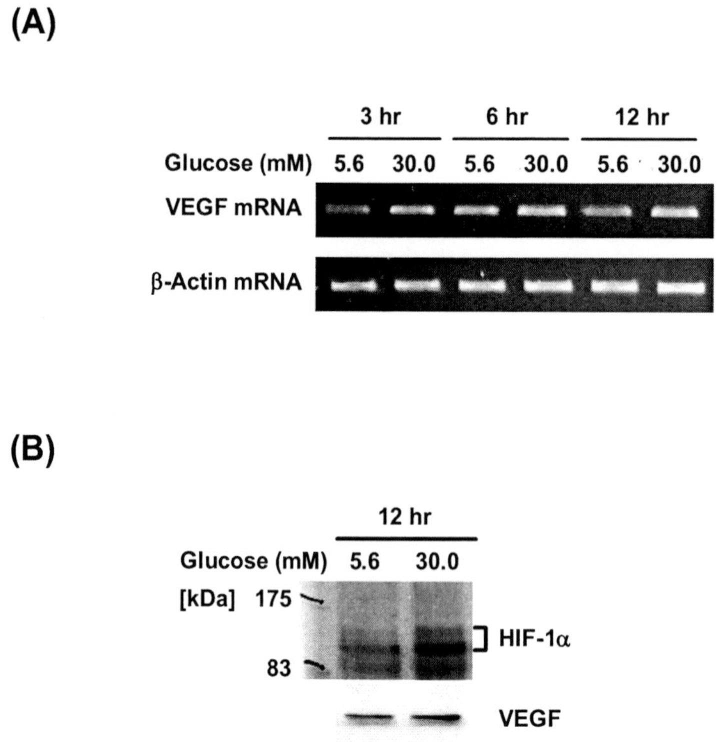

Fig. 7. A high concentration of glucose induces the expression of VEGF mRNA through the up-regulation of HIF-1α in human mesangial cells. (A) Human mesangial cells were exposed to a normal (5.6 mM) or high (30.0 mM) concentration of glucose for 3, 6, or 12 hours. The expression of VEGF mRNA was analyzed by RT-PCR. (B) Human mesangial cells were exposed to a normal (5.6 mM) or high (30.0 mM) concentration of glucose for 12 hours. Whole cell lysates were subjected to immunoblotting with antibodies against HIF-1α or VEGF.

5-3. Tumor

Many human tumors contain hypoxic microenvironments, due to uncontrolled cell growth and insufficient vascularization. Oxygen tension in several cancer types measured using the Eppendorf histograph electrode was significantly lower than that in normal tissues [107]. Actually, tissue oxygen in tumors had 10 to 30 mmHg of partial pressure, whereas in normal tissues the range was 24 to 66 mmHg [108]. In human tumors, HIF-1α is overexpressed as a result of intratumoral hypoxia [109], and HIF-induced angiogenesis contributes to cell growth in solid tumors, malignant progression and metastasis [110]. Thus, HIF-1 plays a critical role in cancer [111-113,109]. VEGF, which is correlated with metastasis and mortality, is believed to be one of the most important mediators of hypoxia in tumor tissues, and its expression is largely controlled by HIF-1α [114]. Recently, we found that NADPN-cytochrome P-450 reductase regulates erythropoietin expression via HIF-1 activation and translocation to the nucleus in Hep3B cells under hypoxic conditions [61]. Furthermore, we examined whether or not NADPN-cytochrome P-450 reductase is overexpressed in human bladder tumors and NADPN-cytochrome P-450 reductase levels are correlated with HIF-1α levels [61]. The results in Fig. 8 showed that high levels of both HIF-1α and NADPN-cytochrome P-450 reductase are observed in human tumor tissue (Fig. 8 A), and the expression of NADPN-cytochrome P-450 reductase mRNA is well correlated with that of VEGF mRNA (Fig. 8 B). We suggest that NADPN-cytochrome P-450 reductase is the protein triggering the signal transduction pathway responsible for tumor angiogenesis.

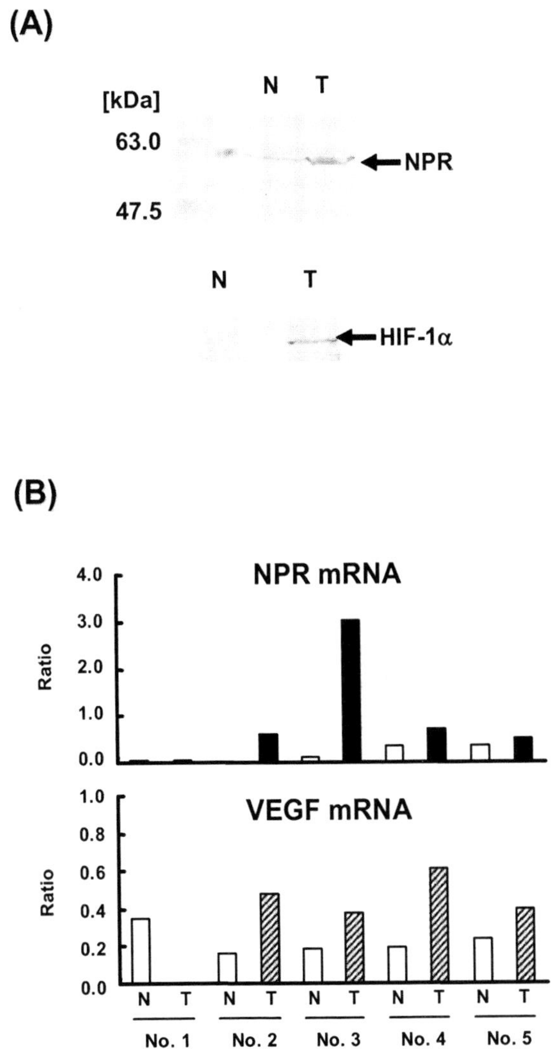

Fig. 8. Expression of NADPH-P450 reductase (NPR) is stronger in tumor regions than in normal regions in human bladder tissues. (A) Proteins of normal (N) or tumor (T) regions were analyzed by immunoblotting with antibodies against HIF-1α or NPR. (B) The normal (N) or tumor (T) regions were obtained from five patients with bladder tumors. The expression of NPR mRNA and VEGF mRNA was analyzed by RT-PCR. Results were expressed as the ratio of NPR mRNA or VEGF mRNA to β-actin mRNA.

On the other hand, tumorigenesis is associated with enhanced cellular glucose uptake and increased metabolism, which is required for rapid proliferation [115]. To facilitate anaerobic glycolysis in tumor neoangiogenesis, the major glucose transporter isoform, GLUT1, is induced by HIF-1α. The result of immunostaining has demonstrated that GLUT1 was expressed at the plasma membrane in renal cell curcinomas [116].

Because of the importance of HIF-1α for tumor development and survival, it is an important target for cancer drug development. The two cancer drugs that inhibit thioredoxin-1 signaling, PX-12 (1-methylpropyl 2-imidazolyl disulfide) and pleurotin, decreased HIF-1α protein levels and HIF-1 transactivation in RCC [117]. HIF-1α and VEGF protein levels in MCF-7 tumor xenografts were also decreased in vivo by PX-12 treatment of mice. Moreover, YC-1 (3-(5'-hydroxymethyl-2'-furyl)-1- benzylinda- zole), which inhibits platelet aggregation, completely blocked HIF-1α expression at the post-transcriptional level and consequently inhibited the transcription factor activity of HIF-1 in hepatoma cells cultured under hypoxic conditions [118].

The activity of the p53 tumor suppresser, including stability and nuclear export through interactions with Mdm2, is regulated by HIF-1α in response to hypoxia (at 0.02% but not at 0.2% O_2) [38,119]. P53 is maintained at low levels in normal cells and strongly stabilized in response to various types of stress including hypoxia. In its active form, p53 can activate the transcription of cycle regulating genes such as $p21^{WAF1/CIP1}$ or *mdm2* and mediate apoptosis. More recently, it was shown that low-level expression of p53 represses HIF-1 transactivation without affecting the amount of HIF-1α protein, whereas high-levels expression of p53 destroys the HIF-1α protein. Thus, the loss of p53 may increase hypoxia-induced HIF-1α levels and augments HIF-1-dependent expression of VEGF in tumor cells [119]. Consistent with the p53 gene being mutated in over 50% of human cancers [120], HIF-1α is frequently overexpressed in common human cancers [109]. Furthermore, hypoxia is also known to induce apoptosis that depends on p53 in mouse embryonic fibroblasts [121]. Apoptosis is an evolutionarily conserved cell death mechanism that is activated by various stimuli. Recently, the diminished apoptotic and the enhanced angiogenic potential of cancer cells have been recognized as key factors in the development of solid tumors, which generally have hypoxic regions. As p53 mediates hypoxia-induced apoptosis and inhibits HIF-1 activity, inactivation of p53 can be a principal mechanism for conferring resistance to apoptosis with an enhanced angiogenic phenotype.

References

[1] Semenza, G. Signal transduction to hypoxia-inducible factor 1. (2002) *Biochem Pharmacol 64*, 993-998

[2] Semenza, G. L., and Wang, G. L. A nuclear factor induced by hypoxia via de novo protein synthesis binds to the human erythropoietin gene enhancer at a site required for transcriptional activation. (1992) *Mol Cell Biol 12*, 5447-5454

[3] Goldberg, M. A., and Schneider, T. J. Similarities between the oxygen-sensing mechanisms regulating the expression of vascular endothelial growth factor and erythropoietin. (1994) *J Biol Chem 269*, 4355-4359

[4] Firth, J. D., Ebert, B. L., Pugh, C. W., and Ratcliffe, P. J. Oxygen-regulated control elements in the phosphoglycerate kinase 1 and lactate dehydrogenase A genes: similarities with the erythropoietin 3' enhancer. (1994) *Proc Natl Acad Sci U S A 91*, 6496-6500

[5] Semenza, G. L., Roth, P. H., Fang, H. M., and Wang, G. L. Transcriptional regulation of genes encoding glycolytic enzymes by hypoxia-inducible factor 1. (1994) *J Biol Chem 269*, 23757-23763

[6] Ebert, B. L., Firth, J. D., and Ratcliffe, P. J. Hypoxia and mitochondrial inhibitors regulate expression of glucose transporter-1 via distinct Cis-acting sequences. (1995) *J Biol Chem 270*, 29083-29089

[7] Murakami, T., Nishiyama, T., Shirotani, T., Shinohara, Y., Kan, M., Ishii, K., Kanai, F., Nakazuru, S., and Ebina, Y. Identification of two enhancer elements in the gene encoding the type 1 glucose transporter from the mouse which are responsive to serum, growth factor, and oncogenes. (1992) *J Biol Chem 267*, 9300-9306

[8] Gerber, H. P., Condorelli, F., Park, J., and Ferrara, N. Differential transcriptional regulation of the two vascular endothelial growth factor receptor genes. Flt-1, but not Flk-1/KDR, is up-regulated by hypoxia. (1997) *J Biol Chem 272*, 23659-23667

[9] Levy, A. P., Levy, N. S., Wegner, S., and Goldberg, M. A. Transcriptional regulation of the rat vascular endothelial growth factor gene by hypoxia. (1995) *J Biol Chem 270*, 13333-13340

[10] Levy, N. S., Chung, S., Furneaux, H., and Levy, A. P. Hypoxic stabilization of vascular endothelial growth factor mRNA by the RNA-binding protein HuR. (1998) *J Biol Chem 273*, 6417-6423

[11] Yamashita, K., Discher, D. J., Hu, J., Bishopric, N. H., and Webster, K. A. Molecular regulation of the endothelin-1 gene by hypoxia. Contributions of hypoxia-inducible factor-1, activator protein-1, GATA-2, AND p300/CBP. (2001) *J Biol Chem 276*, 12645-12653

[12] Zhang, S. X., Gozal, D., Sachleben, L. R., Jr., Rane, M., Klein, J. B., and Gozal, E. Hypoxia induces an autocrine-paracrine survival pathway via platelet-derived growth factor (PDGF)-B/PDGF-beta receptor/phosphatidylinositol 3-kinase/Akt signaling in RN46A neuronal cells. (2003) *Faseb J 17*, 1709-1711

[13] Sakaki, T., Yamada, K., Otsuki, H., Yuguchi, T., Kohmura, E., and Hayakawa, T. Brief exposure to hypoxia induces bFGF mRNA and protein and protects rat cortical neurons from prolonged hypoxic stress. (1995) *Neurosci Res 23*, 289-296

[14] Wang, G. L., Jiang, B. H., Rue, E. A., and Semenza, G. L. Hypoxia-inducible factor 1 is a basic-helix-loop-helix-PAS heterodimer regulated by cellular O2 tension. (1995) *Proc Natl Acad Sci U S A 92*, 5510-5514

[15] Huang, L. E., Arany, Z., Livingston, D. M., and Bunn, H. F. Activation of hypoxia-inducible transcription factor depends primarily upon redox-sensitive stabilization of its alpha subunit. (1996) *J Biol Chem 271*, 32253-32259

[16] Ema, M., Hirota, K., Mimura, J., Abe, H., Yodoi, J., Sogawa, K., Poellinger, L., and Fujii-Kuriyama, Y. Molecular mechanisms of transcription activation by HLF and HIF1alpha in response to hypoxia: their stabilization and redox signal-induced interaction with CBP/p300. (1999) *Embo J 18*, 1905-1914

[17] Maxwell, P. H., Wiesener, M. S., Chang, G. W., Clifford, S. C., Vaux, E. C., Cockman, M. E., Wykoff, C. C., Pugh, C. W., Maher, E. R., and Ratcliffe, P. J. The tumour suppressor protein VHL targets hypoxia-inducible factors for oxygen-dependent proteolysis. (1999) *Nature 399*, 271-275

[18] Huang, L. E., Pete, E. A., Schau, M., Milligan, J., and Gu, J. Leu-574 of HIF-1alpha is essential for the von Hippel-Lindau (VHL)-mediated degradation pathway. (2002) *J Biol Chem 277*, 41750-41755

[19] Lando, D., Peet, D. J., Whelan, D. A., Gorman, J. J., and Whitelaw, M. L. Asparagine hydroxylation of the HIF transactivation domain a hypoxic switch. (2002) *Science 295*, 858-861

[20] Tian, H., McKnight, S. L., and Russell, D. W. Endothelial PAS domain protein 1 (EPAS1), a transcription factor selectively expressed in endothelial cells. (1997) *Genes Dev 11*, 72-82

[21] Cockman, M. E., Masson, N., Mole, D. R., Jaakkola, P., Chang, G. W., Clifford, S. C., Maher, E. R., Pugh, C. W., Ratcliffe, P. J., and Maxwell, P. H. Hypoxia inducible factor-alpha binding and ubiquitylation by the von Hippel-Lindau tumor suppressor protein. (2000) *J Biol Chem 275*, 25733-25741

[22] Huang, L. E., Gu, J., Schau, M., and Bunn, H. F. Regulation of hypoxia-inducible factor 1alpha is mediated by an O2-dependent degradation domain via the ubiquitin-proteasome pathway. (1998) *Proc Natl Acad Sci U S A 95*, 7987-7992

[23] O'Rourke, J. F., Tian, Y. M., Ratcliffe, P. J., and Pugh, C. W. Oxygen-regulated and transactivating domains in endothelial PAS protein 1: comparison with hypoxia-inducible factor-1alpha. (1999) *J Biol Chem 274*, 2060-2071

[24] Blancher, C., Moore, J. W., Talks, K. L., Houlbrook, S., and Harris, A. L. Relationship of hypoxia-inducible factor (HIF)-1alpha and HIF-2alpha expression to vascular endothelial growth factor induction and hypoxia survival in human breast cancer cell lines. (2000) *Cancer Res 60*, 7106-7113

[25] Gothie, E., Richard, D. E., Berra, E., Pages, G., and Pouyssegur, J. Identification of alternative spliced variants of human hypoxia-inducible factor-1alpha. (2000) *J Biol Chem 275*, 6922-6927

[26] Chun, Y. S., Choi, E., Kim, T. Y., Kim, M. S., and Park, J. W. A dominant-negative isoform lacking exons 11 and 12 of the human hypoxia-inducible factor-1alpha gene. (2002) *Biochem J 362*, 71-79

[27] Gu, Y. Z., Moran, S. M., Hogenesch, J. B., Wartman, L., and Bradfield, C. A. Molecular characterization and chromosomal localization of a third alpha-class hypoxia inducible factor subunit, HIF3alpha. (1998) *Gene Expr 7*, 205-213

[28] Maynard, M. A., Qi, H., Chung, J., Lee, E. H., Kondo, Y., Hara, S., Conaway, R. C., Conaway, J. W., and Ohh, M. Multiple splice variants of the human HIF-3 alpha locus are targets of the von Hippel-Lindau E3 ubiquitin ligase complex. (2003) *J Biol Chem 278*, 11032-11040

[29] Makino, Y., Cao, R., Svensson, K., Bertilsson, G., Asman, M., Tanaka, H., Cao, Y., Berkenstam, A., and Poellinger, L. Inhibitory PAS domain protein is a negative regulator of hypoxia-inducible gene expression. (2001) *Nature 414*, 550-554

[30] Epstein, A. C., Gleadle, J. M., McNeill, L. A., Hewitson, K. S., O'Rourke, J., Mole, D. R., Mukherji, M., Metzen, E., Wilson, M. I., Dhanda, A., Tian, Y. M., Masson, N., Hamilton, D. L., Jaakkola, P., Barstead, R., Hodgkin, J., Maxwell, P. H., Pugh, C. W., Schofield, C. J., and Ratcliffe, P. J. C. elegans EGL-9 and mammalian homologs define a family of dioxygenases that regulate HIF by prolyl hydroxylation. (2001) *Cell 107*, 43-54

[31] Ivan, M., Kondo, K., Yang, H., Kim, W., Valiando, J., Ohh, M., Salic, A., Asara, J. M., Lane, W. S., and Kaelin, W. G., Jr. HIFalpha targeted for VHL-mediated destruction by proline hydroxylation: implications for O2 sensing. (2001) *Science 292*, 464-468

[32] Jaakkola, P., Mole, D. R., Tian, Y. M., Wilson, M. I., Gielbert, J., Gaskell, S. J., Kriegsheim, A., Hebestreit, H. F., Mukherji, M., Schofield, C. J., Maxwell, P. H., Pugh, C. W., and Ratcliffe, P. J. Targeting of HIF-alpha to the von Hippel-Lindau ubiquitylation complex by O2-regulated prolyl hydroxylation. (2001) *Science 292*, 468-472

[33] Bruick, R. K., and McKnight, S. L. A conserved family of prolyl-4-hydroxylases that modify HIF. (2001) *Science 294*, 1337-1340

[34] Oehme, F., Ellinghaus, P., Kolkhof, P., Smith, T. J., Ramakrishnan, S., Hutter, J., Schramm, M., and Flamme, I. Overexpression of PH-4, a novel putative proline 4-hydroxylase, modulates activity of hypoxia-inducible transcription factors. (2002) *Biochem Biophys Res Commun 296*, 343-349

[35] Metzen, E., Berchner-Pfannschmidt, U., Stengel, P., Marxsen, J. H., Stolze, I., Klinger, M., Huang, W. Q., Wotzlaw, C., Hellwig-Burgel, T., Jelkmann, W., Acker, H., and Fandrey, J. Intracellular localisation of human HIF-1 alpha hydroxylases: implications for oxygen sensing. (2003) *J Cell Sci 116*, 1319-1326

[36] Berra, E., Roux, D., Richard, D. E., and Pouyssegur, J. Hypoxia-inducible factor-1 alpha (HIF-1 alpha) escapes O(2)-driven proteasomal degradation irrespective of its subcellular localization: nucleus or cytoplasm. (2001) *EMBO Rep 2*, 615-620

[37] Chen, D., Li, M., Luo, J., and Gu, W. Direct interactions between HIF-1 alpha and Mdm2 modulate p53 function. (2003) *J Biol Chem 278*, 13595-13598

[38] Ravi, R., Mookerjee, B., Bhujwalla, Z. M., Sutter, C. H., Artemov, D., Zeng, Q., Dillehay, L. E., Madan, A., Semenza, G. L., and Bedi, A. Regulation of tumor angiogenesis by p53-induced degradation of hypoxia-inducible factor 1alpha. (2000) *Genes Dev 14*, 34-44

[39] Giaccia, A. J., and Kastan, M. B. The complexity of p53 modulation: emerging patterns from divergent signals. (1998) *Genes Dev 12*, 2973-2983

[40] Koumenis, C., Alarcon, R., Hammond, E., Sutphin, P., Hoffman, W., Murphy, M., Derr, J., Taya, Y., Lowe, S. W., Kastan, M., and Giaccia, A. Regulation of p53 by hypoxia: dissociation of transcriptional repression and apoptosis from p53-dependent transactivation. (2001) *Mol Cell Biol 21*, 1297-1310

[41] Suzuki, H., Tomida, A., and Tsuruo, T. Dephosphorylated hypoxia-inducible factor 1alpha as a mediator of p53-dependent apoptosis during hypoxia. (2001) *Oncogene 20*, 5779-5788

[42] Gradin, K., Toftgard, R., Poellinger, L., and Berghard, A. Repression of dioxin signal transduction in fibroblasts. Identification Of a putative repressor associated with Arnt. (1999) *J Biol Chem 274*, 13511-13518

[43] Katschinski, D. M., Le, L., Heinrich, D., Wagner, K. F., Hofer, T., Schindler, S. G., and Wenger, R. H. Heat induction of the unphosphorylated form of hypoxia-inducible factor-1alpha is dependent on heat shock protein-90 activity. (2002) *J Biol Chem 277*, 9262-9267

[44] Isaacs, J. S., Jung, Y. J., Mimnaugh, E. G., Martinez, A., Cuttitta, F., and Neckers, L. M. Hsp90 regulates a von Hippel Lindau-independent hypoxia-inducible factor-1 alpha-degradative pathway. (2002) *J Biol Chem 277*, 29936-29944

[45] Osada, M., Imaoka, S., and Funae, Y. Apigenin suppresses the expression of VEGF, an important factor for angiogenesis, in endothelial cells via degradation of HIF-1alpha protein. (2004) *FEBS Lett 575*, 59-63

[46] Yuan, Y., Hilliard, G., Ferguson, T., and Millhorn, D. E. Cobalt inhibits the interaction between hypoxia-inducible factor-alpha and von Hippel-Lindau protein by direct binding to hypoxia-inducible factor-alpha. (2003) *J Biol Chem 278*, 15911-15916

[47] Ryan, H. E., Lo, J., and Johnson, R. S. HIF-1 alpha is required for solid tumor formation and embryonic vascularization. (1998) *Embo J 17*, 3005-3015

[48] Minet, E., Michel, G., Mottet, D., Raes, M., and Michiels, C. Transduction pathways involved in Hypoxia-Inducible Factor-1 phosphorylation and activation. (2001) *Free Radic Biol Med 31*, 847-855

[49] Richard, D. E., Berra, E., Gothie, E., Roux, D., and Pouyssegur, J. p42/p44 mitogen-activated protein kinases phosphorylate hypoxia-inducible factor 1alpha (HIF-1alpha) and enhance the transcriptional activity of HIF-1. (1999) *J Biol Chem 274*, 32631-32637

[50] Berra, E., Milanini, J., Richard, D. E., Le Gall, M., Vinals, F., Gothie, E., Roux, D., Pages, G., and Pouyssegur, J. Signaling angiogenesis via p42/p44 MAP kinase and hypoxia. (2000) *Biochem Pharmacol 60*, 1171-1178

[51] Minet, E., Arnould, T., Michel, G., Roland, I., Mottet, D., Raes, M., Remacle, J., and Michiels, C. ERK activation upon hypoxia: involvement in HIF-1 activation. (2000) *FEBS Lett 468*, 53-58

[52] Blaschke, F., Stawowy, P., Goetze, S., Hintz, O., Grafe, M., Kintscher, U., Fleck, E., and Graf, K. Hypoxia activates beta(1)-integrin via ERK 1/2 and p38 MAP kinase in human vascular smooth muscle cells. (2002) *Biochem Biophys Res Commun 296*, 890-896.

[53] Liu, C., Shi, Y., Han, Z., Pan, Y., Liu, N., Han, S., Chen, Y., Lan, M., Qiao, T., and Fan, D. Suppression of the dual-specificity phosphatase MKP-1 enhances HIF-1 trans-activation and increases expression of EPO. (2003) *Biochem Biophys Res Commun 312*, 780-786

[54] Huang, L. E., and Bunn, H. F. Hypoxia-inducible factor and its biomedical relevance. (2003) *J Biol Chem 278*, 19575-19578

[55] Mazure, N. M., Chen, E. Y., Laderoute, K. R., and Giaccia, A. J. Induction of vascular endothelial growth factor by hypoxia is modulated by a phosphatidylinositol 3-kinase/Akt signaling pathway in Ha-ras-transformed cells through a hypoxia inducible factor-1 transcriptional element. (1997) *Blood 90*, 3322-3331

[56] Chen, C., Pore, N., Behrooz, A., Ismail-Beigi, F., and Maity, A. Regulation of glut1 mRNA by hypoxia-inducible factor-1. Interaction between H-ras and hypoxia. (2001) *J Biol Chem 276*, 9519-9525

[57] Figueroa, Y. G., Chan, A. K., Ibrahim, R., Tang, Y., Burow, M. E., Alam, J., Scandurro, A. B., and Beckman, B. S. NF-kappaB plays a key role in hypoxia-inducible factor-1-regulated erythropoietin gene expression. (2002) *Exp Hematol 30*, 1419-1427

[58] Hirota, K., and Semenza, G. L. Rac1 activity is required for the activation of hypoxia-inducible factor 1. (2001) *J Biol Chem 276*, 21166-21172

[59] Zhou, J., Schmid, T., Frank, R., and Brune, B. PI3K/Akt is required for heat shock proteins to protect hypoxia-inducible factor 1alpha from pVHL-independent degradation. (2004) *J Biol Chem 279*, 13506-13513

[60] Chandel, N. S., McClintock, D. S., Feliciano, C. E., Wood, T. M., Melendez, J. A., Rodriguez, A. M., and Schumacker, P. T. Reactive oxygen species generated at mitochondrial complex III stabilize hypoxia-inducible factor-1alpha during hypoxia: a mechanism of O2 sensing. (2000) *J Biol Chem 275*, 25130-25138

[61] Osada, M., Imaoka, S., Sugimoto, T., Hiroi, T., and Funae, Y. NADPH-cytochrome P-450 reductase in the plasma membrane modulates the activation of hypoxia-inducible factor 1. (2002) *J Biol Chem 277*, 23367-23373

[62] Goldberg, M., Zhang, H. L., and Steinberg, S. F. Hypoxia alters the subcellular distribution of protein kinase C isoforms in neonatal rat ventricular myocytes. (1997) *J Clin Invest 99*, 55-61

[63] Ohigashi, T., Mallia, C. S., McGary, E., Scandurro, A. B., Rondon, I., Fisher, J. W., and Beckman, B. S. Protein kinase C alpha protein expression is necessary for sustained erythropoietin production in human hepatocellular carcinoma (Hep3B) cells exposed to hypoxia. (1999) *Biochim Biophys Acta 1450*, 109-118

[64] Yuan, H. T., Yang, S. P., and Woolf, A. S. Hypoxia up-regulates angiopoietin-2, a Tie-2 ligand, in mouse mesangial cells. (2000) *Kidney Int 58*, 1912-1919

[65] Mukhopadhyay, D., and Akbarali, H. I. Depletion of [Ca2+]i inhibits hypoxia-induced vascular permeability factor (vascular endothelial growth factor) gene expression. (1996) *Biochem Biophys Res Commun 229*, 733-738

[66] Liu, Q., Moller, U., Flugel, D., and Kietzmann, T. Induction of plasminogen activator inhibitor I gene expression by intracellular calcium via hypoxia-inducible factor-1. (2004) *Blood 104*, 3993-4001

[67] Mottet, D., Michel, G., Renard, P., Ninane, N., Raes, M., and Michiels, C. ERK and calcium in activation of HIF-1. (2002) *Ann N Y Acad Sci 973*, 448-453

[68] Sahai, A., Mei, C., Pattison, T. A., and Tannen, R. L. Chronic hypoxia induces proliferation of cultured mesangial cells: role of calcium and protein kinase C. (1997) *Am J Physiol 273*, F954-960

[69] Sodhi, C. P., Batlle, D., and Sahai, A. Osteopontin mediates hypoxia-induced proliferation of cultured mesangial cells: role of PKC and p38 MAPK. (2000) *Kidney Int 58*, 691-700

[70] Yoshioka, K., Clejan, S., and Fisher, J. W. Activation of protein kinase C in human hepatocellular carcinoma (HEP3B) cells increases erythropoietin production. (1998) *Life Sci 63*, 523-535

[71] Aragones, J., Jones, D. R., Martin, S., San Juan, M. A., Alfranca, A., Vidal, F., Vara, A., Merida, I., and Landazuri, M. O. Evidence for the involvement of diacylglycerol kinase in the activation of hypoxia-inducible transcription factor 1 by low oxygen tension. (2001) *J Biol Chem 276*, 10548-10555

[72] Temes, E., Martin-Puig, S., Aragones, J., Jones, D. R., Olmos, G., Merida, I., and Landazuri, M. O. Role of diacylglycerol induced by hypoxia in the regulation of HIF-1alpha activity. (2004) *Biochem Biophys Res Commun 315*, 44-50

[73] Schmedtje, J. F., Jr., Ji, Y. S., Liu, W. L., DuBois, R. N., and Runge, M. S. Hypoxia induces cyclooxygenase-2 via the NF-kappaB p65 transcription factor in human vascular endothelial cells. (1997) *J Biol Chem 272*, 601-608

[74] Xu, Q., Ji, Y. S., and Schmedtje, J. F., Jr. Sp1 increases expression of cyclooxygenase-2 in hypoxic vascular endothelium. Implications for the mechanisms of aortic aneurysm and heart failure. (2000) *J Biol Chem 275*, 24583-24589

[75] Bradbury, D. A., Newton, R., Zhu, Y. M., Stocks, J., Corbett, L., Holland, E. D., Pang, L. H., and Knox, A. J. Effect of bradykinin, TGF-beta1, IL-1beta, and hypoxia on COX-2 expression in pulmonary artery smooth muscle cells. (2002) *Am J Physiol Lung Cell Mol Physiol 283*, L717-725

[76] Haddad, J. J., Olver, R. E., and Land, S. C. Antioxidant/pro-oxidant equilibrium regulates HIF-1alpha and NF-kappa B redox sensitivity. Evidence for inhibition by glutathione oxidation in alveolar epithelial cells. (2000) *J Biol Chem 275*, 21130-21139

[77] Mottet, D., Dumont, V., Deccache, Y., Demazy, C., Ninane, N., Raes, M., and Michiels, C. Regulation of hypoxia-inducible factor-1alpha protein level during hypoxic conditions by the phosphatidylinositol 3-kinase/Akt/glycogen synthase kinase 3beta pathway in HepG2 cells. (2003) *J Biol Chem 278*, 31277-31285

[78] Schnitzer, S. E., Schmid, T., Zhou, J., Eisenbrand, G., and Brune, B. Inhibition of GSK3beta by indirubins restores HIF-1alpha accumulation under prolonged periods of hypoxia/anoxia. (2005) *FEBS Lett 579*, 529-533

[79] Alfranca, A., Gutierrez, M. D., Vara, A., Aragones, J., Vidal, F., and Landazuri, M. O. c-Jun and hypoxia-inducible factor 1 functionally cooperate in hypoxia-induced gene transcription. (2002) *Mol Cell Biol 22*, 12-22

[80] Laderoute, K. R., Calaoagan, J. M., Gustafson-Brown, C., Knapp, A. M., Li, G. C., Mendonca, H. L., Ryan, H. E., Wang, Z., and Johnson, R. S. The response of c-jun/AP-1 to chronic hypoxia is hypoxia-inducible factor 1 alpha dependent. (2002) *Mol Cell Biol 22*, 2515-2523

[81] Braun, L., Kardon, T., Reisz-Porszasz, Z. S., Banhegyi, G., and Mandl, J. The regulation of the induction of vascular endothelial growth factor at the onset of diabetes in spontaneously diabetic rats. (2001) *Life Sci 69*, 2533-2542

[82] Sodhi, A., Montaner, S., Patel, V., Zohar, M., Bais, C., Mesri, E. A., and Gutkind, J. S. The Kaposi's sarcoma-associated herpes virus G protein-coupled receptor up-regulates vascular endothelial growth factor expression and secretion through mitogen-activated protein kinase and p38 pathways acting on hypoxia-inducible factor 1alpha. (2000) *Cancer Res 60*, 4873-4880

[83] Treins, C., Giorgetti-Peraldi, S., Murdaca, J., and Van Obberghen, E. Regulation of vascular endothelial growth factor expression by advanced glycation end products. (2001) *J Biol Chem 276*, 43836-43841

[84] Yoo, Y. G., Oh, S. H., Park, E. S., Cho, H., Lee, N., Park, H., Kim, D. K., Yu, D. Y., Seong, J. K., and Lee, M. O. Hepatitis B virus X protein enhances transcriptional activity of hypoxia-inducible factor-1alpha through activation of mitogen-activated protein kinase pathway. (2003) *J Biol Chem 278*, 39076-39084

[85] Laughner, E., Taghavi, P., Chiles, K., Mahon, P. C., and Semenza, G. L. HER2 (neu) signaling increases the rate of hypoxia-inducible factor 1alpha (HIF-1alpha) synthesis: novel mechanism for HIF-1-mediated vascular endothelial growth factor expression. (2001) *Mol Cell Biol 21*, 3995-4004

[86] Rak, J., Mitsuhashi, Y., Sheehan, C., Tamir, A., Viloria-Petit, A., Filmus, J., Mansour, S. J., Ahn, N. G., and Kerbel, R. S. Oncogenes and tumor angiogenesis: differential modes of vascular endothelial growth factor up-regulation in ras-transformed epithelial cells and fibroblasts. (2000) *Cancer Res 60*, 490-498

[87] Fukuda, R., Hirota, K., Fan, F., Jung, Y. D., Ellis, L. M., and Semenza, G. L. Insulin-like growth factor 1 induces hypoxia-inducible factor 1-mediated vascular endothelial

growth factor expression, which is dependent on MAP kinase and phosphatidylinositol 3-kinase signaling in colon cancer cells. (2002) *J Biol Chem 277*, 38205-38211

[88] Gorlach, A., Diebold, I., Schini-Kerth, V. B., Berchner-Pfannschmidt, U., Roth, U., Brandes, R. P., Kietzmann, T., and Busse, R. Thrombin activates the hypoxia-inducible factor-1 signaling pathway in vascular smooth muscle cells: Role of the p22(phox)-containing NADPH oxidase. (2001) *Circ Res 89*, 47-54

[89] Page, E. L., Robitaille, G. A., Pouyssegur, J., and Richard, D. E. Induction of hypoxia-inducible factor-1alpha by transcriptional and translational mechanisms. (2002) *J Biol Chem 277*, 48403-48409

[90] Richard, D. E., Berra, E., and Pouyssegur, J. Nonhypoxic pathway mediates the induction of hypoxia-inducible factor 1alpha in vascular smooth muscle cells. (2000) *J Biol Chem 275*, 26765-26771

[91] Shatrov, V. A., Sumbayev, V. V., Zhou, J., and Brune, B. Oxidized low-density lipoprotein (oxLDL) triggers hypoxia-inducible factor-1alpha (HIF-1alpha) accumulation via redox-dependent mechanisms. (2003) *Blood 101*, 4847-4849

[92] Zelzer, E., Levy, Y., Kahana, C., Shilo, B. Z., Rubinstein, M., and Cohen, B. Insulin induces transcription of target genes through the hypoxia-inducible factor HIF-1alpha/ARNT. (1998) *Embo J 17*, 5085-5094

[93] Hellwig-Burgel, T., Rutkowski, K., Metzen, E., Fandrey, J., and Jelkmann, W. Interleukin-1beta and tumor necrosis factor-alpha stimulate DNA binding of hypoxia-inducible factor-1. (1999) *Blood 94*, 1561-1567

[94] Thornton, R. D., Lane, P., Borghaei, R. C., Pease, E. A., Caro, J., and Mochan, E. Interleukin 1 induces hypoxia-inducible factor 1 in human gingival and synovial fibroblasts. (2000) *Biochem J 350 Pt 1*, 307-312

[95] Spinella, F., Rosano, L., Di Castro, V., Natali, P. G., and Bagnato, A. Endothelin-1 induces vascular endothelial growth factor by increasing hypoxia-inducible factor-1alpha in ovarian carcinoma cells. (2002) *J Biol Chem 277*, 27850-27855

[96] Bjornheden, T., Levin, M., Evaldsson, M., and Wiklund, O. Evidence of hypoxic areas within the arterial wall in vivo. (1999) *Arterioscler Thromb Vasc Biol 19*, 870-876

[97] Rydberg, E. K., Salomonsson, L., Hulten, L. M., Noren, K., Bondjers, G., Wiklund, O., Bjornheden, T., and Ohlsson, B. G. Hypoxia increases 25-hydroxycholesterol-induced interleukin-8 protein secretion in human macrophages. (2003) *Atherosclerosis 170*, 245-252

[98] Wada, Y., Sugiyama, A., Yamamoto, T., Naito, M., Noguchi, N., Yokoyama, S., Tsujita, M., Kawabe, Y., Kobayashi, M., Izumi, A., Kohro, T., Tanaka, T., Taniguchi, H., Koyama, H., Hirano, K., Yamashita, S., Matsuzawa, Y., Niki, E., Hamakubo, T., and Kodama, T. Lipid accumulation in smooth muscle cells under LDL loading is independent of LDL receptor pathway and enhanced by hypoxic conditions. (2002) *Arterioscler Thromb Vasc Biol 22*, 1712-1719

[99] Wight, T. N., Raugi, G. J., Mumby, S. M., and Bornstein, P. Light microscopic immunolocation of thrombospondin in human tissues. (1985) *J Histochem Cytochem 33*, 295-302

[100] Brody, J. I., Pickering, N. J., and Fink, G. B. Immunocytochemical features of obstructed saphenous vein coronary artery bypass grafts. (1989) *J Clin Pathol 42*, 477-482

[101] Raugi, G. J., Mullen, J. S., Bark, D. H., Okada, T., and Mayberg, M. R. Thrombospondin deposition in rat carotid artery injury. (1990) *Am J Pathol 137*, 179-185

[102] Norman, J. T., Clark, I. M., and Garcia, P. L. Hypoxia promotes fibrogenesis in human renal fibroblasts. (2000) *Kidney Int 58*, 2351-2366

[103] Manotham, K., Tanaka, T., Matsumoto, M., Ohse, T., Inagi, R., Miyata, T., Kurokawa, K., Fujita, T., Ingelfinger, J. R., and Nangaku, M. Transdifferentiation of cultured tubular cells induced by hypoxia. (2004) *Kidney Int 65*, 871-880

[104] Higgins, D. F., Biju, M. P., Akai, Y., Wutz, A., Johnson, R. S., and Haase, V. H. Hypoxic induction of Ctgf is directly mediated by Hif-1. (2004) *Am J Physiol Renal Physiol 287*, F1223-1232

[105] Eickelberg, O., Seebach, F., Riordan, M., Thulin, G., Mann, A., Reidy, K. H., Van Why, S. K., Kashgarian, M., and Siegel, N. Functional activation of heat shock factor and hypoxia-inducible factor in the kidney. (2002) *J Am Soc Nephrol 13*, 2094-2101

[106] Kim, N. H., Jung, H. H., Cha, D. R., and Choi, D. S. Expression of vascular endothelial growth factor in response to high glucose in rat mesangial cells. (2000) *J Endocrinol 165*, 617-624

[107] Bennett Kaufman, O. S., Jeffrey Arbeit, Margaret Ashcroft, J. Martin Brown,. Proceeding of the Oxygen Homeostasis/Hypoxia Meeting. (2004) *Cancer Research 64*, 3350-3356

[108] Dachs, G. U., Patterson, A. V., Firth, J. D., Ratcliffe, P. J., Townsend, K. M., Stratford, I. J., and Harris, A. L. Targeting gene expression to hypoxic tumor cells. (1997) *Nat Med 3*, 515-520

[109] Zhong, H., De Marzo, A. M., Laughner, E., Lim, M., Hilton, D. A., Zagzag, D., Buechler, P., Isaacs, W. B., Semenza, G. L., and Simons, J. W. Overexpression of hypoxia-inducible factor 1alpha in common human cancers and their metastases. (1999) *Cancer Res 59*, 5830-5835

[110] Melnyk, O., Zimmerman, M., Kim, K. J., and Shuman, M. Neutralizing anti-vascular endothelial growth factor antibody inhibits further growth of established prostate cancer and metastases in a pre-clinical model. (1999) *J Urol 161*, 960-963

[111] Kondo, K., Klco, J., Nakamura, E., Lechpammer, M., and Kaelin, W. G., Jr. Inhibition of HIF is necessary for tumor suppression by the von Hippel-Lindau protein. (2002) *Cancer Cell 1*, 237-246

[112] Mandriota, S. J., Turner, K. J., Davies, D. R., Murray, P. G., Morgan, N. V., Sowter, H. M., Wykoff, C. C., Maher, E. R., Harris, A. L., Ratcliffe, P. J., and Maxwell, P. H. HIF activation identifies early lesions in VHL kidneys: evidence for site-specific tumor suppressor function in the nephron. (2002) *Cancer Cell 1*, 459-468

[113] Maranchie, J. K., Vasselli, J. R., Riss, J., Bonifacino, J. S., Linehan, W. M., and Klausner, R. D. The contribution of VHL substrate binding and HIF1-alpha to the phenotype of VHL loss in renal cell carcinoma. (2002) *Cancer Cell 1*, 247-255

[114] Carmeliet, P., Dor, Y., Herbert, J. M., Fukumura, D., Brusselmans, K., Dewerchin, M., Neeman, M., Bono, F., Abramovitch, R., Maxwell, P., Koch, C. J., Ratcliffe, P., Moons, L., Jain, R. K., Collen, D., and Keshert, E. Role of HIF-1alpha in hypoxia-mediated apoptosis, cell proliferation and tumour angiogenesis. (1998) *Nature 394*, 485-490

[115] Schwartzenberg-Bar-Yoseph, F., Armoni, M., and Karnieli, E. The tumor suppressor p53 down-regulates glucose transporters GLUT1 and GLUT4 gene expression. (2004) *Cancer Res 64*, 2627-2633

[116] Nagase, Y., Takata, K., Moriyama, N., Aso, Y., Murakami, T., and Hirano, H. Immunohistochemical localization of glucose transporters in human renal cell carcinoma. (1995) *J Urol 153*, 798-801

[117] Welsh, S. J., Williams, R. R., Birmingham, A., Newman, D. J., Kirkpatrick, D. L., and Powis, G. The thioredoxin redox inhibitors 1-methylpropyl 2-imidazolyl disulfide and pleurotin inhibit hypoxia-induced factor 1alpha and vascular endothelial growth factor formation. (2003) *Mol Cancer Ther 2*, 235-243

[118] Yeo, E. J., Chun, Y. S., Cho, Y. S., Kim, J., Lee, J. C., Kim, M. S., and Park, J. W. YC-1: a potential anticancer drug targeting hypoxia-inducible factor 1. (2003) *J Natl Cancer Inst 95*, 516-525

[119] Schmid, T., Zhou, J., Kohl, R., and Brune, B. p300 relieves p53-evoked transcriptional repression of hypoxia-inducible factor-1 (HIF-1). (2004) *Biochem J 380*, 289-295

[120] Levine, A. J. p53, the cellular gatekeeper for growth and division. (1997) *Cell 88*, 323-331

[121] Graeber, T. G., Osmanian, C., Jacks, T., Housman, D. E., Koch, C. J., Lowe, S. W., and Giaccia, A. J. Hypoxia-mediated selection of cells with diminished apoptotic potential in solid tumours. (1996) *Nature 379*, 88-91

In: Trends in Cellular Signaling
Editor: Dave E. Caplin, pp. 143-227

ISBN 1-59454-633-9

Chapter V

WNT SIGNALLING IN NORMAL DEVELOPMENT AND CANCER PROGRESSION

Nathalie Le Floch[1], Patricia Forgez[1], Frédérique Souazé[1], Erik Bruyneel[2], Marc Bracke[2], Michèle Sabbah, Gérard Redeuilh, Amazia Zimber[3], Jan Mester[1] and Christian Gespach[1,*]

[1]INSERM U673, Cancérologie Fondamentale et Clinique des Tumeurs Solides;
[1]Paris VI Université Pierre et Marie Curie, Hôpital Saint-Antoine, 75571 Paris Cedex 12 France,
[2]The Laboratory of Experimental Cancerology, Ghent University Hospital, B-9000 Ghent, Belgium, and
[3]The Hebrew University of Jerusalem, POB 12, Rehovot 76100, Israel.

ABSTRACT

The secreted proteins of the multimember Wnt family activate Frizzled, G-protein-coupled serpentine receptors (GPCR), and regulate a multitude of cellular functions implicated in normal development, stem cell fate determination, proliferation of embryonic stem cells and progenitor cells in adult life, postnatal woundhealing and tissue regeneration, as well as neoplastic transformation. Wnt family signalling factors evolved and diversified early in multicellular animals about 650 million years ago. In molecular terms, multiple signalling pathways are transduced downstream of the Fz receptors, resulting in regulation of protein kinases and transcription of specific Wnt target genes. These canonical β-catenin -dependent and noncanonical Wnt pathways involve numerous intracellular signalling cascades, molecular scaffolds at cell-cell contacts, as well as secreted and membrane-bound proteins which reinforce or impede Wnt signals. The intricacy of the Wnt functions is finely tuned through the appropriate expression of Wnt ligands (19 genes identified in man), Wnt receptors (9 human Frizzled genes), Wnt cofactors, antagonists, and other proteins that participate in signal transduction. Multiple

* Correspondence concerning this article should be addressed to Dr. Christian Gespach, INSERM U673, Hôpital Saint-Antoine, 75571 Paris Cedex 12 France. E-mail: gespach@st-antoine.inserm.fr.

crosstalks between Wnt and Hedgehog, Notch, bone morphogenic proteins (BMP) and transforming growth factor-β (TGFβ) signalling pathways, as well as receptor and nonreceptor kinases are integrated with critical networks involved in cell growth and differentiation, survival, migration, angiogenesis, and invasion. Dysregulation of the activities of the Wnt signalling elements by activating /inactivating mutations and epigenetic changes is linked to several pathologies. Several molecular alterations targeting Wnt ligands and internal elements of the Wnt and APC/ β-catenin pathways are frequently detected in sporadic cancers of the digestive and urogenital systems. In the present review, we summarize the molecular mechanisms involved in the transduction of the Wnt signals, and their widespread implication in development and malignancy, with special emphasis on stem cell biology and cancer progession.

Keywords: Hedgehog, Notch, BMP/ TGFβ, EGF-R, ERB-B2, AP-1, Cdx/Hox, Stem Cells, Human Solid Tumors, Colon and breast cancer, Leukemia, Therapy, Iressa, Trastuzumab.

1. Introduction: Signalling Pathways Initiated by Wnt Proteins

Wnt proteins are a large family of highly conserved, secreted protein factors involved in normal development and cancer. For their activity, these factors rely on binding to the serpentine G-protein-coupled receptors (GPCR) Frizzled (Fz) that are connected to different intracellular signalling pathways through heterotrimeric G-proteins using Gα and Gβγ subunits as functional elements (Fig. 1):

a) The canonical Wnt/ GSK-3β pathways leading to the transcription of β-catenin target genes;
b) the planar polarity pathways that involves the c-Jun C-terminal kinase (JNK) and cytoskeletal rearrangements;
c) noncanonical Wnt pathways also use other GPCR effectors such as RhoGTPAses and ROK, PKC, calcium translocations, activation of phospholipase C (PLC) and of its downstream target protein kinase C (PKC);
d) Wnt pathways are also controlled by the dual-specific transcriptional repressor Kaiso, and signalling crosstalks using other pools of β-catenins engaged with molecular scaffolds comprising receptor kinases, or the tumor suppressor gene E-cadherin/ β-catenin/ APC / γ-catenin/ p120 / p85 subunit of PI3-kinase involved in cancer cell survival, cell-cell adhesion and invasion (Birchmeier, 1994; Kotelevets, 1998; Muller, 1999b; Woodfield, 2001; Brembeck, 2004; Kim, 2001 and 2004).

1. a. The Canonical Wnt Pathway

In the absence of Wnt, β-catenin is a component of a multiprotein complex comprising the scaffold protein axin, APC (the product of the *APC* tumor suppressor gene), and glycogen synthase kinase-3β (GSK-3β). GSK-3β, a constitutively active kinase, phosphorylates Ser/Thr residues in the N-terminal region of β-catenin. Phosphorylated β-catenin at ser /thr residues (not true for Tyr-phosphorylated β-CTN) is then ubiquitinylated *via* the F-box protein β-TrCp (β-transducin repeat -containing protein), an E3 ubiquitin ligase component, and directed to degradation by the proteasome.

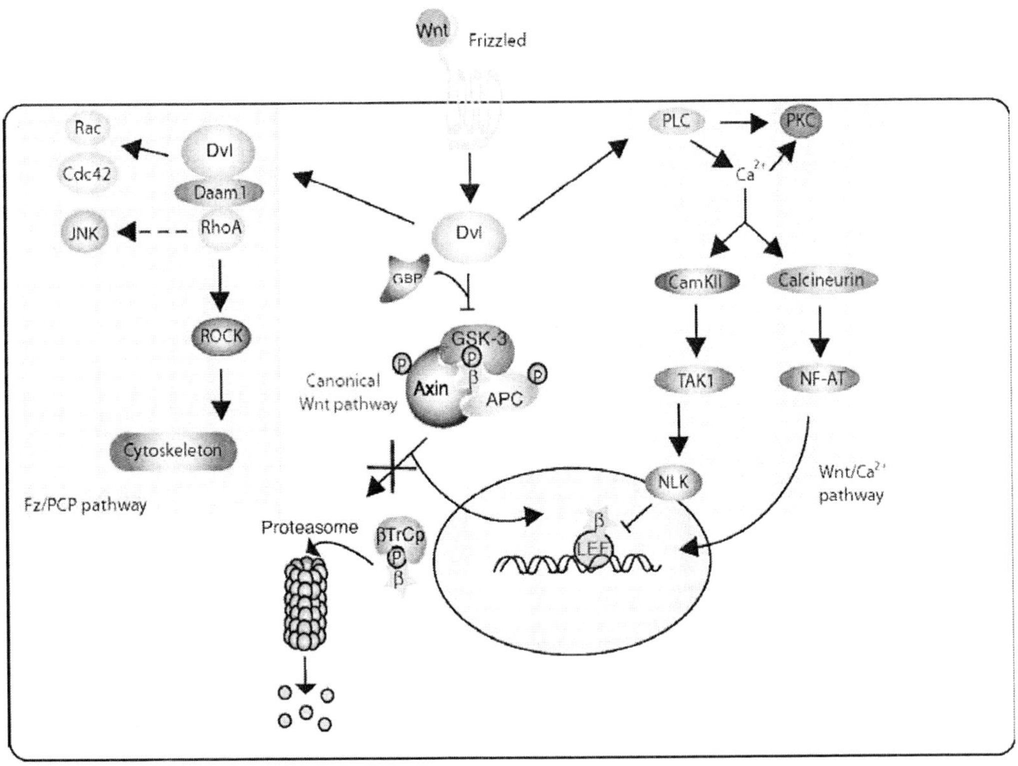

Figure 1. Wnt signaling pathways

After binding Wnt, Fz receptors associated to LRP co-receptors activate Dvl proteins that recruit GBP/Frat -1and inhibit GSK-3β. The stabilised β-catenin accumulates in the cytoplasm and becomes available for translocation to the nucleus, where it associates with the transcription factors LEF/TCF lacking a transactivation domain, leading to the formation of a functional bipartite transcription factor. Thus, the canonical Wnt pathway, through the protection of β-catenin, leads to the expression of a large set of target genes. Upon cessation of the Wnt signal, APC exports β-catenin from the nucleus to the cytoplasm where it is targeted for degradation, while LEF/TCF are converted to the inactive state and resume their initial function as repressors of transcription.

The canonical Wnt pathway participates in the development of different organisms ranging from *Drosophila* to mammals (Fig. 2). In *Drosophila*, the product of the *wingless* gene (*wg*), a Wnt-1 homologue, is involved in the establishment of segment polarity. At this stage of embryonal development, several segment polarity genes encode secreted proteins that transmit signals to neighboring cells. These segment polarity genes are responsible for cellular organization, and positioning within the segments of the larva. The cuticle of the *Drosophila* embryo displays alternated bands with and without protrusions (denticles). When the *wg* gene is invalidated by mutation, the cuticle is entirely covered with denticles. A similar phenotype is seen in mutants of the genes *dishevelled* (*dsh*), *armadillo* (*arm,* encoding a β-catenin homologue) and *pangolin* (*pan,* a LEF homologue), whereas mutations in *zeste-white 3* (*zw3*, GSK3 homologue) display an opposite phenotype characterised by a « naked » cuticle. Double mutant strategy showed that the products of these genes all participate at the Wg signalling pathways (Noordermeer, 1994; Peifer, 1994; Siegfried, 1994; Brunner, 1997; van de Wetering, 1997).

In *Xenopus*, the canonical Wnt pathway is involved in the establishment of the dorsal axis of the embryo. Injection of mRNA encoding Wnt-1 or XWnt-8A into ventral blastomers of early *Xenopus* embryo induces duplication of the embryonal axis, resulting in a tadpole with two heads (McMahon, 1989; Christian, 1993). The expression of downstream mediators of the Wnt pathway can short-circuit the Wnt signal. For instance, overexpression of Dvl (a *disheveled* homologue) or β-catenin (the vertebrate homologue of Armadillo), or introduction of a dominant -negative GSK-3 variant (ZW3 homologue) results in the formation of an additional embryonal axis (Dominguez, 1995; Guger, 1995; He, 1995; Sokol, 1995).

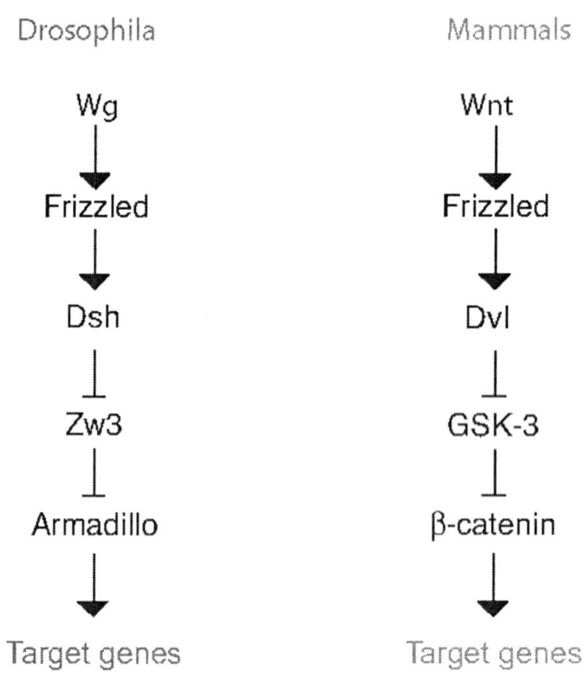

Figure 2. The canonical Wnt pathways

Although the injected Wnt mRNA induces the duplication of the dorsal axis, the role of the endogenous Wnt protein in the establishment of the primary axis in *Xenopus* has not been clearly demonstrated. Dominant negative variants of XWnt-8A and Dvl impair the induction of a supplementary axis by Wnt, but do not prevent the formation of the primary axis (Hoppler, 1996; Sokol, 1996). In contrast, depletion of β-catenin by antisense RNA injection, overexpression of GSK-3, as well as introduction of a dominant negative variant of the Xtcf-3 transcription factor (homolog of Pangolin) or cadherins, induce the ventralisation of embryos as a consequence of the absence of the primary dorsal axis (Heasman, 1994; Dominguez,

1995; He, 1995; Molenaar, 1996). Besides its role in embryonal development, deregulation of the canonical Wnt pathway is associated with cancerogenesis. This point will be discussed in Paragraph 7 of this review.

1. b. The Fz /PCP Planar Polarity Pathway

Planar cell polarity signalling (PCP, also known as tissue polarity) controls the polarity of epithelial cells within a plane orthogonal to their apical-basal axis. In *Drosophila*, the Fz/PCP pathway controls the establishment of the epithelial cell polarity, needed for the oriented organization of hairs, bristles, and ommatidia (Shulman, 1998). This pathway is activated by Fz receptors, and relayed by the Dsh proteins. For 5 out of the 7 *Drosophila* Wnt genes, their absence or overexpression did not affect the Fz/ PCP signalling (Lawrence, 2002). In the nematode *C. elegans*, polarization of the endoderm potential by non-canonical Wnt signalling has been seen in a single embryonic cell that divides to produce one endoderm and one mesoderm precursor (Thorpe, 2000). The rotation of the mitotic spindle that takes place during this polarized division does not depend on the TCF homolog of the worm, and can be induced even when transcription is blocked (Schlesinger, 1999; Moon, 2002).

Recent research has demonstrated a similarity between the Fz/Dsh pathway in *Drosophila* and the molecular mechanisms involved in the convergent extension movements that determine the elongation of the embryo during gastrulation in *Xenopus* and zebrafish. While the implication of Wnt has not been demonstrated during the activation of the Fz/PCP pathway in *Drosophila* (see above), the parallel morphogenetic movements pathways in vertebrates are activated by Wnt-11/ Silberblick and Wnt-5a /pipetail (Tada, 2000; Heisenberg, 2000; Moon, 1993; Du, 1995). Wnt genes and signalling elements probably had an ancestral function in the development of early multicellular animals, such as sea anemone *Nematostella vectensis* and freshwater polyp *Hydra magnipapillata* (Kusserow, 2005). Genetic controls exerted by Wnt signalling diversified during the evolution and development of nematodes, as shown in *Caenorhabditis elegans* and *Pristionchus pacificus* (Zheng, 2005).

Similar to the Wnt/β-catenin pathway, the Fz/PCP pathway also utilizes Dsh/Dvl as relays (Figs.1 and 2). Thus, mutations of *dsh* induce anomalies of cell orientation in the wings, eyes and thorax. On the other hand, genetic and biochemical studies showed that PCP signalling does not rely on Zw-3 (GSK-3), Arm (β-catenin), or Pan (Tcf-3) that act downstream of Dsh in the canonical Wnt pathway. The distinction between the two pathways exists also in vertebrates. The Wnt-11/ Silberblick loss of function phenotype in *Xenopus* and zebrafish is compensated by a truncated form of Dsh that is incapable to transduce the canonical Wnt signals (Tada, 2000; Heisenberg, 2000).

Receptors and Mediators Involved in the Fz/PCP Pathway

Five Frizzled receptors have been identified in *Drosophila*: Frizzled (termed DFz1 for simplicity in this review), DFz2, DFz3, DFz4, and Smoothened. It has been suggested that DFz1 signals mainly towards the PCP pathway, whereas Dfz2 signal is preferentially directed to Arm (Boutros, 2000). In vertebrates, Wnt-11 and Wnt-5a are involved in convergent extension movements respectively transduced through Fz7 and Fz2 (Djiane, 2000; Kilian, 2003). The PCP signals in *Drosophila* are transduced *via* small G-proteins RhoA and Rac, as well as the stress -activated serine/threonine kinase SAPK-JNK (Strutt, 1997; Boutros, 1998). The Rho-associated kinase (Drok) is also required for PCP signalling downstream of RhoA (Winter, 2001). In vertebrates, RhoA activation by Wnt11 requires the formation of the complex Dsh-RhoA and interaction with Daam1 (Habas, 2001). Moreover, the Rok2 kinase downstream of RhoA appears to be involved as a morphogen in *Xenopus* (Marlow, 2002). It has been demonstrated that Rho and Rac are activated in parallel by Wnt/Fz signals. This activation requires two independent domains of Dsh, while Daam1 intervenes only in the activation of Rho by Dsh (Habas, 2003). Downstream Rac, the Wnt /Fz signal induces JNK activation; the deregulation of this kinase affects the morphogenetic movements (Habas, 2003; Yamanaka, 2002). Finally, the small GTPase Cdc42, whose activation by Wnt /Fz requires Gβγ dimers as well as PKC, regulates the convergent extension movements (Djiane, 2000; Choi, 2002; Penzo-Mendèz, 2003).

1. c. The Wnt/ Ca^{2+} pathways

The first indication that Wnt pathways are connected with calcium signalling came from the observation that ectopic expression of Xwnt-5a and rat Fz2 (Rfz2, but not of Xwnt-8 and Rfz1) specifically activates intracellular calcium release and phosphatidylinositol signalling (Slusarski, 1997a; Slusarski, 1997b). Similar studies in Xenopus (Sheldahl, 1999; Kühl, 2000; Saneyoshi; 2002) identified PKC, Ca^{2+} /Calmodulin-dependent protein Kinase II (CamKII) and calcineurine, as elements in the signalling pathway induced by Xwnt-5a, Xwnt-11, rat Rfz2, and intracellular Ca^{2+} translocations (Fig.1). Because CamKII is activated within minutes following receptor stimulation, this calcium/ calmodulin -dependent kinase is probably a direct Xwnt11 target (Kühl, 2000). A recent report suggested that downstream of Wnt-5a/Rfz2 -induced Ca^{2+} flux, CamKII activated the MAPKs TAK1/NLK cascade (Ishitani, 2003), as shown in Figure 3. Heterotrimeric G-proteins are also involved in Wnt/Ca^{2+} signal transduction. Accumulation of intracellular Ca^{2+} as well as activation of PKC and CamKII are blocked by (i) Pertussis toxin (PTx) which through ADP-ribosylation of Gαo/i subunits inhibits signal transduction using both Gα subunits and Gβγ heterodimers, and (ii) Gα-transducin, which sequesters Gβγ subunits (Slusarski, 1997b; Sheldahl, 1999; Kühl, 2000; Le Floch, 2005). Dsh mediates the Wnt/β-catenin and Wnt /Ca^{2+} pathways *via* Fz and PCP signals. A truncated form of Dsh failed to mediate the Wnt /β-catenin signal but increased intracellular Ca^{2+}levels, activation of CamKII, and PKC translocation. Depletion of Dsh by antisense RNA inhibited Xfz7-dependent translocation of PKC (Sheldahl, 2003). Similar to the Fz/PCP pathway, the Wnt/Ca^{2+} is involved in the regulation of morphogenetic movements during gastrulation.

1. d. Crosstalk Between Wnt Pathways

Inhibition of the canonical Wnt signalling by the Wnt/Ca2+ pathway

On the basis of their transforming activities in mammary epithelial cells and their capacity to induce a secondary axis in *Xenopus*, Wnt ligands have been sorted in two classes

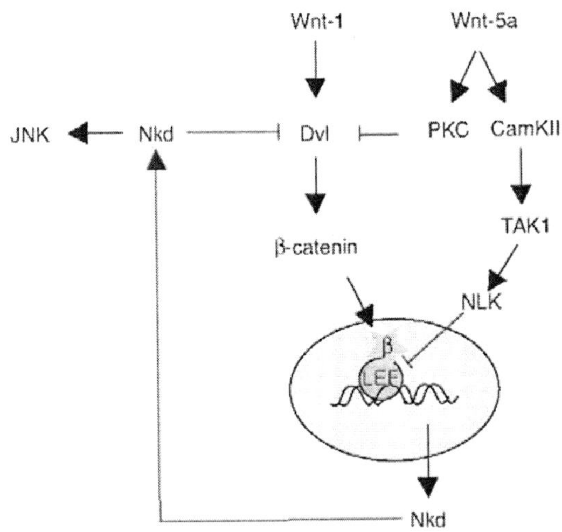

Figure 3. Inhibition of the canonical Wnt-1 pathway by Nkd and the Wnt-5a/PKC/Ca2+ pathways

(see also Paragraph 2 below): (1) a class of Wnt ligands, including Wnt-1 and Wnt-8, which display these two biological functions through β-catenin; (2) a class of Wnt ligands lacking these activities, such as Wnt-5a and Wnt-11, both signalling through the Fz /PCP and Wnt/Ca^{2+} pathways.

Interactions between several signalling cascades take place in cells exposed to Wnt factors of different classes. Wnt-5a can partly reverse the morphogen activity of Wnt-1 in C57mg mammary epithelial cells (Olson, 1998). In *Xenopus*, Wnt-5a blocks Wnt-8 -induced formation of an additional axis. Activation of the Wnt /Ca^{2+} pathway inhibits the transduction of the Wnt-8 /β-catenin signals (Torres, 1996; Slusarski, 1997a). In *Drosophila*, DWnt4 and Dfz3 display an antagonistic action on Wg signalling, as shown by the reduced accumulation of Arm /β-catenin (Sato, 1999; Gieseler; 1999).

A recent study of Kühl (2001) indicated the participation of Wnt-8/ β-catenin signalling in the elongation of the Xenopus embryo. These authors showed that Wnt-8 acted as a positive regulator of the elongation *via* the stabilisation of β-catenin and the ensuing transcription of the *Xnr3* gene. On the other hand, the Wnt /Ca^{2+} signals induced by Wnt-5a exerted an inhibitory action, possibly through the inhibition of the Wnt/ β-catenin pathway on two levels targeting respectively Dvl and LEF (Fig.3). Interestingly, Wnt-2 is overexpressed in human colon cancer (Vider, 1996). The fact that the inhibition of the β-catenin signalling by PKC was reversed by overexpression of β-catenin or of a dominant -negative variant of

GSK3 suggests that the negative control by PKC operates upstream of β-catenin degradation through Dsh phosphorylation. In contrast, the negative control of the canonical Wnt pathway by CamKII occurs downstream of β-catenin, through LEF-1 phosphorylation (Fig. 3). In agreement, Ishitani (1999, 2003) demonstrated that expression of Wnt- 5a or of a dominant constitutive variant of CamKII, as well as the activation of TAK1 and overexpression of NLK, all inhibited the transcriptional activity of β-catenin, without altering its cellular levels or localisation. Moreover, the injection of mRNA encoding NLK on the dorsal side of *Xenopus* embryo induced its ventralisation, a phenotype similar to that induced by GSK-3. This intervention also blocked the formation of a secondary embryonal axis induced by the ventral injection of β-catenin mRNA, but not the transcription of *Siamois* and *Twin*, two genes induced by β-catenin. These observations suggest that TAK1 /NLK acts downstream of β-catenin and upstream of Siamois and Twin. Thus, these data pinpoint LEF factors as targets of the NLK kinase. NLK could associate with and phosphorylate LEF/TCF coupled to β-catenin, causing the dissociation of the complex between LEF /TCF / β-catenin and target DNA sequences.

Inhibition of the Canonical Wnt Pathway by Nkd Proteins

The transcription of the genes encoding Naked cuticle (Nkd) proteins is induced by Wnt signalling. In turn, these proteins act as antagonists of the Wnt/β-catenin pathway, both in *Drosophila* and vertebrates. For instance, Nkd inhibits the XWnt-8 -induced axis duplication in the ventral blastomere of *Xenopus* embryo (Zeng, 2000; Yan, 2001a). The target of Nkd is upstream of β-catenin as Nkd inhibits the transcriptional effects of Wnt-1, but not of β-catenin. The so-called "EF-hand" domain of Nkd (bearing a Ca^{2+}-binding helix-loop-helix motif commonly present in Ca^{2+}-binding proteins) appears to directly interact with the central region of Dsh that contains a protein/protein interaction domain PDZ and an adjacent basic region at its N-terminal terminus (Yan, 2001a; Wharton, 2001; Rousset; 2001). The same domain of Dsh also binds GBP /Frat and Casein Kinase 1ε, two positive regulators of the canonical Wnt pathway (Li, 1999; Peters, 1999). Thus the binding of Nkd prevents the interaction between Dsh and positive regulatory proteins (Fig.3). Nkd may also be involved in Fz/PCP signalling. Firstly, Nkd induces the phosphorylation of c-Jun, suggesting that it can activate the JNK pathway. Secondly, overexpression of Nkd perturbs the normal elongation of *Xenopus* embryo in a manner independent of the Wnt /β-catenin pathway since GSK-3β is unable to reverse this effect (Yan, 2001a). Therefore, Nkd appears to interfere with the convergent extension by stimulating Fz/PCP signalling. It is possible that this protein acts as a molecular switch, directing the Wnt signal from the canonical to the PCP pathway. In addition, the presence of the Ca^{2+} binding domain in Nkd suggests that its activity is regulated by the Wnt / Ca^{2+} pathway.

Other Wnt -inducible Pathways

In spite of much information on Wnt signalling, many questions remain unanswered. For example, it is not quite clear whether the Wnt/Ca^{2+} and Fz/PCP pathways are really distinct.

Prickle, a specific member of the PCP pathway, involved in the localisation of the Fz-Dvl complex, also stimulates Ca^{2+} release in zebrafish. An atypical form of Dvl targeting the PCP pathway can also activate the Wnt/Ca^{2+} in *Xenopus* (Sheldahl, 2003; Veeman, 2003). Altogether, Dsh /Dvl controls at least three major Wnt pathways during the integration of the signals initiated by Fz receptors. To date, 18 partners of Dsh /Dvl have been identified (Axelrod, 1998; Wharton, 2003; Cong, 2004a). A recent study demonstrated that the Wnt/Ca^{2+} activator Rfz2 down -regulates cGMP levels. Rfz2 activates heterotrimeric G-proteins and phosphodiesterases (PDE) involved in cGMP destruction. Inhibition of these PDE affects both Ca^{2+} flux and gastrulation in Xenopus (Ahumada, 2002; Wang, 2003).

These pathways may not entirely cover the modalities of Wnt-dependent signal transduction. For instance, a recent report (Chen, 2005) suggests that embryonal myogenesis may be induced by Wnt-activated Frizzled receptors coupled to trimeric G-proteins that activate adenylyl cyclase, the PKA- CREB pathways, and induce activation of the myogenic MYF5 and PAX3 promoter genes (Chen, 2005).

2. Canonical Wnt Pathways: Molecular Elements and Mechanisms

2. a. Wnt Ligands

The *Wnt* genes are defined by sequence homology with *Wnt-1* and *Wg* (Wingless), as the first members of this family identified in mouse and *Drosophila*, respectively. Subsequently, highly conserved Wnt homologues were found in a large variety of vertebrate and invertebrate species. In the human genome, 19 *Wnt* genes are known. They encode secreted glycoproteins of 350-400 aminoacids, characterized by a conserved motif of 21 - 23 Cys (Fig. 4). In Drosophila, these Cys have been shown to form S-S bridges that direct protein folding. Wnt proteins are monomeric *in vivo* (Tanaka, 2002).

Figure 4. Structure of Wnt ligands.

Post- translational Modifications of Wnt Proteins

Early genetic evidence suggested post-translational modification of Wnt proteins. The genes encoding *porcupine* (*porc*) in *Drosophila* and *mom-1* in *Caenorhabditis elegans* are required for Wnt signalling, and display homologies with acyltransferases (Kadowaki, 1996; Rocheleau, 1997). In the case of *porc*, these homologies concern membrane-bound acyltransferases which are localised at the ER and acylate various substrates (Hofmann, 2000). Due to their physicochemical nature, the Wnt proteins are difficult to isolate, and their purification has only recently been successful. The polypeptide chain has indeed been found to be modified by covalent attachment of a lipidic palmitoyl-type group to a cysteine residue: Cys77 in the mouse Wnt-3a, Cys 58 in Dwnt-8 (Fig.4). The palmitoylated Cys is present in all Wnt family members, and is situated at the N-terminal position of the conserved Cys motif (Willert, 2003). Experiments with *Drosophila* and *C.elegans* suggest that it is essential for Wnt function. Indeed, its replacement by a serine in one allele of *egl-20* in *C.elegans* or by a tyrosine in one allele of *Wg* in Drosophila inactivate the Wnt/Wg proteins (Maloof, 1999; Couso, 1994). Moreover, enzymatic de-palmitoylation by APT-1 (acyl-protein thioesterase-1) inhibits β-catenin accumulation induced by Wnt-3a (Willert, 2003).

The similarities between the *wg* and *porc /mom-1* phenotypes indicate that Porc and MOM-1 are part of Wnt signalling, and underline the importance of the lipid group for Wnt activity. Overexpression of Wnt-3a mutated at the site of attachment to the palmitoyl group leads to β-catenin accumulation but less efficiently than wild-type Wnt-3a (Willert, 2003). The lipid is apparently not indispensable but may serve to concentrate Wnt at the membrane.

In embryos bearing mutated *porc,* the Wg protein is retained in the intracellular space (Kadowaki, 1996). Wnt secretion depends on post-translational mechanisms that are decisive for the formation of S-S links ensuring the appropriate tridimensional structure of the protein (Nusse, 2003). Besides its role in the acylation of Wnt, Porc has been shown to physically interact with Wg and stimulate its N-glycosylation (Tanaka, 2002). The covalent attachment of a lipid residue and the resulting association with the ER membrane may also favor Wnt glycosylation.

Classification of Wnt Ligands

Ectopic expression in *Xenopus* and zebrafish revealed two types of Wnt activity. One type can be mimicked by lithium, an inhibitor of GSK-3β. The induced phenotype is characterized by the duplication of the dorsal axis in the *Xenopus* embryo. First observed with Wnt-1, this class includes Wnt-2b, 3a, 8a and 8b (McMahon, 1989; Landesman, 1997; Wolda, 1993; Christian, 1993; Cui, 1995). Wnt factors of the second class (Wnt-5a, 4 and 11) interfere with the morphogenetic movements in the embryo (Moon, 1993; Du, 1995, Ungar, 1995). In agreement with this classification, not all of the mouse Wnt factors can transform mammary epithelial cells C57mg. Wnt-1, -3a and -7a display a strong transforming activity, whereas Wnt-4, 5a and 6 are not transforming in this model. Wnt-2, -5b and -6 have a transforming activity but weaker than Wnt-1, and may define a third Wnt class (Wong, 1994). In fact, the distinction of two classes may be too simplistic. The study of Wnt-5a reveals additional complexities in Wnt signalling. Not only does this factor inhibit the Wnt-8-induced canonical pathway (Torres, 1996), but in the presence of the Hfz4, Xfz5, and Xfz7 receptors, Wnt-5a induces axis duplication. This effect, as in the case of the Wnt-1 class is mediated by the inhibition of GSK3 (He, 1997, Umbhauer, 2000). One can postulate that all Wnt factors are potentially compentent to induce an ectopic axis, depending on the expression of appropriate Fz receptors at appropriate time during development (Wodarz, 1998).

Apart from their essential role in the development of invertebrates and lower vertebrates (*Xenopus*, zebrafish), *Wnt* genes intervene also at numerous sites during the development of the mouse embryo, including the central nervous system and limbs. Several Wnt factors are often co-expressed, suggesting that their functions may be overlapping in part or redundant. Studies based on loss-of-function mutations confirmed that Wnt are essential for the normal embryonic development. For instance, Wnt-1 deficiency causes malfunction of the cerebellum and mesencephalic part of the brain, and few animals survive after birth (McMahon, 1990; Thomas, 1990; McMahon, 1992). Wnt-2 is expressed in the heart and placenta and is required for the correct placental vascularisation (Monkley, 1996). Loss of function of Wnt-3a affects the formation of somites and of the hippocampus (Takada, 1994, Greco, 1996; Yoshikawa, 1997, Lee, 2000). Wnt-3a in association with Wnt-1 also regulates the expansion of the progenitors of the neural crest (Ikeya, 1997).

2.b. Wnt Receptors

Five members of the family Frizzled (Fz) have been identified in Drosophila, and 9 in man. They display an extracellular N-terminal region, followed by the 7 transmembrane domains typical for the G-protein coupled receptors (GPCR), and a cytoplasmic C-terminal tail. The N-terminal region contains a putative signal sequence followed by 120 aminoacid sequence that includes 10 conserved Cys residues (CRD, for cysteine-rich domain) and a highly divergent hydrophilic sequence (40 to 100 aminoacids) that constitutes the flexible junction between the extracellular and transmembrane domains (Fig.5). The CRD domain interacts with Wnt ligands (Cadigan, 1998; Lyons, 2004). A majority of the Fz family members contain the conserved Lys-Thr- X-X-X-Trp element situated 2 aminoacids downstream of the last transmembrane segment. This element appears to be necessary for the signal transduction to the canonical Wnt/β-catenin pathway. Following the activation of the Fz receptor, Dvl proteins are translocated from the cytoplasm to the plasma membrane. The conserved sequence element may be necessary for the translocation and subsequent phosphorylation of Dvl (Umbhauer, 2000). Besides, a direct interaction between Dvl and the C-terminal portion of Fz has recently been demonstrated (Wong, 2003). Certain Fz receptors also possess also a Ser/Thr-X-Val motif that may constitute a binding site for scaffolding proteins with multiple PDZ protein interaction domains. The length of the intracellular Fz portion is variable. In Drosophila, Fz1 and Fz2 display 45% identity over the first 29 aminoacids, while Fz2 has 61 additional aminoacid residues. Both signal to the canonical pathway, but only Fz1 activates the PCP pathway. They are differentially localized in the imaginal disk epithelium. The C-terminal tail is decisive for this localisation and signalling: Fz1 is preferentially localized in the apical region with members of the PCP pathway. In contrast, Fz2 receptors are uniformly distributed over the basolateral side where they preferentially bind Wg, an agonist of the canonical pathway (Wu, 2004).

Specificity of Wnt-Fz Interactions

Given the large number of combinations between the Wnt ligands and Fz receptors, the specificity of interactions remains obscure. Table I shows the results of a study of the binding of 5 out of 7 Wnt with CRD domains to the Drosophila Fz receptors.

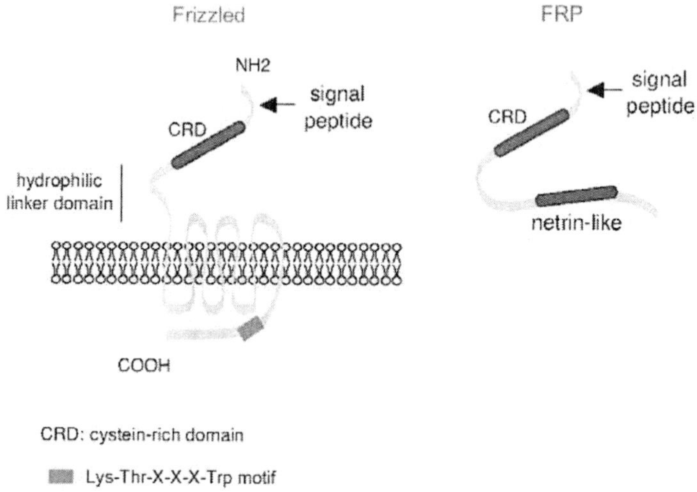

Figure 5. Structural homologies between Frizzled receptors and soluble FRP inhibitors.

As in other signalling systems, a given ligand can bind to different receptors, as observed in the glucagon-secretin receptors superfamily. Wg binds to DFz1, DFz2, and DFz3 (Gazit, 1999; Wu, 2004). Fz can act in a redundant manner since the Wingless phenotype of *Drosophila* is observed only when both DFz1 and DFz2 are mutated (Bhanot, 1999; Chen, 1999). Fz seem to act in a redundant manner since the Wingless phenotype of *Drosophila* is observed only when both Fz1 and Fz2 are mutated (Bhanot, 1999; Chen, 1999).

Table I: Binding specificity between Drosophila Wg/Wnts and Fz receptors.

	Fz (Fz1)	DFz2	DFz3	DFz4	Smo
Wg	+	+++	+	-	-
DWnt2	+	+	+	-	-
DWnt3	-	-	-	-	-
DWnt4	++	++	-	+	-
DWnt8	-	-	-	+	-

Transduction of Wnt Signals to Heterotrimeric G-Proteins

In agreement with their serpentine receptor structure, experimental data obtained with the rat RFz1 and RFz2 receptors support the model in which the Wnt signal is transduced by coupling between Fz receptors and heterotrimeric G-proteins. For instance, the differentiation of F9 teratocarcinoma cells is induced by XWnt-8 via RFz1, and this effect is blocked by

PTx which impedes specifically Gαo/i subunits. Neutralisation of Fz signalling is also observed by oligonucleotides directed against Gαo et Gαq (Liu, 1999b). The isoproterenol - induced signalling by a chimeric RFz1/ β-adrenergic receptor is also blocked by the same inhibitors (Liu, 2001). Conversely, GTPase-deficient, constitutively active Gαo and Gαq subunits, activate LEF/TCF-dependent transcription (Liu, 2001). In *Xenopus*, the activation of the Ca2+ flux by RFz2 and Wnt-5a is abolished by PTx, as well as by a non-hydrolyzable GDP analog (which irreversibly arrests protein G-dependent processes), and by the Gβγ sequestering agent Gα-transducin (Slusarski, 1997b). Finally, PTx as well as antisense RNA oligonucleotides targeting Gαo, Gαt2, and Gβ2 block the RFz2/ -β-adrenergic receptor -mediated signal (Liu, 1999c). Thus, Fz receptors are, at least in part, coupled to G-proteins, as illustrated for Rfz1 and Gαo/Gαq, RFz2 and Gαo, as well as Gαt2 and Gβ2.

Oligomerization of Frizzled

Fz, like other GPCR, form homo- or hetero-oligomers (Carron, 2003; Kaykas, 2004). By complementation test in 293Y cells, Kaykas and collegues showed that Fz1, 2, 4, 7 and 9 could oligomerize with each other, and that the closer the phylogenetic proximity the stronger was the interaction. Interestingly, Fz7 interacted strongly with all other forms. The Fz oligomers may form in the ER, during or after the synthesis of the receptor polypeptide, and independently of ligand binding. The oligomers may then be transported to the plasma membrane. In the early *Xenopus* embryos, the oligomerization of Xfz3 is sufficient to activate the canonical signalling in the absence of the Wnt ligand (Carron, 2003). Thus, oligomerization of the Fz GPCR contributes to signal transduction, in the sense that different combinations of isoforms may favor specific pathways downstream.

Similar to the Wnt ligands, Fz expression varies according to the tissue and cell type, often in overlapping fashion, so that a given tissue can express several Fz (Table II).

Table II : Expression of the GPCR Fz in human brain and peripheral tisues.

Receptors	Fz expression in human tissues	References
Fz1	Abundant in heart, placenta, lung, kidney, pancreas, prostate and ovary in adults, as well as in fetal lung and kidney.	Sagara, 1998
Fz2	Adult heart; fetal brain, lung and kidney.	Sagara, 1998
Fz4	Adult heart (+++), skeletal muscle (+++), ovary (+++), fetal kidney (+++), adult liver (++), kidney (++), pancreas (++), spleen (++), fetal lung (++), placenta (+), adult lung (+), prostate (+), testes (+), colon (+), fetal brain (+) and liver (+).	Kirikoshi, 1999
Fz5	Fetal liver and lung. Adult liver, kidney, pancreas.	Saitoh, 2001a
Fz6	Fetal brain, lung, liver, kidney. Adult brain, placenta, lung, liver, skeletal muscle, kidney, pancreas, thymus, prostate, testes, ovary, small intestine, and colon.	Tokuhara, 1998
Fz7	Adult skeletal muscle, heart, brain. Fetal development: kidney and lung.	Sagara, 1998

Table II : Expression of the GPCR Fz in human brain and peripheral tisues. (cont.)

Receptors	Fz expression in human tissues	References
Fz8	Fetal kidney (+++), brain and lung. Adult: kidney, heart, pancreas and skeletal muscle.	Saitoh, 2001b
Fz9 (Fz3)	Predominant in: fetal brain, testes, skeletal muscle; adult kidney.	Sala, 2000
Fz10	Strong expression in adult placenta. Fetal kidney, lung and brain.	Koike, 1999

2. c. LRP Coreceptor Family

The distinction between the different signalling pathways is essential to ensure an adequate transduction of the Wnt signals. How does this distinction operate? As debated above, the differential expression of the Wnt and Fz isoforms can preferentially trigger a given signalling pathway. At overlapping expression sites, features such as differential affinities characteristic of the ligand/receptor pair, as well as the subcellular localisation of the receptor, and presence of Fz oligomers, or heterodimers with other GPCR, may contribute to the specification of the signals. In addition, the presence of Fz co-receptors is yet another crucial element involved in the process of Wnt signal transduction.

Genetic studies have established that Wnt/β-catenin signal tranduction requires Fz co-receptors that belong to the Low density lipoprotein-receptors Related Proteins (LRP) family (Fig.6), i.e. Arrow in Drosophila, and LRP5 / LRP6 in mammals. LRP co-receptors are transmembrane proteins comprising some 1,600 aminoacids. The extracellular domains of LRP5 and LRP6 show 64-73 % identity with Arrow. Overall, there is about 40% identity between Arrow and LRP5/6 (He, 2004). LRP proteins are characterized by four scattered "EGF" motifs in their extracellular part, followed by 3 groups of LDL receptor motifs (LDL-R repeats) adjacent to the transmembrane domain, and a proline /serine-rich-cytoplasmic domain (Bejsovec, 2000). Drosophila embryos bearing mutations in arrow display the wingless phenotype (Wehrli, 2000). The phenotype of the LRP6 -/- mice is similar to that produced by inactivation of Wnt-1, Wnt-3a and Wnt-7a. In contrast, the LRP6-/- mice do not resemble those bearing inactivated wnt5a, suggesting that LRP6 is implicated in a sub-class of Wnt-Fz signals (Pinson, 2000). It is possible that the LRP co-receptors only participate in the canonical Wnt/β-catenin pathway. Mutation in arrow does not prevent the establishment of planar polarity, and LRP6 does not interfere with the XWnt-11-dependent morphogenetic movements (Wehrli, 2000; Semenov, 2001).

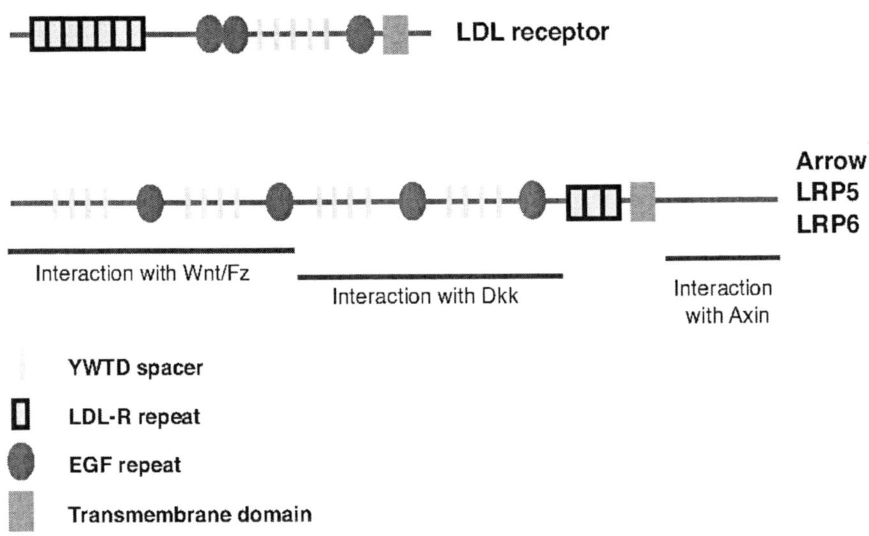

Figure 6. Structure of LRP co-receptors.

Interactions of LRP with Wnt and Fz.

LRP intervene in numerous developmental mechanisms that rely on the canonical Wnt/β-catenin signalling. For instance, LRP6 in Xenopus activates the transcription of β-catenin target genes and induces the duplication of the embryonal axis. It also synergizes with Wnt-5a et HFz5 (Tamai, 2000). LRP5 potentiates the actions of Wnt-1 and Wnt-3a (Mao, 2001b). What is the mechanism by which LRPs exert their activity? In fact, they intervene downstream of Wnt and Fz, and upstream of Dvl (Tamai, 2000; Wehrli, 2000). The LRP5/6 co-receptors form a complex with Fz in the presence of Wnt. The extracellular domain of LRP5/6 binds Wnt-1 and forms a complex with the CRD domain of mFz8, in a Wnt-1 -dependent manner (Tamai, 2000). Both LRP5 and LRP6 bind Wnt-1 (Mao, 2001a). The region of LRP6 that is required for the interaction with the Wnt/Fz complex is located at the extracellular portion of the molecule encompassing the EGF repeats 1 and 2 (Fig.6), (Mao, 2001b). LRP5 lacking the cytoplasmic and transmembrane domains act as dominant- negative variants, probably by sequestering Wnt. Conversely, LRP lacking the extracellular domain behave as dominant constitutive variants capable to activate the canonical Wnt pathway in the absence of ligand (Mao, 2001b). The intracellular domain of LRP5/6 is therefore implicated in signalling downstream of Wnt receptors.

Interaction of LRP with Axin

A possible mechanism underlying the implication of LRP in Wnt signalling relies on the destabilisation of the β-catenin degradation complexes by recruitment of axin (Fig. 6). LRP5 /6 directly interact with the C-terminal region of axin, and this interaction is critical for their function in the canonical Wnt pathway (Mao, 2001b; Tamai, 2004). In the presence of Wnt-3a, axin rapidly associates with the plasma membrane, via LRP. The axin /LRP complex has been detected as early as 4 min after stimulation, suggesting that this process is involved at the initiation of the Wnt cascade. The phosphorylation of Dvl and dephosphorylation of axin occurs later after 30 min, while axin is degraded at 2h (Willert, 1999). It has been proposed that axin recruitment favors its degradation (Mao, 2001b).

The C-terminal region of LRP5/6 contains 5 PPP(S/T)P repeats involved in Wnt signalling. When phosphorylated, these motifs are acting as axin binding sites (Tamai, 2004). Remarkably, the kinase activity of GSK-3β is required for the axin /LRP intercation (Mao, 2001b). It is plausible that LRP5/6 only associate with axin after their phosphorylation by GSK-3β. The phosphorylation of axin increases its capacity to bind β-catenin and GSK-3β (Ikeda, 1998; Willert, 1999). Thus, LRP may recruit the axin molecules that are engaged in β-catenin degradation. Alternatively, GSK-3β may phoshorylate another hypothetical kinase that in turn phosphorylates the PPP(S/T)P motifs of LRP.

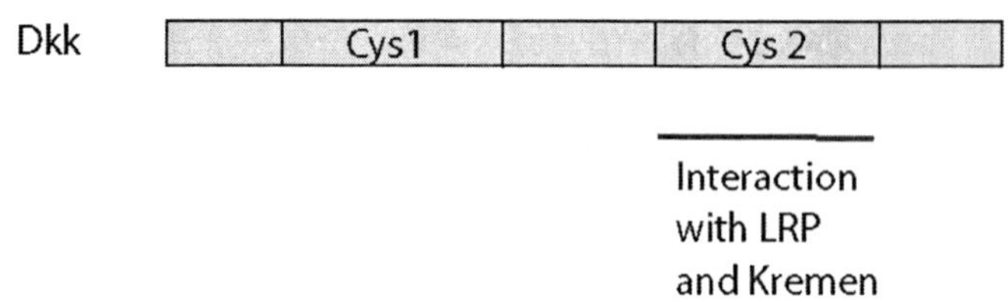

Figure 7. Structure of Dickkopf proteins.

2. d. Wnt Signalling Inhibitors

Two classes of secreted Wnt antagonists have been described: sFRP (soluble Frizzled-Related Protein) and Dickkopf (Fig.7). The sFRP family members, sFRP, WIF-1 (Wnt inhibitory factor) and Cerberus, bind Wnt factors and alter their capacity to interact with Fz. The Dickkopf proteins interfere with Wnt signalling by binding to the LRP5/6 co-receptors. Thus, in theory, sFRP inhibit canonical as well as non-canonical pathways, whereas Dickkopf proteins are specific for the Wnt /β-catenin pathway (Kawano, 2003). Eight sFRP proteins have been described. A simplified classification of the sFRP (1 to 5 in mammals) is presented in Table III.

Table III: Soluble Wnt inhibitors FRP (sFRP) targeting Fz receptors and signaling.

sFRP-1	SARP-2, FrzA, Frp
sFRP-2	SARP-1, SDF-5
sFRP-3	FrzB, Frzb-1, Fritz
sFRP-4	FrpAP, frpHE
sFRP-5	SARP-3

Three additional sFRP have been identified in *Xenopus* and chicken, namely Sizzled, Sizzled2 and Crescent (Kawano, 2003). Similar to Fz receptors, sFRP possess a CRD domain comprising 10 conserved Cys residues and display 30 to 50% homology with the CRD of Fz. The C-terminal region of sFRP contains a netrin-like sequence (Fig. 5), a protein involved in axonal guidance and apoptosis controlled by the tumor suppressor gene DCC (deleted in colon cancer).

Interactions Between sFRP, Wnt and Fz

A direct interaction between sFRP and Wnt has been clearly established, as well as inhibition of Wnt signals by sFRP. Like in Fz, the interaction domain of sFRP with Wnt is the CRD region (Fig. 5). Deletion of CRD in sFRP-3 abolished the interaction with Wnt-1 (Lin, 1997). In addition, the CRD of sFRP-1 is able to bind Wnt-1 and -2 (Bafico, 1999). However, the binding to Wnt may not be sufficient for the function of sFRP: indeed, sFRP-3 binds Wnt-5a but does not inhibit its activity (Lin, 1997). Additional facts lead to the conclusion that the actions of sFRP are Wnt-specific. In particular, sFRP-3 inhibits the actions of Wnt-1 and Wnt-8, but is inactive against Wnt-3a, Wnt-5a, and Wnt-11 signalling (Lin, 1997, Wang, 1997). The inhibitory effect of sFRP-1 on the duplication of axis induced by Wnt-3a is much stronger than in the case of Wnt-1, Wnt-8 or Wg (Finch, 1997). Besides Wnt, sFRP have been shown to interact also with Fz, and in particular, the CRD domains of sFRP-1 and HFz6 bind to each other in vitro (Bafico, 1999). This interaction can explain the inhibitory activity of sFRP in certain physiological situations. Remarkably, low sFRP-1 concentrations are associated with the Wg-induced β-catenin accumulation, whereas higher concentrations are inhibitory (Uren, 2000). During normal development, gradients of diffusible factors specify the differentiation of embryonic cells. Overlapping gradients of Wnt and sFRP modulate Wnt signalling.

Dickkopf Proteins

The Dickkopf (Dkk) family include Dkk-1 to -4, as well as Soggy, a protein closely related to Dkk-3 (Krupnik, 1999). In Dkk, two cysteine -rich domains Cys1 and Cys2 are present, each containing 10 conserved Cys residues (Fig.7). The topology of Cys2 resembles that of co-lipases that facilitate the interactions between pancreatic lipases and lipid micelles. Soggy lacks both Cys1 and Cys2 domains. Like sFRP, Dkk were identified as secreted Wnt antagonists at the duplication of the Xenopus embryonic axis, accumulation of β-catenin and transcription of Wnt target genes. In contrast with FRP, Dkk act specifically on the canonical pathway (Semenov, 2001).

The two classes of secreted Wnt antagonists act through different mechanisms. Unlike sFRP, Dkk do not interfere with the Wnt/Fz interaction (Fig.8), but instead interact with the LRP co-receptors (Bafico, 2001; Mao, 2001a; Semenov, 2001). The C-terminal end of Dkk interacts with LRP6 (Brott, 2002). The region of LRP involved in the interaction with Dkk is the extracellular domain containing the EGF-like motifs 3 and 4 (Mao, 2001a). Dkk also interact with another transmembrane receptor family, Kremen (Krm-1 et -2). Again, the Cys2 repeats of Dkk interact with the extracellular domain of Krm to form a LRP-Dkk-Krm complex. This ternary complex is internalized by endocytosis and degraded (Mao, 2002).

This leads to a reduction of the number of LRP molecules at the cell surface and possibly to a less efficient formation of Wnt/LRP complexes (Fig. 8).

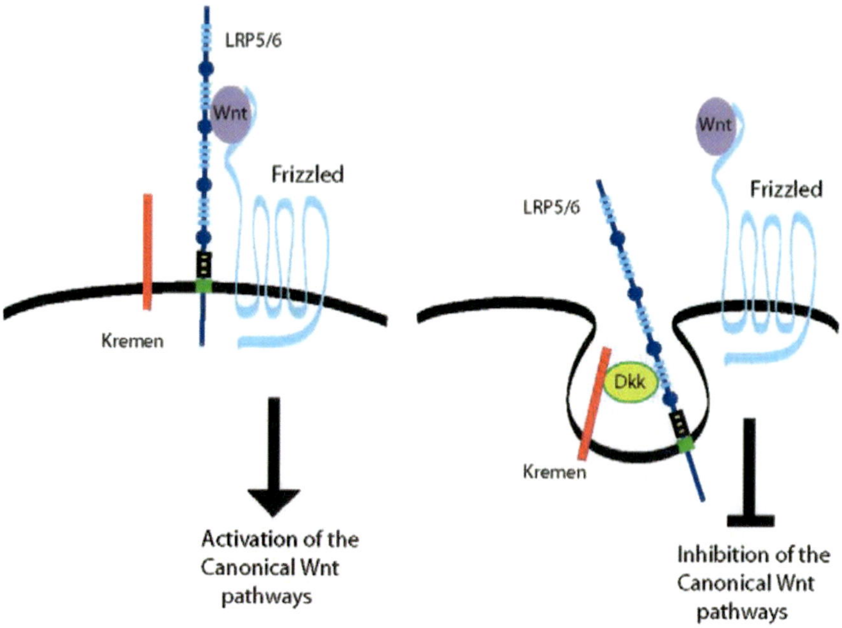

Figure 8. Inhibition of the canonical Wnt signaling pathways by Dkk proteins.

Dkk-1 has been identified in Xenopus where it inhibits the XWnt-8 -induced axis duplication (Glinka, 1998). While Dkk-1 and Dkk-4 clearly antagonize the canonical Wnt pathway, Dkk-3 has no effect, and does not interact with either LRP5/6 or with Kremen1/2 (Wu, 2000; Mao, 2001a; Mao, 2002). Dkk-2 acts as Wnt agonist or antagonist, according to the cellular context (Krupnik, 1999; Wu, 2000; Brott, 2002). Although both Dkk-1 and Dkk-2 interact with LRP5/6 and Krm, and inhibit β-catenin -dependent transcription (Wu, 2000; Brott, 2002), Dkk-2 activates Xenopus Wnt/ β-catenin pathway in synergy with LRP6 (Brott, 2002). Studies with deletions and chimeric constructs indicate that the functional differences between Dkk-1 and Dkk-2 C-terminal sequences are determined by the N-terminal domain. When the N-terminal domain of Dkk-1 was fused to the C-terminal domain of Dkk-2, the resulting protein behaved as Dkk-1, suggesting that the N-terminal Dkk-1 domain inhibited the function of the C-terminal domain (Brott, 2002) by abrogating its capacity to bind LRP6. The C-terminal region of Dkk-2 is also involved in the interaction with Krm-2. Co-expression of Krm-2 and Dkk-2 is sufficient to convert Dkk-2 from agonist to antagonist of LRP6 (Mao, 2003), through internalisation and degradation of LRP6.

3. Transmission of Wnt Signals to the β-Catenin Degradation Complex

The activation of Fz and its co-receptors LRP stabilizes β-catenin. In the basal state, the half-life of cytoplasmic β-catenin is regulated by the β-catenin degradation complex containing APC, GSK-3β and axin (Fig.9). Activation of the canonical Wnt pathway destabilizes this complex. In the following Paragraph, we will review the mechanisms by which the Wnt signal is transduced to the degradation complex.

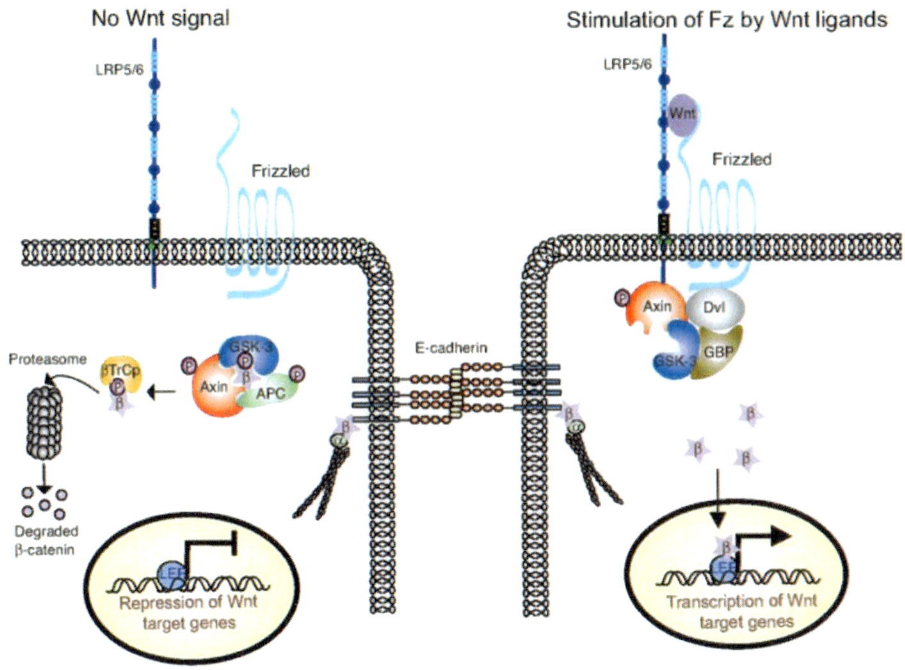

Figure 9.The canonical Wnt pathways.

β : β-catenin

3. a. Dishevelled Proteins (Dvl).

Although the function of Dvl proteins is not yet sufficiently elucidated, it is clear that they are active elements in the transduction of the Wnt signal downstream of Fz receptors. All organisms which use Wnt/ β-catenin signalling express Dvls: a single one in Drosophila (Dsh), and three in mammals (Dvl-1, -2, -3). Genetic evidence places Dvl between Fz and β-catenin. In addition, Dvl overexpression induces the accumulation of cytoplasmic β-catenin and activates LEF/TCF-dependent transcription (Lee, 1999; Smalley, 1999).

Structure of Dvl

Dvl proteins possess three conserved domains, DIX, PDZ and DEP. In addition, Dvl display regions with conserved basic aminoacid residues, and numerous potential phosphorylation sites (Wharton, 2003). The roles of these different regions in Dvl signalling have been the subject of extensive research. There is consensus that all three conserved domains are involved in Dvl functions at the Wnt/β-catenin and Fz/PCP pathways. The DIX domain, of approximately 80 aminoacid residues at the N-terminal end is found in both dishevelled and axin (DIX) and is involved in homodimer formation (Kishida, 1999b). The central portion of Dvl contains another conserved domain of approximately 90 aminoacid residues, the PDZ signature (***P***ost Synaptic Density-95, **d**iscs-large and **Z**onula occludens-1). This domain is involved in the interaction with the Lys-Thr-X-X-X-Trp motif present in Fz receptors (Wong, 2003). Finally, the DEP (***D***ishevelled, ***e***gl-10, ***p***leckstrin) domain of approximately 90 aminoacid residues at the C-terminal portion of Dvl is also present in RGS proteins which participate in the negative regulation of Gα subunits of the heterotrimeric G-proteins. RGS proteins, like GAP (GTPase Activating Proteins) of Ras -like small GTPases enhance the intrinsic GTPase activity of the Gα subunit, but paradoxically may enhance GPCR activation.

Subcellular Localization of Dvl

Dvl is partitioned in two pools: one associated with stress fibers, the other with cytoplasmic vesicles: endosomes, Golgi-derived vesicles, or transport vesicles (Krylova, 2000). This differential localisation is directed by motifs in the DIX domain that target actin or phospholipids. The phospholipids -targeting motif is critical for the Wnt/β-catenin pathway, suggesting that the Dvl pool involved in the canonical Wnt pathway is associated with vesicles (Capelluto, 2002).

Phosphorylation of Dvl

Stimulation of Fz receptors leads to hyperphosphorylation of Dvl. These proteins contain about 15% of Ser and Thr residues and present multiple phosphorylation sites (Wharton, 2003; Willert, 1997). Wnt-1, Wnt-3a and Wnt-5a induce phosphorylation of Dvl (Lee, 1999; Gonzalez-Sancho, 2004) during activation of both canonical and non-canonical Wnt pathways, independently of β-catenin stabilisation. Dvl phosphorylation is abolished by a dominant -negative Fz mutant, but not when the canonical pathway is inhibited at the LRP co-receptor level. Thus, the phosphorylation status of Dvl depends on the activation of Fz receptors. Deletion of the C-terminal 143 aminoacid residues of Dvl prevents phosphorylation, suggesting that the phosphorylation sites are located in this region (Gonzalez-Sancho, 2004). The region between aminoacid residues 178 and 254 also harbors potential phosphorylation sites but their mutation does not affect the phenotypic rescue function of *dsh* -deficient Drosophila embryo (Penton, 2002).

Dvl phosphorylation is observed 20 to 30 min following stimulation by Wnt, far after the axin-LRP interaction. It is possible that this phosphorylation step may not be required for the Wnt signal transduction but would intervene in secondary processes such as degradation or recycling of the Fz receptors. One of the functions of Dvl could be the removal of the activated Fz from the plasma membrane. Indeed, Dvl-2 appears to allow the recruitment of β-Arrestin 2 to the Fz4 receptor. In the presence of Wnt-5a, β-Arrestin 2 favors the internalisation of the Fz4-Dvl complex by a clathrin-dependent endocytosis (Chen, 2003). This event takes place some 30 min after stimulation by the ligand and Dvl phosphorylation. PKC activity is required for the recruitment of β-arrestin by Dvl-2. Thus, the Wnt signal may induce the phosphorylation of Dvl by PKC or another PKC-associated kinase.

3.b. Dvl-associated Proteins

About twenty interacting Dvl proteins have been described so far (Fig.10). Some appear to intervene specifically at the Wnt /β-catenin pathway, others at the Fz /PCP pathway, and yet others at non-Wnt pathways. Several protein kinases including PKC, casein kinase 1 (CK1), casein kinase 2 (CK2) and Par1, are among proteins that associate with Dvl.

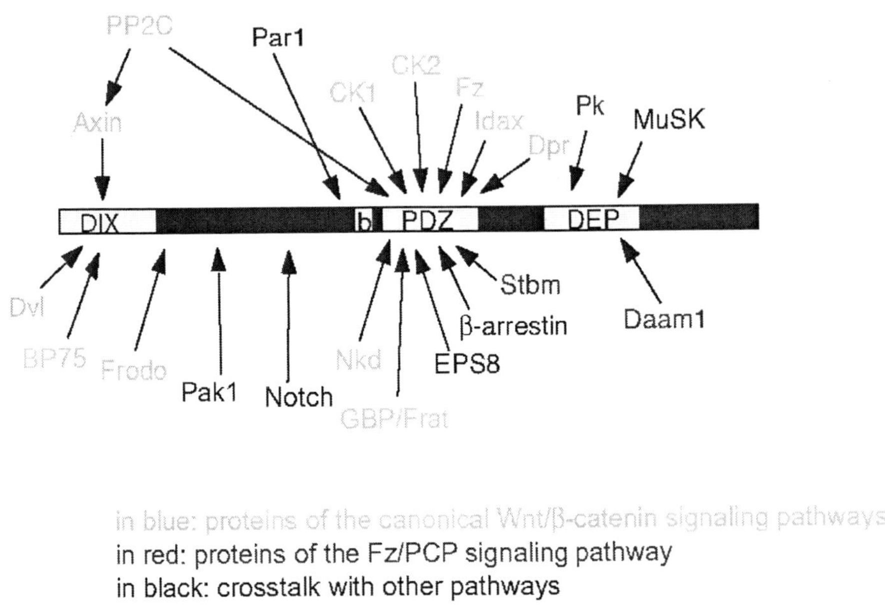

Figure 10. Dvl structure and interacting partners

Different CK1 isoforms are involved in the phosphorylation of other Wnt pathway components: axin, APC, β-catenin and TCF3. CK2 also phosphorylates β-catenin (Wharton, 2003). According to a recent report, the CK1ε isoform acts in synergy with Dvl to activate the Wnt/ β-catenin pathway, while inhibiting JNK/PCP (Cong, 2004a). In HeLa S3 cells, depletion of CK1ε by RNA interference reduced the Dvl-1 phosphorylation in the presence of

Wnt-3a, while CK1α depletion had no effect (Hino, 2003). CK1α silencing in Drosophila S2R+ cells abrogated Dvl phosphorylation during Wg stimulation, and depletion of CK1ε, CK2 and Par-1 was ineffective (Matsubayashi, 2004). It is possible that different protein kinases are implicated in Dvl phosphorylation in mammals and invertebrates.

The PDZ domain of Dvl interacts with the PP2C protein phosphatase. It seems plausible that this interaction may lead to a topologic proximity between PP2C and axin, one of its substrates. Dephosphorylation of axin by PP2C shortens its half-life and reduces its affinity for β-catenin (Strovel, 2000). The PDZ domain of Dvl also binds GBP/ Frat, identified as positive regulators of Wnt/β-catenin signalling. These proteins interact with GSK-3β, and, after recruitment by Dvl, they destabilize the β-catenin degradation complex by interfering with the axin/ GSK-3β association (Wharton, 2003). GBP/Frat binding may be favored by the phosphorylation of Dvl by CK1ε (Hino, 2003). A double-hybrid approach using the DIX domain of Dvl led to the identification of BP-75, a Dvl interacting partner which down-regulates GSK-3β activity by Tyr 216 phosphorylation (Kim, 2003). Protein phosphatase PP2A C-subunit interacts with the axin C-terminus and dephosphorylates axin *in vitro* (Hsu, 1999; Willert, 1999).

3.c. Axins as Scaffold Platforms for the Wnt Signalling Elements

There is a general consensus that the complex which directs the degradation of β-catenin in the absence of the Wnt signal is assembled on the axin "scaffold" containing β-catenin, GSK-3β and APC (Fig. 9). The tumor suppressor axin appears to bring together the Ser/Thr kinase GSK-3β and β-catenin since these two elements do not interact directly in the complex. Axin is about 1,000 times more abundant than the other elements of the degradation complex and constitutes a critical element that controls the canonical Wnt pathway (Lee, 2003). Two key mechanisms regulate axin levels: in the absence of Wnt, axin is stabilised by GSK-3β; in the presence of Wnt, relocalisation by Arrow /LRP directs axin to degradation. It appears that the control of axin stability is more important in Wnt signalling than the regulation of β-catenin phosphorylation by GSK-3β. In Drosophila, the ZW-3 homologue of GSK-3 can be abrogated without disrupting the Wg signal (Tolwinski, 2003).

Both axin and its homologue axin/conductin display an RGS domain and a DIX domain (Fig.11). The RGS motif present in the N-terminal region of axin, is homologous to RGS4 and is involved in the interaction with APC. The DIX domain of Dvl is situated at its C-terminal end, and is required for the interaction with Dvl and axin homodimerisation. The central region of the scaffold protein axin is required for binding to β-catenin and GSK-3β. The minimal portion of the murine axin molecule sufficient to bind GSK-3β has been located between residues 501 and 560 (Smalley, 1999).

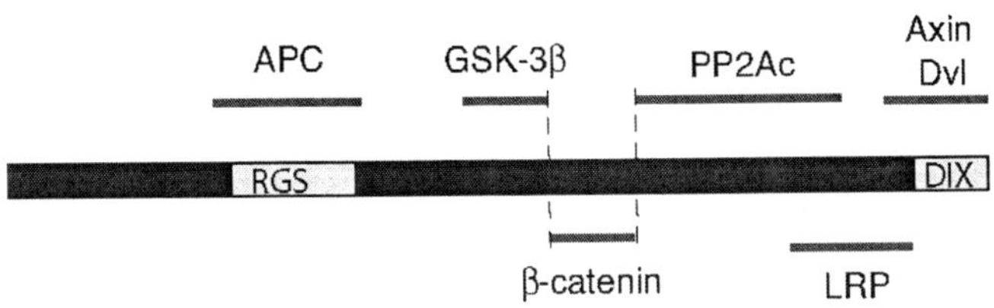

Figure 11. Axin structure and interaction domains with other signaling elements.

Degradation Complex

In the absence of Wnt signal, the formation of the quaternary complex favors the phosphorylation of APC, axin, and β-catenin by GSK-3β (Ikeda, 1998; Kishida, 1998). The phosphorylation of axin prolongs its half-life, whilst that of β-catenin accelerates its degradation. In addition, the phosphorylation of APC and axin, two negative regulators of Wnt signalling, increases their affinity for β-catenin (Rubinfeld, 1996; Willert, 1999).

Although it is generally accepted that the liberation of β-catenin, linked to the inhibition of its phosphorylation by Dvl is a critical event in the canonical Wnt pathway (Kishida, 1999a), the mechanisms involved in the stability of the quaternary complex remain unclear. It has been shown that the stimulation of the Fz/LRP pathway leads to a rapid translocation of axin, a cytoplasmic protein with a vesicular type of distribution, to the plasma membrane. This translocation requires Dvl that interacts with axin via the DIX motif (Cliffe, 2003). Prior to the Fz/LRP stimulation, Dvl is also localized in vesicles and may transport axin at the receptor site, where Dvl binds Fz while axin associates with LRP. In this manner, Dvl appears to recruit the degradation complex. This model is in agreement with the fact that GSK-3β activity is required for the association of axin with LRP (see above and Fig. 9).

The GBP/Frat proteins, which interact with Dvl, are positive co-regulators of the Wnt/β-catenin pathway (Yost, 1998; Yuan, 1999). The binding of Frat to Dvl is increased by CK1ε (Hino, 2003). In contrast, this kinase reduces the formation of the Dvl/axin complexes. This may be a transient structure as Wnt-1 causes the dissociation of Dvl and axin from Frat but not from GSK-3β (Li, 1999). CK1ε can thus replace Frat by axin. The binding of GBP/Frat and axin to GSK-3β is mutually exclusive as their binding domains overlap (Farr, 2000; Ferkey, 2002; Fraser, 2002). Thus, Frat can cause the release of axin - and of β-catenin - from GSK-3β. At the same time, the dephosphorylated axin is degraded. The β-catenin molecule, released from the degradation complex, is now made available at high levels for translocation to the nucleus, to activate the transcription of its target genes, in a LEF /TCF- dependent manner.

Additional Functions for Axin?

Recent reports suggest that, in addition to its role as an anchor to retain β-catenin in the cytoplasm, axin may also shuttle between the nuclear and cytoplasmic compartments (Wiechens, 2004; Cong, 2004b). These authors have identified a nuclear import (NLS) and export sequences (NES) in axin. In the presence of axin, nuclear β-catenin is transported to the cytoplasm. Mutation of the NLS motif inhibits this relocalisation of β-catenin, in favor of the model in which axin commutes between the nucleus and cytoplasm in order to remove β-catenin from the nuclear compartment, to allow its degradation.

3. d. APC

The APC gene product is a critical component of the degradation complex and functions as a negative regulator of the Wnt /β-catenin pathway (Polakis, 1997). A coiled-coil structure of 7 armadillo-type repeats at the N-terminal portion of this large protein (more than 300 kDa) is involved in its oligomerisation (Fig.12). The central part of APC contains domains of interaction with axin and β-catenin. Three 15 aminoacid repeats and 7 motifs, each containing 20 aminoacid residues, participate in the interactions with β-catenin. The 20 aminoacid motifs are characterized by the presence of a consensus site for phosphorylation by GSK-3. Three SAMP domains (Ser-Ala-Met-Pro signature) responsible for the interaction with axin are scattered within the 20 aminoacid repeat region (Behrens, 1998; von Kries, 2000). Conductin's and axin's RGS domains bind the SAMP repeats of APC. The C-terminal APC region contains motifs involved in the interaction with structural proteins: a basic domain involved in the binding to microtubules, a domain of interaction with the microtubule - associated EB1 proteins, and a potential binding site for PDZ -type proteins (Dikovskaya, 2001).

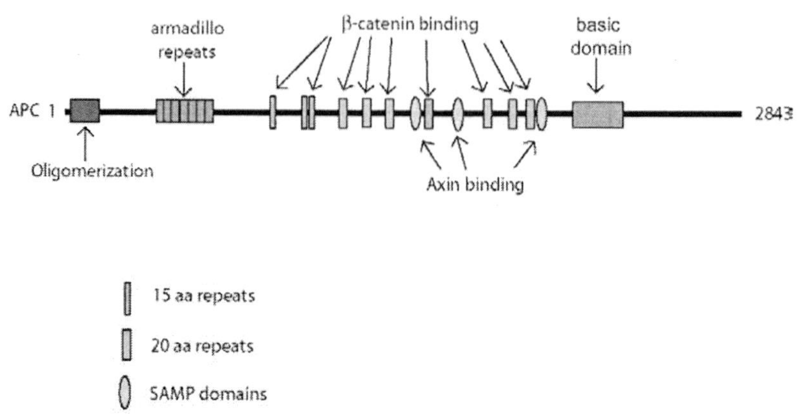

Figure 12. APC protein structure and interaction domains with other signaling elements.

It is generally accepted that APC needs to bind axin in order to favor β-catenin degradation. Although its role remains unclear, APC may regulate the activity of axin. Overexpression of axin is sufficient to counteract the accumulation of β-catenin in APC-deficient cells (Hart, 1998). APC can be detected in different cellular compartments and structures, including cytoplasm, nucleus, plasma membrane at cell-cell junctions, microtubule ends, mitotic spindles, and F-actin filaments (Bienz, 2002; Mogensen, 2002; Näthke, 2004). Actin -dependent pools of APC have been identified at the apical plasma membrane, at lateral membrane interface and near the basal membrane in microtubule-dependent clusters. As axin, APC has nuclear import and export motifs. APC may facilitate the association of β-catenin and axin at the plasma membrane and E-cadherin/β-catenin interactions (Henderson, 2002), suggesting that cell -cell contact associations and dissociations during cell scattering and invasion can supply a new pool of free β-catenin available for nuclear import and canonical Wnt transcription signals. Accordingly, restoration of wt-APC in APC-deficient human colon cancer cells causes an increase in the translocation of E-cadherin to the plasma membrane, resulting in cell-cell adhesion and maintenance of the epithelial phenotype (Faux, 2004). Similar to β-catenin, APC is subject to degradation by the proteasome (Choi, 2004). Wnt signals slow down APC degradation and favor the nuclear accumulation of APC and β-catenin. The stabilisation of APC by Wnt signalling may constitute a limiting step in which APC exported from the nucleus reduces LEF/TCF-dependent transcription.

3. e. GSK-3β

In Wnt signalling, GSK-3β controls the stability of β-catenin within the degradation complex. Interestingly, GSK-3β is a central regulatory element of several signalling pathways regulating glucose metabolism, cell division cycle, stem cell renewal and differentiation, epithelial-mesenchymal transition, cell survival and invasion, and transcription.

Structure of GSK-3

Two genes encoding the GSK-3α and -3β isoforms (respectively 51 and 47 kDa) have been identified in mammals (Fig. 13a). A glycin-rich extension at the N-terminus is inherent to GSK-3α. The kinase domain is highly conserved (98% identity) but there are important differences in primary structure between the α /β isoforms (Woodgett, 1990). GSK-3α and -3β isoforms exhibit different functions since GSK-3β knock -out causes death of mouse embryos and this lethality is not compensated by the α isoform (Hoeflich, 2000). A minor splice variant GSK-3β2 containing a 13 aminoacid insert in kinase domain showed decreased kinase activity towards two phosphorylation sites on tau compared with the conventional isoform (Mukai, 2002).

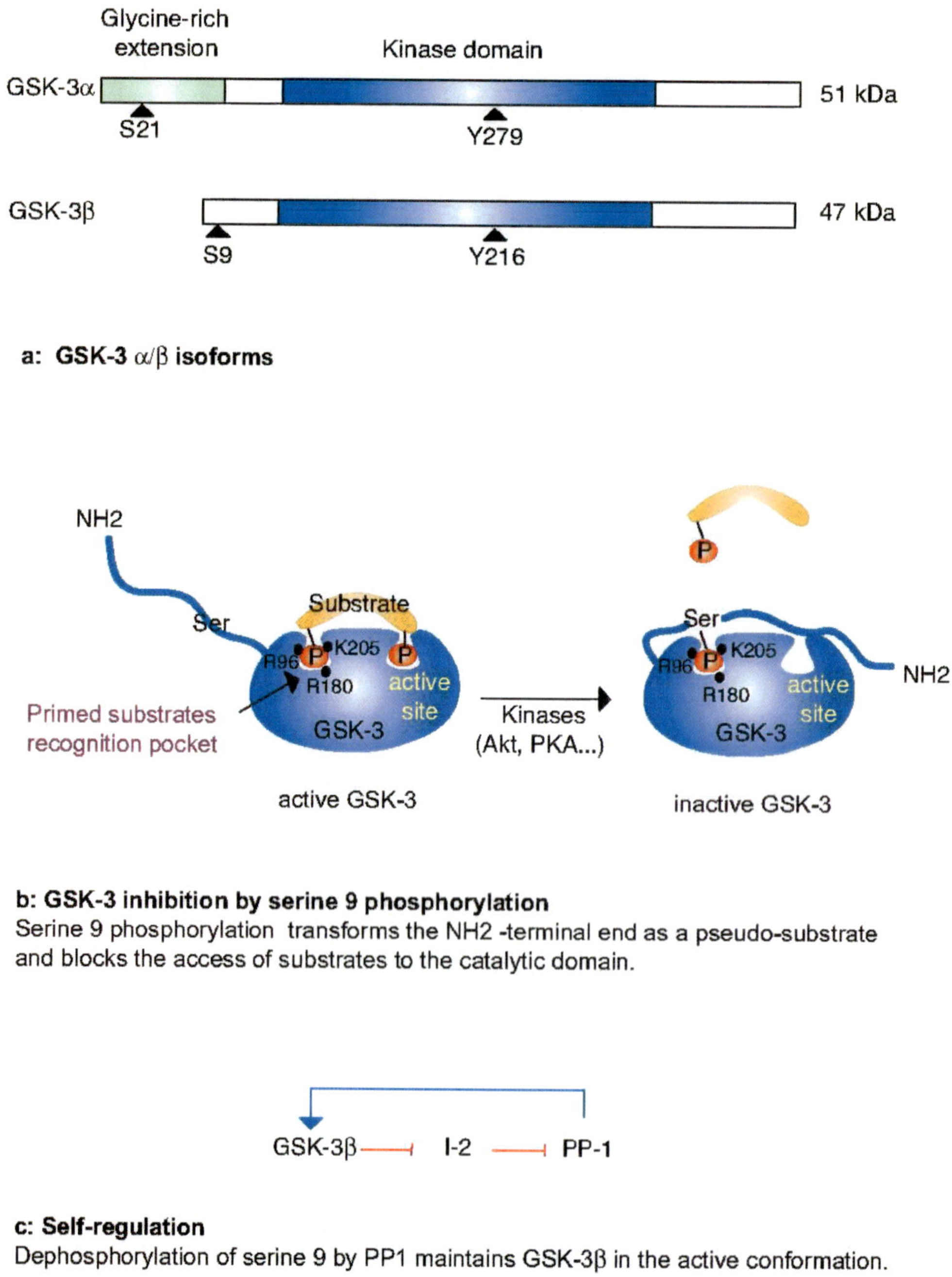

Figure 13. Structure and regulation of the GSK-3α/β serine-threonine kinases.

Regulation of GSK-3β Activity

This kinase shows a rather unusual type of regulation: it is constitutively active and can be functionally repressed by Wnt ligands and several signalling pathays. GSK-3 -related kinases such as CDK-2, p38γ and ERK2 require their phosphorylation at the so-called "activation loop". This phosphorylation leads to a conformational change that allows the access of the substrate to the catalytic site. GSK-3 β and α are autophosphorylated at Tyr216 and Tyr279, respectively (Cole, 2004). Mutation of Tyr276 perturbs the kinase activity, suggesting that this residue also needs to be phosphorylated to produce a fully active GSK-3 (Hagen, 2002). GSK-3 preferentially phosphorylates "primed substrates" pre-phosphorylated by another kinase, on a residue 4 aminoacids distant from its target at the C-terminal terminus. The pre-phosphorylated substrate interacts with GSK-3 at the positively charged pocket constituted of Arg96, Arg180 and Lys205 (Frame, 2001; Dajani, 2001; ter Haar, 2001). This interaction positions the substrate within the catalytic groove in the appropriate manner to allow the transfer of the phosphate group.

GSK-3 activity is obliterated by several proinvasive signalling pathways, including PI3K-α /Akt and MAPK (Eldar-Finkelman, 1995; Kotelevets, 1998; van Weeren, 1998; Fukumoto, 2001; Kim, 2001; Le Floch, 2005). GSK-3 negatively regulates Akt-1 motor protein kinesin-based motility (Morfini, 2002). Accordingly, Akt-1 induces MMP-2 activity and invasion (Park, 2001), and regulates recycling of cell-matrix adhesion proteins integrins (Roberts, 2004). Inhibition of GSK-3 by these kinases is a consequence of its phosphorylation on Ser9 or Ser21 in the β/α isoforms, respectively. The phosphorylated N-terminal end of GSK-3 re-folds and interacts with the pre-phosphorylated substrate recognition site, thus occupying the catalytic site and acting as a "pseudo-substrate" (Fig.13b). So, the catalytic site is blocked by the pseudo-substrate and has no access to pre-phosphorylated substrates. Experiments with synthetic peptides confirmed the role of the N-terminal sequence as a "pseudo-substrate". The activity of the GSK-3 mutated at Arg96 towards unprimed substrates is resistant to the inhibition by the phosphorylation of Ser9 (Frame, 2001). GSK-3 autoregulates its activity by phosphorylation of I-2, an inhibitory subunit of PP-1 protein phosphatase. This enzyme removes the phosphate group from the Ser 9/21, as shown in Figure 13c (Zhang, 2003). In the absence of GSK-3 activity, I-2 blocks PP-1 so that Ser 9/21 remains phosphorylated and GSK-3 inactive. This mechanism reinforces the GSK-3 auto-regulation loop. The canonical Wnt pathway does not inactivate GSK-3 through Ser9/21 phosphorylation (Ding, 2000). GSK-3 activity is not directly down -regulated following the activation of the canonical pathway (Thomas, 1999; Ding, 2000; Farr, 2000). Instead, protein/protein interactions regulate the phosphorylation of GSK-3 substrates within the β-catenin degradation complex.

Subcellular Distribution of GSK-3

Although GSK-3 is generally considered as a cytosolic enzyme, it has been also found in the nucleus and mitochondria (Bijur, 2003). The nuclear pool of GSK-3 is particularly interesting since many substrates of this kinase are transcription factors. The level of nuclear GSK-3 is not constant but vary for instance as a function of the cell cycle progression. Highest levels of nuclear GSK-3 coincide with the S-phase, favoring the phosphorylation of cyclin D1 (Diehl, 1998). GSK-3 also accumulates in the nucleus in the course of apoptosis (Bijur, 2001).

Table IV: GSK-3 substrates and their implication in molecular biology.

Activity and Signal transduction	Targets	Transcription factors
AcetylCoA carboxylase	DF3/MUC1	AP-1
Amyloid precursor protein	Dynamin-like protein	β-catenin
APC	Kinesin light chain	C/EBP
ATP-citrate lyase	MAP1B	CREB
Axin	MAP2	GATA4
Cubitus interruptus	Neural cell-adhesion protein	Glucocorticoid receptor
Cyclic-AMP-dependent protein kinase	Neurofilaments	HIF-1
Cyclin D1	Ninein	HSF-1
Cyclin E	Tau	Mash1
eIF2B	Telokin (KRP)	MITF
Glycogen synthase		c-Myb
hnRNP		c-Myc
Insulin receptor substrate-1		NeuroD
Myelin basic protein		NFAT
NGF receptor		NF-κB (p65 et p105)
Nucleoporin p62		Notch
p21		p53
Presenilin-1		TCF
Protein kinase A (sous-unité RII)		
Protein phosphatase 1		
Protein phosphatase inhibitor-2		
Pyruvate dehydrogenase		

GSK-3 Substrates

More than 40 potential GSK-3 substrates are known (Jope, 2004). They include transcription factors, as well as other proteins involved in a plethora of mechanisms and biological functions, including cell structure, signal transduction and metabolism, as shown in Table IV. As observed for β-catenin, inhibition of GSK-3β by Wnt-1 results in the stabilization of the transcription factor Snail involved in E-cadherin down-regulation and control of the epithelial-mesenchyme transition (Zhou, 2004; Yook, 2005).

GSK-3 in Health and Disease

GSK-3 has been initially identified as the glycogen synthase (GS) kinase inducing phosphorylation of the insulin receptor substrate -1 (IRS-1), a potential inhibitory mechanism for insulin resistance in type 2 diabetes. Serine phosphorylation of IRS-1 by GSK-3 attenuates insulin signalling. GSK-3 activates GS and favors the conversion of glucose to glycogen. Inhibition of GSK-3 downstream of insulin action might improve glucose utilization and restore insulin sensitivity. GSK-3 is also involved in neurodegenerative diseases, including the Alzheimer syndrome characterized by the presence of filamentous protein inclusions in the central nervous system. These inclusions contain tau protein, a target of GSK-3 associated with microtubules (Liu et al, 2004). Such deposits containing hyperphosphorylated tau have been also observed in other neurodegenerative diseases (Kaytor, 2002). GSK-3 also enhances the formation of β-amyloid sheet deposits, another characteristic of Alzheimer disease. The formation of amyloid fibrils is also associated with major human diseases, including prion diseases and type 2 diabetes -related human islet amyloid polypeptide (hIAPP).

Pharmacological Inhibitors of GSK-3

GSK-3β requires for its activity the presence of Mg^{2+}/ATP, and lithium acts as a direct and reversible inhibitor by competition with Mg^{2+}. In addition, lithium inhibits GSK-3 indirectly through Ser9/21 phosphorylation (Zhang, 2003). Lithium has alternative targets, including inositol-phosphate phosphatase, phosphoglucomutase and histone deacetylases involved in transcriptional regulations. Maleimide-based compounds such as SB216763 and SB415286, which compete with ATP binding to the active site of the enzyme are relatively more selective for GSK-3 inhibition but they also inhibit related kinases, including CDK. The development of selective inhibitors of GSK-3 remains a challenge for the therapeutic applications in major human pathologies related to the central nervous system, diabetes, regenerative diseases, stem cell biology and maintenance of embryonic stem cells (Coghlan, 2000; Cohen, 2004; Meijer, 2004).

4. β-CATENIN

Structure of *β*-Catenin

Three functional domains can be distinguished in the primary structure of β-catenin (Fig.14a). The 150 aminoacid N-terminal domain of β-catenin contains phosphorylation sites implicated in its stability, as well as sequences involved in α-catenin binding. The central domain (about 550 aminoacid residues) harbors twelve 42-aminoacid *armadillo* motifs binding axin, APC, E-cadherin, and TCF. It has been shown that APC, E-cadherin and TCF compete for β-catenin binding and that APC and E-cadherin phosphorylation increases their

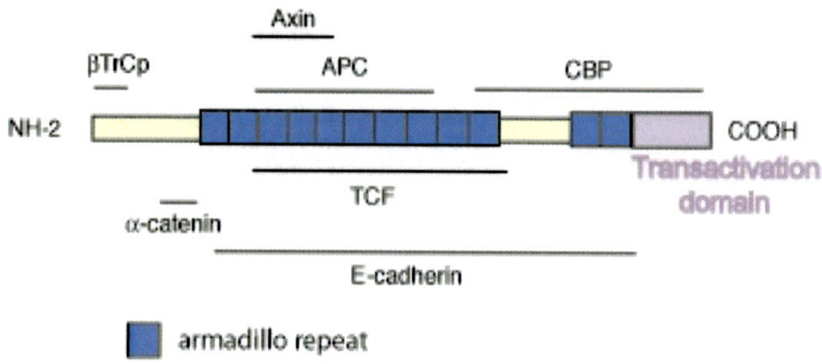

a) β-catenin structure and interacting proteins

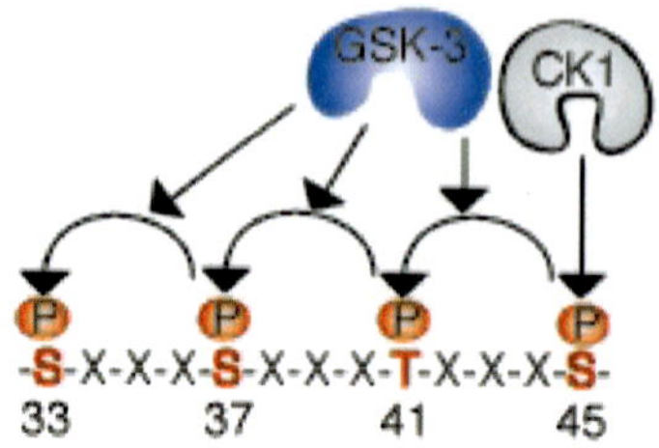

b) Sequential phosphorylation of β-catenin by CK and GSK-3

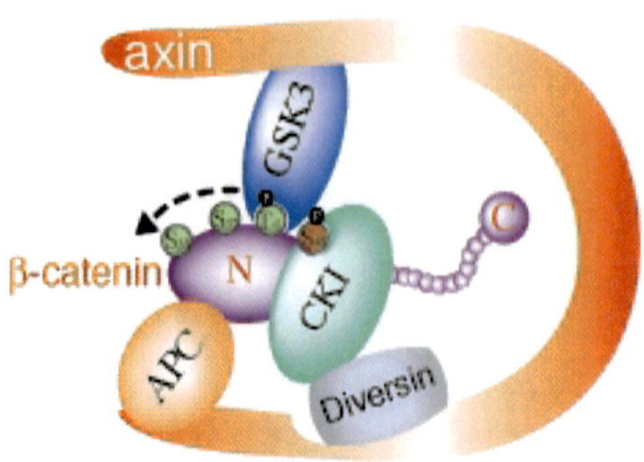

c) Mechanism of β-catenin phosphorylation inside the degradation complex

Figure 14. β-catenin structure and phosphorylation.

affinity (Gottardi, 2001; Daniels, 2001). The C-terminal end of the molecule carries the transactivation function, active in association with LEF/TCF. There is no DNA binding sequence in the β-catenin molecule.

Sequential Phosphorylation Steps control β-Catenin Stability

The rapid ubiquitinylation and degradation of β-catenin in the absence of Wnt activation relies on the recognition of phosphorylated Ser 33 and 37 residues in the N-terminal region of β-catenin by βTrCp, a subunit of the E3 ubiquitin-ligase complex (Liu, 1999a). The priming phosphorylation reaction on Ser 45 is carried out by casein kinase 1, CK1, probably recruited by another protein named diversin (Schwarz-Romond, 2002) which interacts with axin and facilitates the access of CK-1 to β-catenin (Fig. 14, b and c). GSK-3β can then phosphorylate Thr 41, Ser 33, and Ser 37 (Liu, 2002).

Crystallographic analysis of the binding of β-catenin to axin and to APC led to the proposition that APC ensures fast liberation of β-catenin from axin (Figure 15, and Xing, 2003). Axin interacts with the Arm 3-4 motifs of β-catenin. While the 15 aminoacid motifs of APC bind the Arm 5-8 sequences of β-catenin, the 20 aminoacid motifs of APC interact with the same sites as axin (motifs 3-4). The phosphorylation of the 20 aminoacid motifs by CK-1 and GSK-3β increases their affinity for β-catenin, disrupting their binding to axin. Thus, the phosphorylation of APC leads to β-catenin release, and the phosphorylated β-catenin is ubiquitinylated and degraded by the proteasome.

4.a. Transcriptional Functions of β-Catenin

Once released by Wnt ligands from the degradation complex, β-catenin is translocated to the nuclear compartment by a mechanism independent of the usual importin/karyopherin mediators (Fagotto, 1998). There are no NLS sequences in the β-catenin molecule and no direct interaction with nuclear pores. Instead, the nuclear localisation of β-catenin may rely on Pygopus and Legless (Townsley, 2004). Pygopus is anchored in the nuclear envelope and Legless shuttles between the nuclear and cytoplasmic compartments. In the nucleus, β-catenin forms a complex with the lymphoid enhancer factor/ T-cell-specific transcription factor (LEF/TCF) family including LEF1, TCF1, TCF3, and TCF4. These HMG-box-containing proteins are devoid of a transactivation function but recognize DNA elements corresponding to the consensus sequence 5' [A/T] [A/T] CAAAG 3'. In the absence of β-catenin, LEF/TCF transcription factors bind to their cognate sequences and act as repressors, in association with p300/CBP, CtBP and members of the TLE/Groucho family (Cavallo, 1998; Roose, 1998; Waltzer, 1998; Brannon, 1999). When complexed with β-catenin, a functional transcription factor complex is formed, due to the powerful transactivation domain at the C-terminal end of the latter that binds TBP (TATA-box Binding Protein) and recruits the general transcription initiation complex (Hecht, 1999). Sumoylation of LEF-1 and TCF-4 is important in the regulation of their transcriptional activity. This modification is affected by the PIASy (protein

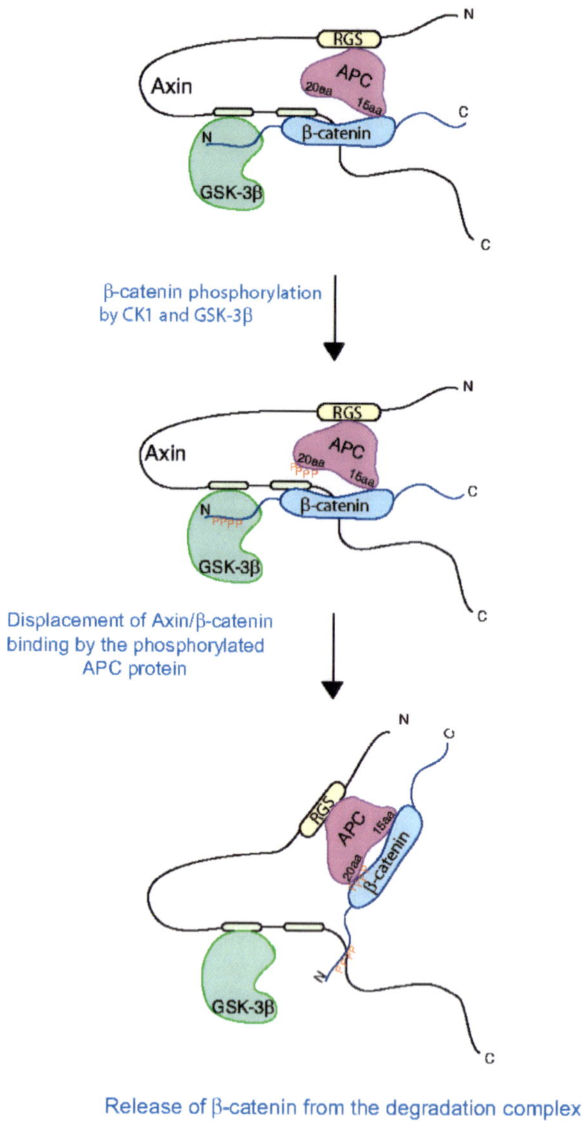

Figure 15. Model of the phosphorylated β-catenin release from the axin/APC degradation complex

inhibitor of activated STAT), a SUMO (small ubiquitin-related modifier) E3 ligase, and results in the sequestration of LEF-1 and TCF-4 in the nucleus. PIASy was shown to repress LEF-1 activity (Sachdev, 2001), and to activate TCF-4 (Yamamoto, 2003). This sequestration involve the co-localization of TCF /LEF, SUMO-1 and PIASy in promyelocytic leukemia (PML) nuclear bodies. This association is interesting in view of the role of PML-NBS in transcription and in chromatin remodelling (Zimber, 2004).

4.b. Role of co-Activators

Several transcription co-activators are required for the expression of the canonical Wnt pathway target genes. Chromatin remodelling has to take place to cancel the repression by LEF/TCF and Groucho/TLE, and to allow access of TBP, the general transcription machinery and RNA polymerase II to the initiation site. While in drosophila CBP/p300 represses the LEF/TCF activity, in mammals it functions as a co-activator of β-catenin-dependent transcription (Takemaru, 2000). CBP/p300 modifies chromatin structure due to its histone acetyltransferase activity, and moreover interacts with general transcription factors such as TBP and TFIIB as well as with RNA polymerase II. The Brg-1 factor participates at chromatin remodelling complexes and associates with β-catenin to favor the transcription of target genes (Barker, 2001). The *Pontin52* ATPase and helicase also interacts with β-catenin and TBP (Bauer, 2000), and facilitates transcription by remodelling the chromatin structure.

4.c. Negative Regulation of β-Catenin-dependent Transcription

Several mediators exert an inhibitory action on β-catenin-induced transcription: The nuclear protein Chibby directly interacts with the C-terminal region of β-catenin and represses the transcriptional activity of β-catenin, probably by interfering with its association with LEF/TCF (Takemaru, 2003). Disabled-2 (also known as Dab2 or DOC-2), a putative tumor suppressor recently identified as an antagonist of the canonical Wnt/ β-catenin pathway and Wnt-3A signaling, associates with Dvl-3 and Axin, increasing β-catenin degradation (Reviewed in Prunier, 2004). As Chibby, the nuclear protein Duplin binds β-catenin and inhibits its association with LEF/TCF (Sakamoto, 2000). ICAT has been identified by double hybrid screening as a partner and co-localizes with β-catenin in the nucleus and cytoplasm, and disrupts β-catenin interactions with LEF/TCF and CBP/p300 transcription factors (Daniels, 2002). Reptin 52, identified as a partner of Pontin 52, forms complexes with β-catenin, TBP, and harbors ATPase and helicase activities. Unlike Pontin 52, Reptin 52 is a negative regulator of β-catenin-dependent transcription (Bauer, 2000). Both Pontin 52 and Reptin 52 interact with the same β-catenin region and compete with each other. NLK represses the canonical Wnt-induced transcription by a mechanism that involves the phosphorylation of LEF/TCF, thereby inhibiting DNA binding (Ishitani, 1999). Wnt receptors Fz6 represses the canonical Wnt cascade activated by Dvl-1, oncogenic β-catenin mutant and lithium (Golan, 2004), suggesting that Fz6 can function as a tumor suppressor gene under certain circumstances.

4. d. β-Catenin Target Genes

A selected list of genes targeted by Wnt signalling in human colorectal cancer is presented in Table V.

Clearly, β-catenin regulates the expression of numerous genes involved in the development and a variety of critical cellular (dys)functions, such as proliferation, cell cycle and survival, differentiation, inflammation and cancer progression (invasion, angiogenesis and metastasis). A few concrete examples are discussed below.

Wnt-dependent genes involved in cell proliferation and differentiation

During the cell division cycle, DNA is replicated during the S-phase and distributed between the daughter cells during mitosis phase M. The "gap" phases G1 and G2 allow the preparation of the S and M phases, respectively, as well as checkpoints for accurate completion of the preceding events. During the G0-G1 transition, noncycling cells respond to signals such as growth factors and oncogene products that induce cell cycle entry and progression. Other signals are linked to exit from the cell cycle, or quiescence at G0. Cyclin-dependent kinases (Cdk) sequentially regulated by cyclins D, E, and A are decisive for the entry into S-phase and its completion, whereas Cdk1 regulated by cyclin B triggers mitosis. Among substrates of Cdk, an important subset are proteins of the Rb family that in turn regulate the activities of the transcription factors E2F needed for the expression of genes involved in DNA replication, cell cycle progression, and upstream components of signal transduction (Sherr, 1996; Chaussepied, 2005). Cdk activities are also regulated by inhibitory proteins such as $p21^{WAF1}$, whose expression is subject to regulation by yet other transcription factors, for instance the protooncogene c-myc (van de Wetering, 2002). Wnt signalling induces the expression of cyclin D1 and c-Myc linked to the G1 phase progression and G1/S transition (He, 1998; Tetsu, 1999; Shtutman, 1999). These proteins are frequently overexpressed in human solid tumors, including colorectal cancer. In the normal intestinal epithelium, the highest expression of β-catenin is observed in the stem cell niche, where the canonical Wnt signalling is essential for the control of the proliferation -differentiation balance (van de Wetering, 2002). Wnt is also essential for the architecture of the intestinal epithelium. TCF-4 knock -out mice lack intestinal proliferative cells and all cells appear to be differentiated (Korinek, 1998). Inhibition of the canonical Wnt pathway by Dkk1 blocks the differentiation of the secretory cells of the intestinal epithelium (Pinto, 2003). Wnt modulates the expression of both ephrin-B and its receptor EphB, which specify the positioning of epithelial cells along the crypt/villus axis (Battle, 2002). The expression of the transcription factor SOX9 depends on the Wnt/ β-catenin signals (Blache, 2004). SOX9 represses several differentiation-related genes, including the homeobox gene CDX-2 and MUC-2 encoding the principal component of the mucus in the digestive mucosa.

Wnt-dependent Genes and Signalling Elements Involved in Cell Migration and Invasion

Wnt induces the expression of the uPA receptor uPAR at the initiation of the MMP activation cascade (Mann, 1999). The invasive behavior is associated with the expression, localisation, and activation of several metalloproteases, including matrilysin /MMP-7, MMP-26, and MT1-MMP (Brabletz, 1999; Marchenko, 2002; Takahashi, 2002). MT1-MMP is anchored in plasma membrane, and its activity is required for the activation of MMP-2 (Sato, 1994). During cell migration, this transmembrane MMP is detected in lamellipodia structures at the invasive front. This localisation, essential for the digestion of the extracellular matrix

(ECM) facing the moving cell, is determined by the interaction with the hyaluronan receptor CD44 (Mori, 2002). Wnt induces the expression of CD44 which facilitates cell invasion in vitro as well as the formation of metastases in vivo (Faassen, 1992; Gunthert, 1991).

Table V. β-catenin targets genes and their functions.

Genes	Functions	Activation/ repression	References
c-myc	Transcription factor, cell cycle regulation	+	He, 1998
Tcf-1	Canonical Wnt pathway transcription factor	+	Roose, 1999
Cyclin D1	Cell cycle regulator	+	Tetsu, 1999 Shtutman, 1999 Civenni, 2003
LEF1	Transcription factor	+	Hovanes, 2001
PPARδ, COX-2	Peroxisome proliferator-activated receptor, nuclear receptors binding retinoic X receptor (RXR). Cyclo-oxygenase-2, prostanoid synthesis. Wound repair, inflammation, invasion	+	He, 1999 Wang, 2004
c-jun	Transcription factor, component of AP-1	+	Mann, 1999
Fra-1	Component of the AP-1 oncogene	+	Mann, 1999
uPAR	Matrix degradation, cell migration, adhesion, apoptosis, and growth	+	Mann, 1999
MMP-7	Matrix degradation, invasion and metastasis	+	Brabletz, 1999 Crawford, 1999 and 2001 Le Floch , 2005
MT1-MMP	Matrix degradation at focal adhesion sites	+	Takahashi, 2002
Axin-2	Scaffold protein, negative regulator of Wnt pathway	+	Yan, 2001b Lustig, 2002 Jho, 2002
Nr-CAM	Cell adhesion molecule, motility	+	Conacci-Sorrell, 2002
ITF-2	Helix-loop-helix transcription factor (HLH)	+	Kolligs, 2002
Gastrin	Gastrointestinal hormone, cell proliferation	+	Koh, 2000
CD44	ECM adhesion factor	+	Wielenga, 1999
Ephrin B	Ligand of Ehp receptor tyrosine kinase: intestinal Paneth cell positioning in crypt-villus axis	+/-	Battle, 2002
SOX9	Transcription factor	+	Blache, 2004
BMP-4/ BMP-7	Bone morphogenic protein-4: dorsal-ventral patterning in *Xenopus*, embryonic dorsal-ventral axis, ectodermal fate decisions in ES cells; dorsal closure in Drosophila; epithelial cell differentiation.	+	Wilson, 2001 Kim, 2002 Haegele, 2003
Claudin-1	Tight junction protein	+	Miwa, 2000
VEGF	Angiogenic factor, cancer cell proinvasive agent	+	Zhang, 2001 Rivat, 2005
FGF-18	Growth and differentiation factor	+	Shimokawa, 2003
Met	HGF receptor, oncogene, tyrosine kinase	+	Boon, 2002
MMP-26	Matrilysin-2, pro-gelatinasse B (MMP-9) activation, matrix degradation, initiation of breast cancer invasion.	+	Marchenko, 2002
Id2	Inhibitor of HLH transcription factors	+	Rockman, 2001

The localisation of MMP-7 and -9 at the cell surface is regulated by the membrane-anchorage CD44 receptor, allowing functional targeting and proteolytic activation of latent proinvasive factors, such as heparin-binding epidermal growth factor (HB-EGF) or latent, pro-TGF-β (Yu, 1999, 2000, and 2002). Interestingly, both HB-EGF and EGF-R signalling are elevated in several human solid tumors and regulate invasive growth and tumor angiogenesis via autocrine and paracrine pathways (Ongusaha, 2004; Rodrigues, 2003).

Cell motility induced by Wnt-3a is not inhibited by disruption of the β-catenin pathway, but is partially blocked by ROCK inhibitor and Dvl-2 down-regulation by RNA interference (Endo, 2005). Similarly, cancer cell invasion responses induced by Wnt-2 and Wnt-3a are blocked by sFRP-3, pharmacological inhibitors and interfering mutants of the Rho-ROK axis (Le Floch, 2005). Both canonical and noncanonical Wnt pathways are involved in the proinvasive activity of the Fz-9 ligand Wnt-2 (Karasawa, 2002) acting through convergent GSK-3β signals at c-jun/AP1 and LEF response elements in the matrix metalloprotease matrilysin promoter (de Groot, 1993; Crawford, 1999 and 2001; Rivat, 2003; Le Floch, 2005). In agreement with these data, we have recently demonstrated that depletion of GSK-3β by RNA interference or blockade of the AP1 oncogene by a c-jun interfering mutant obliterates collagen type I invasion by cancer cells treated by Wnt-2 and other proinvasive factors (Rivat, 2003, Le Floch, 2005).

5. Positive and Negative Wnt Auto-Regulatory Loops

β-catenin regulates the expression of proteins involved in the transduction of the canonical Wnt signal, suggesting feedback mechanisms. Axin homolog axin-2, also called conductin or axil, is regulated by Wnt-1, Dvl-2 and β-catenin, *via* LEF/TCF (Jho, 2002; Leung, 2002; Lustig, 2002). Like axin, axin-2 facilitates the assembly of the β-catenin degradation complex, inhibits its transcriptional activity (Behrens, 1998), and participates in the negative Wnt auto-regulatory loop to limit the duration and intensity of the Wnt signals.

Table VI: Positive and negative feedback control by Wnt pathways of genes encoding Wnt signalling elements.

Target genes	Effects of Wnt on the target gene expression	Effects of target gene products on Wnt signalling	References
Fz	Repression	Inactivation	Muller, 1999a
Dfz2	Repression	Inactivation	Cadigan, 1998
Dfz3	Induction	Activation	Sato, 1999
Arrow/ LRP	Repression	Inactivation	Wehrli, 2000
Naked	Induction	Inactivation	Rousset, 2001
Axin 2	Induction	Inactivation	Yan, 2001b Lustig, 2002 Jho, 2002
β-TrCP	Induction	Inactivation	Spiegelman, 2000
TCF1	Induction	Inactivation	Roose, 1999
LEF1	Repression	Activation	Hovanes, 2001

Canonical Wnt pathways may interact with the adhesion, invasion and transcription regulators HGF receptor Met, EGF-R, c-erbB-2, src, and intestinal trefoil factor targeting β-catenins at cell-cell contacts (Hoschuetzky, 1994; Kanai, 1995; Liu, 1997; Efstathiou, 1998; Hiscox, 1999; Monga, 2002; Lu, 2003; Rivat, 2003; Emami, 2004; Le Floch, 2005). Physical interactions between LEF-1 and c-Jun contribute to a synergistic cooperation at the matrilysin promoter in human colorectal cancer cells transformed by the src oncogene (Rivat, 2003). E-cadherin mutations frequently observed in human gastric and lobular breast carcinoma (Oda, 1994) may play a role in Wnt pathways through increased levels of free cytoplasmic /nuclear β-catenins. A general role of the cAMP- protein kinase A (PKA) /CREB-dependent transcription in Wnt-directed transcription programs during embryonic patterning, stem cell renewal and cancer has been recently proposed (Chen, 2005; Le Floch, 2005). Positive regulatory loops use phosphorylation and inactivation of GSK-3 by PKA (Fang, 2000).

Table VI summarizes Wnt target genes whose products participate at Wnt signalling.

6. Crosstalk Between Wnt and other Developmental Signalling Pathways

6.a. Wnt, Notch/ Hedgehog and TGFβ/ BMP

A prominent interactive network of information between Wnt and Notch and/or Hedgehog signalling pathways is operating in the so called "molecular clock" that determines the fate of presomitic mesoderm and the vertebrate segmentation process during development (Aerne, 2004; Aulehla, 2003 and 2004). A transcriptional oscillator comprising several feedback loops was often suggested, presumably driven by components of the Wnt and Notch signalling pathways. These include the Wnt antagonist Nkd1 (Ishikawa, 2004), Hes7 and axin 2 (Giudicelli, 2004), the Lrp6 co-receptor for Wnt ligands (Kokubu, 2004), and cross-talk between Wnt and Notch/Delta that is necessary for the normal formation and patterning of somites and bones (Aulehla, 2004; Zamurovic, 2004). LEF-1 was shown to act on Dll1 promoter to affect Notch pathways and somitogenesis (Galceran, 2004; Hofmann, 2004). The segmentation clock of vertebrates, certainly not yet fully deciphered, is therefore often discussed and reviewed (Aulehla, 2004; Giudicelli, 2004; Pires-daSilva, 2003).

The development of different brain regions, including tissue patterning/ lateral inhibition operating in the development of rhombomers, including the maintenance of boundaries, is controlled by several Wnts (-1, -3a, -8b, -10b); and Delta genes (A, B, D), acting in transverse stripes flanking rhombomere boundaries (Riley, 2004; Amoyel, 2005). FGF8 gradients act to control somite size, together with Wnt-3a and the "Oscillator" (Cossu, 1996). Wnt, Notch/ Delta, together with FGF and BMP, determine also somite polarity and segmental patterning of the peripheral nervous system, as shown in the chick embryo.

Concomitantly with the activation of Wnt, signalling pathways regulated by Notch, TGFβ and BMP-2 were also shown to control bone morphogenesis, both positively and negatively. This was deduced from data obtained with Affimetrix GeneChip Oligonucleotide arrays performed on MC3T3 murine cells (Zamurovic, 2004). Chick telencephalon development, the acquisition of dorsal versus ventral, or intermediate characters, also depend of Wnt and FGF versus Shh signals (Gunhaga, 2003).

The retinal determination gene network (RDGN) also integrates information from Notch, TGFβ, Wnt, Hh and EGF signalling pathways (Silver, 2005). Both differentiation of ES cells to form dorsal interneurons, and early neural cell death were shown to be controlled by crosstalk between Wnt, Shh, BMP and FGF members. These factors also play an essential role in epithelial-mesenchymal interactions that induce lung and kidney development (Shannon, 2004; Stuart, 2003; Davies, 2002). A role for distinct Wnt ligands, Lrps, Ihh or Shh, and BMP-2, -4 has been demonstrated in chondrogenesis and osteogenesis (Church, 2002; Shea, CM, 2003), and in tooth development (Theslef, 2003). Among several interactions implicated, FGF-4 was shown to be a direct target of LEF-1/Wnt, where reciprocal communication between epithelium and mesenchyme is critical (Kratochwill, 2002). Analysis of patterning mechanisms that control limb development in vertebrates, a paradigm for the analysis of pattern formation in many other structures and organs, revealed the existence of complex interactions between Hh, TGFβ /BMP, Wnt and FGF gene superfamilies encoding secreted factors and other networks in the control of limb positioning, outgrowth and patterning (Capdevila, 2001).

In the colonic epithelium, positive co-operation between Hh and Wnt pathways was previously reported. However, Hh has recently been shown to antagonize Wnt signalling, described as a new "Wnt-Hh axis", where Ihh is expressed in mature colonocytes (Van den Brink, 2004). Interactions between Hh, BMP, Notch, Sox, and Wnt in the embryonic and adult mammalian intestinal development (and relationship to cancer) were recently reviewed (Sancho, 2004; de Santa Barbara, 2003).

The mammary epithelium, including the stem cell niche, and its plasticity evident not only during embryonic development but also in postnatal and adult life, demonstrating periodic growth and regression, wound-healing, and tumorigenesis (as studied in the mouse, rat, ruminants, and man) is an exciting model-system for exploring the role of Wnt pathway and its interactions with other major signalling pathways in the control of pivotal dynamic biological processes (Chepko, 2003). Wnt-1 and Notch-4 display opposing roles in the regulation of branching morphogenesis in the mammary epithelium (Uyttaendalel,1998). As a matter of fact, progesterone activity in the development of the mammary gland involves a paracrine mechanism *via* the induction of Wnt-4 (Robinson, 2000). Interplay between different signalling pathways, including Wnt, Hh, Notch, ERBB, and TGF-β, is essential in mammary development and tumorigenesis (Muller, 2001; Smalley, 2001).

6.b. Wnt, Cdx, and Hox Interactions

Intricate connections between Wnt signalling and several homeobox genes, including *Cdx-1* (and -2), and several Hox family members have been demonstrated in vertebrate embryogenesis and in adult regenerating tissues, such as the intestinal epithelium, and in cancer (Lohnes, 2003). Similar observations were made in invertebrates *e.g. C. elegans* and *Drosophila.* Wnt /β-catenin signals regulate the expression of the homeobox *Cdx-1* gene in the embryonic intestine, *Cdx-1* being a direct transcriptional target, *via* Tcf-4 promoter /binding motifs (Lickert, 2000). Activation of β-catenin pathway stimulates endogenous expression of Cdx-1 mRNA and the activity of Cdx-1 promoter in human colon cancer cells. Cdx-2 exerts an inhibitory effect on the basal and β-catenin-stimulated activity of the Cdx-1 promoter (Dommon-Dell, 2002; Moucadel, 2002). Cdx-2 transcription is induced by APC in colon cancer cells (da Costa, 1999).

In zebrafish gastrula, Spemann organizer induces neuroectoderm with anterior character, followed by posteriorization by a transforming signal. Experiments with mutants of the homeobox gene bozozok suggested that boz promotes neural induction, promotes organizer formation by limiting BMP -2b/ -4 activity, and limits posteriorization of the neuroectoderm in the late gastrula by negatively regulating Wnt-8 signalling (Fekany-Lee, 2000).

As for invertebrates, activation and inhibition of Wnt signalling in *C. elegans* by activating Hox gene expression in the migratory neuroblast QL cells controls the migration of its descendants. Also, inhibition of Wnt signalling is important in restricting Hox gene expression to regional domains thereby affecting the specification of body pattern along the anteroposterior body axis (Maloof, 1998). In *Drosophila* embryos, homeobox gene were shown to maintain Wingless expression (Jagla, 1997). Interactions between Wnt and Notch, and other signalling pathways are important in hematopoiesis and operate in stem cells (discussed below).

6.c. Stem Cells

The interplay between Wnt signalling and major signalling pathways implicated in the survival and differentiation of embryonic and adult stem cells (Tsai, 2002; Li, 2003), has recently attracted much interest, in the perspective of the therapeutic potential of stem cells (Sato, 2004). In addition, the emerging concept of cancer as a disease of progenitor stem cells provides a connection between these two fields of research. A role of distinct Wnt ligands in the control of survival and proliferation of hematopoietic progenitor cells has been demonstrated (Austin, 1997; van den Berg, 1998). Interactions between Notch and Wnt signalling pathways controlled by β-catenin and Wnt-3a were shown to be essential for the maintenance of undifferentiated hematopoietic stem cells (Bonde, 2004; Brandenberger, 2004; Moore, 2004 ; Duncan, 2005). It has been reported that the Wnt antagonist DKK1 is required for adult bone marrow steem cells to reenter the cell cycle (Gregory, 2003). Understanding the biology of the hematopoietic stem cell niche and the role of Wnt ligands is now enhanced by the development of techniques for the isolation of hydrophobic active Wnt molecules (Willert, 2003). The possibility to expand the pools of HSCs grown in culture by over-expression of β-catenin (resulting in elevated expression of Notch 1 and HOXB4)) stresses the interest of the manipulation of Wnt signalling pathways for clinical applications (Reya T, 2003).

More recently, hedgehog and Wnt transcription pathways have been implicated in the maintenance of stem cells or progenitor cells in the gut, prostate, blood, muscle and nervous system (Beachy, 2004). Interestingly, Wnt pathways play a major role in hair growth, skin epidermis stem cells and their progenitors. TCF-3 is expressed in the bulge of adult hair follicles, presumably acting as a repressor protein in most phases of the hair cycle. It maintains the stem cell character, or directs stem cell progeny along the outer root sheath program of differentiation. A switch from TCF-3 to LEF-1 expression, in conjunction with reception of a Wnt signal, is essential in the commitment of stem cells to a hair lineage fate (Fuchs, 2001). Lineage commitment in the post-natal epidermis, and the role of Wnt, Notch, and other signalling molecules was recently shown (Nieman, 2002; Alonso, 2003), as well as the transcriptional profile of the stem cell niche in the skin epidermal basal layer (Tumbar, 2004). β-catenin mutations were found in hair tumor (Gat, 1998).

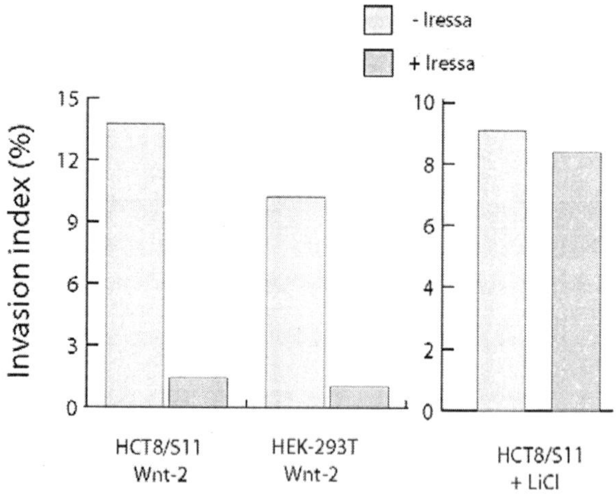

Figure 16. Wnt-2 proinvasive activity is dependent on EGF-R. Collagen type I invasion induced by the Wnt-2 expression vector in stably transfected human colorectal and kidney cancer cells HCT8/S11-Wnt-2 and HEK-293T-Wnt-2 (LeFloch, 2005) was obliterated by the EGF-R tyrosine kinase inhibitor Iressa (Gefitinib, AstraZeneca). Accordingly, invasion promoted by external addition of Wnt-3a was blocked by the dominant negative mutant of the EGF-R (HER-CD533). In contrast, Iressa was ineffective against invasion induced by the direct GSK-3 beta inhibitor lithium.

6.d. Other Signalling Connections with Wnt Pathways

Several signalling elements are connected with the canonical and noncanonical Wnt cascades (*vide supra*, Introduction), including EGF-R and ErbB2, as shown in Figures 16 and 17. Similarly, Tiam 1, a GEF for Rac1 is regulated by the invasion promoter heregulin-β1 involved in LEF/ β-catenin nuclear signalling transactivation *via* PI3-kinase (Adam, 2001).

7. Wnt Signalling and Cancer

The discovery of the Wnt signalling pathways is linked to their role in tumorigenesis: Wnt1 activation by enhancer integration is a mechanism of tumorigenesis induced by the mouse mammary tumor virus MMTV (Nusse & Varmus, 1982; Nusse, 1984 and 1991). The fact that the canonical Wnt pathway leads to the expression of the protooncogenes c-myc and *CCND* (encoding growth-and survival-promoting genes cyclin D1), and invasion-promoting genes MMP7 (matrilysin) account, at least in part, for the role of Wnt in cancer progression (Tetsu 1999; Lin 2000). Inappropriate activation of the Wnt pathways is associated with the developmement of cancer from human solid tumors and leukemia (Staal, 2005). Canonical and non-canonical Wnt signals are involved in a series of cellular dysfunctions linked to cancer progression, including cell proliferation /survival and differentiation, inflammation, invasion and angiogenesis (see Table V above, and Moon, 2004). Oncogenic β-catenin mutations are detected in numerous types of cancers (Polakis, 2000). Moreover, autocrine Wnt signalling has been shown to contribute to the malignant phenotype, in the absence of β-catenin mutations (Bafico, 2004). Here we outline new insights into the role of the Wnt pathways in the cancereous progression of human solid tumors.

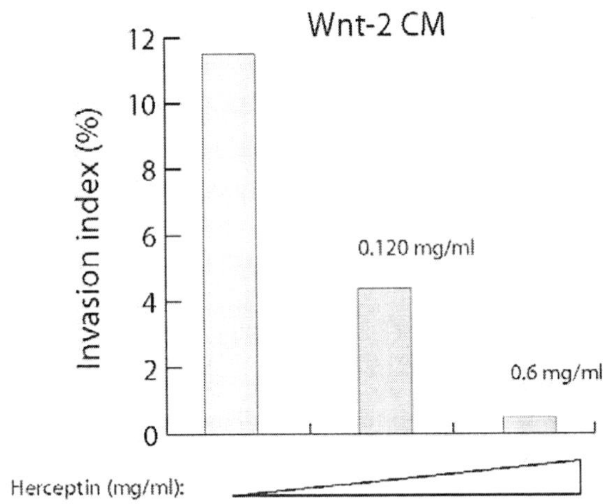

Figure 17. Wnt-2 proinvasive activity is dependent on ErbB2. Collagen type I invasion induced by external addition of Wnt-2 in parental human colorectal cancer cells HCT8/S11 (LeFloch, 2005) was reversed by the ErbB2 neutralizing antibody Herceptin (Trastuzumab, Roche) pre-incubated for 1 h at 37°C before Wnt-2 addition. Herceptin mAb was ineffective against invasion induced by EGF-R agonist amphiregulin or the proinvasive factor spasmolytic polypeptide, i.e. trefoil factor TFF2.

7. a. Colon Cancer

The elucidation of the canonical Wnt pathways provided the link between Wnt signalling and colorectal cancer, mutations of β-catenin and of other genes whose products participate at this pathway. The emergence of colorectal adenocarcinomas (ADK) is a complex multistep process linked to genomic and chromosomal instability (CIN), DNA aneuploidy, and generalized deregulation of gene expression and signal transduction pathways (Figure 18). A role for APC in the origin of CIN and aneuploidy has been suggested (Fodde, 2001). Wild - type APC may be involved in maintaining proper connections of microtubules with chromosomes. These deregulations target proliferation cell cycle checkpoints, loss of the functional identity of the epithelial cancer cells and of tumor stromal cells. Such genetic and molecular changes lead to the formation of aberrant crypt foci (ACF), which precede the appearance of adenomas anchored in the colon mucosal wall. The next stage is the evolution of the adenoma towards more aggressive lesions and irreversible acquisition of dominant and anarchic functions, chronic inflammation, oxidative DNA damage, autocrine and paracrine regulatory loops linked to IGF-R, VEGF-R, EGF-R ligands, induction of thrombin protease - activated receptors PAR-1 and trefoil factor pS2 (TFF-1) and the immediate response gene cyclooxygenase *COX-2* (Rodrigues, 2001; Nguyen, 2002; Emami, 2004). Loss of imprinting (LOI) at the *IGF2* gene or activation of the normally silent maternally inherited allele was discovered in embryonal tumors of childhood, such as Wilms' tumor, and occurs in many common cancers, including colorectal tumors. Genomic imprinting is regulated by epigenetic modifications, such as CpG methylations coordinated by impinting centers containing differentially methylated regions. Epigenetic LOI can be detected in peripheral blood lymphocytes in 10% of the normal population, corresponding to 3.5- 5-fold increased risk of neoplastic disease. LOI affecting the *IGF2* locus contributes to overexpression of the autocrine factor IGF2, tumorigenesis and dedifferentiation in intestinal epithelia of APC +/min mice and may be a valuable predictive marker of colorectal cancer (Cui, 2002 and 2003; Sakatani, 2005). By this model, *IGF2* and *H19* promoters compete at the same chromosome 11p15 region for a shared enhancer located distal of *H19*. Increased or ectopic activity of DNA methyltransferase might lead to aberrant methylation of the maternal differentially methylated region of the *H19* promoter, leading to reciprocal cross activation of the maternal *IGF2* gene (Cui, 2002). Access to these enhancers is regulated by binding of the insulator protein CTCF, a CCCTC-binding transcription factor, to the unmethylated maternal imprinting center-1 (IC-1) allele, thus creating a functional chromatin boundary and blocking the interaction of *IGF2* promoters and enhancers. The methylation of IC-1 on the paternal chromosome prevents CTCF binding, thus allowing *IFG2* expression. Human IC-1 contains seven CTCF target sites. Microdeletions at the *H19 /IGF2*- imprinting center are necessary for manifestation of Wilms' tumors (Prawitt, 2005). In aging -related cancers, altered imprinting patterns of *IGF2* may contribute to the development of replicative senescence in human prostate and urothelial epithelial cells (Fu, 2004). Paternal expression of the positive Wnt pathway regulator Frat3 has been identified in the 7C imprinted region of the mouse genome (Kobayashi, 2002).

Familial adenomatous polyposis (FAP) is induced by mutations of the APC gene, a defect that contributes to CIN and appearance of more than 100 colorectal adenomas. Non-polyposis form of the hereditary colon cancer (HNPCC) is more frequent than FAP, and is caused by germ cell mutations that invalidate the DNA repair systems. DNA mismatch repair is deficient in 90% of the HNPCC patients. The mutations concern mostly the hMSH2 and hMLH1 DNA repair enzymes, less frequently hPMS1 and hPMS2. Genetic factors predisposing to familial colorectal cancer are also involved in sporadic cancers in families at risk, defined either by cancers in siblings, or by adenomas appearing before the age of sixty. Relatively recently, attenuated form of FAP (MHY-associated polyposis) has been found to result from germ-line mutations in MHY involved in in BER and oxidative DNA damage repair (Al-Tassan, 2002). Accordingly, MAP patients show p53 overexpression, and point mutations at guanine nucleotides in APC and Ki-ras, because of the MYH failure to repair lesions induced by the variant base 8-oxo-guanine (Lipton, 2003; Sieber, 2004).

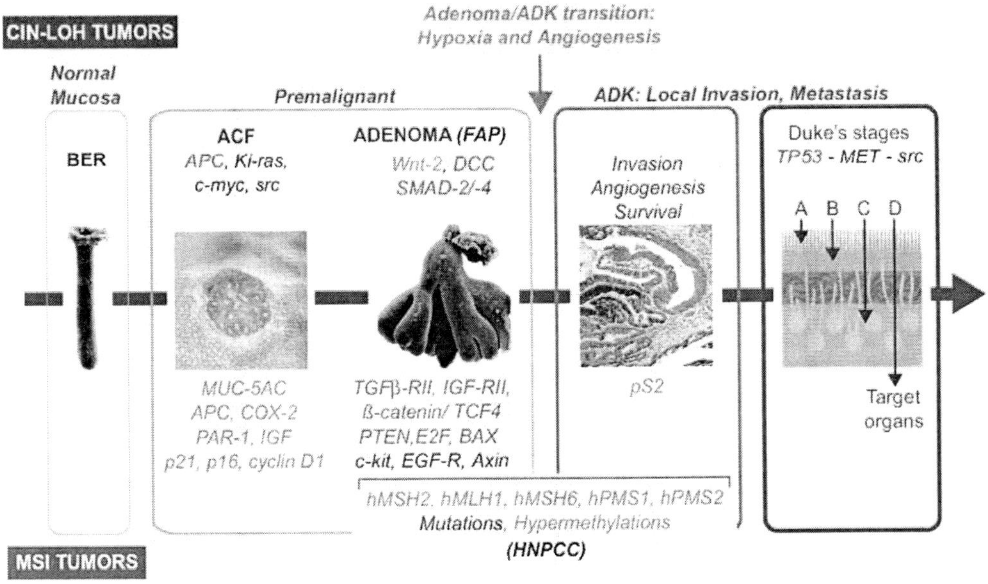

Figure 18. Genetic and molecular alteractions linked to the multistep progression of familial and sporadic human colorectal cancers.

A large majority of colorectal cancers are sporadic (70%), and are strongly associated with inflammatory situations, ulcerative colitis and Crohn's disease. Tumor progression in the digestive tract is controlled by key environmental factors and lifestyle of patients: dietary factors, high caloric intake, obesity, sedentary life, tobacco smoking inducing epigenetic events such as promoter gene methylation and COX-2 induction. Sporadic cancers can be divided in two groups according to the initiating mechanisms, generally a loss of tumour suppressors and constitutive, dominant activation of proto -oncogenes.

The first mechanism, which accounts for 80% of sporadic cases, is connected with chromosomal instability and loss of heterozygosity (LOH) targeting the tumor suppressors APC (5q), TP53 (17p), DCC (18q), and TGFβ pathway signalling elements SMAD2 and SMAD4 at 18q (Prunier, 1999). During the adenoma formation, loss and inversion in polarity gradients of proliferating and apoptotic epithelial cells occurs along the crypt-villus axis together with delocalization of regulatory cytokines such as TGFβ, and progressive disorganisation of the mucosal architecture, leading to inappropriate and persistent release of paracrine and autocrine agents acting at illegitimate sites in the nascent tumor. Early dysregulations in the proliferation/ differentiation balance occur in premalignant lesions in the colon, as shown by induction of the gastric MUC-5AC mucin gene in aberrant crypt foci ACF (Bara, 2003 and Figure 18). Loss-of-function events in the TGFβ pathways at early steps of the neoplastic progression (inhibition of cell proliferation, induction of apoptosis) are switched to TGFβ transforming functions at later stages: induction of cell proliferation, survival, complex epithelial-mesenchyme and fibroblast-myofibroblast transitions, angiogenesis, and invasion at the adenoma-adenocarcinma transition (De Wever, 2004a and b). Activation of Wnt and TGFβ signalling occurs together with other transforming factors related to the transcriptional machinery, cell-cell communication, ECM-basement membrane interaction, and cytokeletal-cell adhesion characteristics.

The second mechanism is characterised by microsatellite instability (MSI), targeting alterations of short repetitive DNA sequences at the immediate vicinity of key (onco) genes. MSI tumors are linked to inactivating mutations or to methylation of gene promoters encoding DNA repair enzymes. These epigenetic changes are observed for instance in the hMLH1 promoter, hypermethylated in certain patients. The hypermethylation of the hMLH1 locus reflects that of numerous genes and the « CpG Island Methylator Phenotype » or CIMP. The hMSH2, hMSH6, hPMS1 and hPMS2 gene promoters are also concerned by hypermethylation. Inactivation of DNA repair genes favours generalized genetic instability and increases the frequency of subsequently acquired mutations ("Mutator phenotype"). Other forms of DNA repair include the short, single nucleotide base excision repair sytem (BER) involved in point mutation repair (Ki-ras activation), and the long patch multinucleotide 2-13 nucleotide repair BER (Wood, 2001). Sporadic MSI tumours are frequently mucinous, predominantly localised in the right colon, and generally diploid. In MSI patients, alterations in type II TGFβand IGF receptors, β-catenin, TCF-4 (Duval, 1999), and E2F transcription factors, as well as loss of the PTEN tumor suppressor, and the apoptosis regulator BAX are frequently reported.

Among somatic mutations in sporadic colorectal cancer, early events concern - for instance – i) the Wnt pathway (including APC, Axin-2), overexpression of the Wnt target (onco) genes (Table V); ii) constitutive activation or overexpression of the protooncogenes Ki-ras, src tyrosine kinase and STAT-3 (Rivat, 2005), EGF-R, c-kit (Attoub, 2002), c-myc and c-Met (gene amplification as well); iii) loss of the tumor suppressors and cell cycle regulators p16 and $p21^{WAF1}$ by promoter hypermethylation (Sancho, 2004). After the adenoma-adenocarcinoma transition and in later stages characterised by metastatic dissemination, other genetic changes become dominant, such as loss of *DCC* (Mazelin, 2004) and *TP53* seen in 50% of patients. Wnt pathway activation is an early event in colorectal cancerogenesis. APC is mutated in familial adenomatous polyposis coli (FAP) as well as in 80% of sporadic cancers. The mutations generally produce a truncated protein deficient for axin binding. Interestingly, several authors described nuclear-cytoplasmic shuttling of APC in human carcinoma cells (Bienz, 2002; Rosin-Arbesfeld, 2003), suggesting that APC controls the β-catenin transcription factor and other mechanisms in transformed cells (Narayan, 2003 and 2005). However, this mechanism has been recently revaluated and questionned (Brocardo, 2005). In the mouse, the expression of oncogenic β-catenin leads to the appearance of adenomatous polyps in the intestine and of microadenomas in the colon (Harada, 1999). Epigenetic down-regulation of the Wnt antagonists encoded by the *SFRP* genes by promoter methylation may be instrumental in the appearance of aberrant crypt foci as early premalignant changes that predispose to colorectal cancer through constitutive Wnt ligand signalling and cell resistance to apoptosis (Deng, 2004; Suzuki, 2004). Overall, mutations in the Wnt signalling elements cause β-catenin stabilisation, induce and enhance malignant transformation in familial and sporadic colorectal tumors. Sporadic colorectal cancers that express normal APC frequently display β-catenin mutations (Morin, 1997), usually localised at the GSK-3 phosphorylation site in the N-terminal. The mutated protein is not recognised by β-TrCP and escapes degradation by the proteasome. In addition, mutations of axin-2 have been identified in colorectal cancers induced by the DNA repair enzymes defects, in absence of APC dysfunction (Liu, 2000). These mutations are also present in cancer-derived cell lines, with mutations in APC or β-catenin, suggesting their additive impact on Wnt pathways. In addition to genetic alterations, overexpression of Wnt-2, -4, -5a, -6 and -7a has been observed in human colorectal tumors (Vider, 1996; Dimitriadis, 2001; Holcombe, 2002). Up-regulation of Frizzled-1 and -2 was also detected at the invasive front (Holcombe, 2002). In contrast, sFRP1 and sFRP2 inhibit Wnt signalling and are frequently down -regulated. Epigenetic inactivation of their genes by promoter methylation has been observed in aberrant crypt foci, an early preneoplastic lesion (Suzuki, 2004). In the Wnt signalling cascade, TCF1 is overexpressed in numerous colorectal cancer cell lines. LEF1 expression has been described in tumoral tissu whereas normal colonic mucosa was negative (Mayer, 1997; Hovanes, 2001). Previously unidentified Wnt target genes were recently obtained from a DNA microarray screen in adenomatous gastrointestinal tumors of the APC Min /+ mouse (Reichling, 2005), including *Igfbp5*, *Lcn2*, *Ly6d*, *Ly6d*, *N4wbp4*, *S100c*, and *Sox4*. Several transgenic animal models indicate that molecular alterations targeting the Wnt pathways and other modifiers are linked to stem-cell /progenitor and Paneth-cell differentiation programmes, as well as priming of adenocarcinoma transitions in the intestinal mucosa (Fodde, 1994; Sonoshita, 2001; Philipp- Staheli, 2002; Boivin, 2003; Yamada, 2003; Gupta, 2004; van Es, 2005).

7.b. Other Digestive Cancers

Gastric Cancer

Gastric cancer has been histologically classified as intestinal-type gastric cacinoma and undifferentiated diffuse-type carcinoma. Defects in E-cadherin tumor suppressor gene are commonly observed in sporadic and hereditary diffuse-type gastric carcinoma. Intestinal metaplasia of the stomach is characterized by the transdifferentiation of gastric epithelial cells to an intestinal phenotype, expression of intestine-specific genes, including mucin MUC2, sucrase-isomaltase, and carbonic anhydrase, and amplification of the ERBB2 and cyclin E genes. Activating β-catenin gene mutations were found in a high percentage of sporadic fundic gland polyps (FGPs), the most common adenomas of the stomach (Abraham, 2001), generally regarded as non-neoplastic lesions. However, a high frequency of somatic, "second hit" alternations in the APC gene on chromosome 5q was present in FGPs associated with FAP, but not in sporadic FGPs (Abraham, 2000). As in colon cancer, deletions within the coding region of TCF-4 gene were shown in 9% of gastric carcinomas with high MSI, but without change in non-MSI tumors (Duval,1999). β-catenin-exon 3 mutations are a frequent cause for Wnt pathway activation in gastric cancer, found in 73 /311 (26%) of tumors examined (Clements, 2002). Also, Wnt-5a mRNA was found to be upregulated in 5/8 primary gastric cancers and in 5 /18 primary colorectal tumors (Saitoh, 2002). Loss of expression of the sFRP Wnt inhibitor family (WIF) has been reported in gastric cancers, breast and cervical carcinoma. In esophageal squamous-cell carcinoma, overexpression of β-catenin was found in 3 /22 cases), but with no mutation in β-catenin gene or in APC (Ninomiya, 2000).

Carcinogenic infections by *Helicobacter pylori*, chronic gastritis, peptic ulcers, adenoma dysplasia, and deregulation of the CDX1 /-2 balance are observed in intestinal-type gastric carcinoma (Bai, 2002; Yuasa, 2003). *H. pylori* is also involved in gastric mucosa-associated lymphoid tissus lymphomas (Peek, 2002). CDX2 in turn inhibits β-catenin -induced CDX1 promoter activation and cell cycle-arrest p21 /WAF-1 gene transcription (Domon-Dell, 2002). CDX1 gene is also down-regulated by oncogenic ras (Lorentz, 1999; Lickert, 2000). Thus, the relative activity of the homeobox CDX-1 and -2 transcription factors monitored by oncogenic ras, Wnt / β--catenin pathways, and CDX2 promoter methylation status may exert a crucial influence on gastric epithelial (trans) differentiation and carcinogenesis.

Barrett's oesophagus (BO) is a precancerous state of oesophageal adenocarcinoma usually induced by gastro-oesophageal reflux disease and *H. pylori* infection. BO is associated intestinal metaplasia, aberrant activation of CDX2 (Eda, 2003), and LOH at multiple tumor suppressor gene loci, including TP53, DCC, RB1, and P16 (McManus, 2004). The normal oesophageal sqamous epithelium is replaced by a specialized intestinal-like metaplasic epithelium. Intestinal metaplasia has been also observed in other digestive organs, inducing cholangiocarcinoma and gallbladder carcinoma.

Hepatocellular Carcinomas and Hepatoblastomas

Wnt signalling elements are also altered in a significant percentage of human hepatocellular carcinomas (HCC). Deletions or missense mutations targeting Wnt signals were detected in 12/35 (34%) of hepatocelular carcinomas (Van Nhieu, 1999). Accumulation of β-catenin was shown already at early stages of HCC, and in less differentiated tumors membrane localization of β-catenin could be related to intra-hepatic dissemination (Suzuki, 2002). β-catenin mutations are frequent in HCC associated with hepatitis C virus infection (Huang, 1999). Besides about 20% of HCCs and 80% of HBS (hepatoblastomas) with β-catenin mutations, AXIN1 and AXIN2 mutations appear to be involved in HCC (Taniguchi, 2002). Nuclear accumulation of β-catenin resulting from mutations were reported in HCC (Terris, 1999; Yau, 2005), plus overexpression of cyclin D1 and fibronectin were found in poorly-differentiated childhood hepatoblastomas (Takayasu, 2001). In another study, nuclear accumulation of β-catenin was found in 7/15 (46%) of hepatic adenomas, indicating activation of the Wnt signalling pathway, in the absence of β-catenin or APC mutations. In patients with normal β-catenin, axin mutations have been detected (de La Coste, 1998; Satoh, 2000). Significant β-catenin expression (25-39%) was demonstrated in grade I, II and III hepatocellular tumors, associated with poor prognosis (Inagawa, 2002). Upregulation of conductin was found in liver and colorectal tumors, and this was shown to result from the loss of the direct regulation of conduction/axin 2 by the Wnt signalling pathway (Lustig, 2002). The causative role of oncogenic β-catenin in hepatocarcinogenesis was also demonstrated in transgenic mice and laboratory animals treated with chemical carcinogens (Cadoret, 2001; Calvisi, 2001). Down-regulation of the Wnt /β-catenin pathway inhibitor HDPR1, the human homolog of the *Xenopus* Dvl inhibitor Dapper, are frequently observed in HCC (Yau, 2005).

7.c. Cancers of Breast and Urogenital Tract

Breast Cancer

As mentioned above, Wnt-1 acts as a protooncogene in the mouse mammary gland. Overexpression of of Wnt-2, -5a, -7a and -10b has been observed in hyperproliferative lesions of the breast as well as in malignant tumors (Huguet, 1994; Lejeune, 1995). A recent study also reports amplification and increased expression of Dvl-1 in breast cancer (Nagahata, 2003). Different studies report up-regulation of Wnt-7b, frizzled 1 and 2 receptors in breast cancer (Huguet 1994, Milovanovic 2004). Somatic mutations of APC and β- catenin are infrequent in breast cancer, in contrast to colorectal cancer. However, a high rate of APC promoter methylation was found in primary breast tumors (36-47%). APC hypermethylation was detected in ductal and lobular carcinoma in situ and in all grades and stages of invasive breast cancer (Jin, 2001; Dulaimi, 2004). In human breast cancer cells MCF-7, estradiol has been reported to up-regulate Wnt-1, -10b and Fzd-10 (Katoh 2003; Kirikoshi 2002; Saitoh 2002). It is relevant to note that the estrogen receptor and β-catenin cooperate to mutually enhance each other's transcriptional activities (Kouzmenko et al, JBC, 2004).The situation is quite different in the cancer of the uterus since Wnt-2, -3, -4 and -5a were often down-regulated in human endometrial sarcoma, whereas Wnt-7a, -7b and -10b were absent from normal endometrium (Bui, 1997).

In human breast carcinoma cells, GSK-3 was shown to be associated with the membrane-bound MUC1 oncogene and β-catenin (Li, 1998). Interestingly, DF3/MUC1 is overexpressed in several human carcinomas and is regulated by proinvasive factors including the ERB receptor family, PKC, and src oncogene (Olayioye, 2000; Li, 2001a and b; Ren, 2002; Li, 2003;Schroeder, 2003). Previous studies reported that the cytoplasmic tail of MUC1 interacts with β-catenin at a serine-rich motif similar to binding sites found in E-cadherin and APC (Li, 1998). Under conditions of overexpression, fragments of the MUC1 cytoplasmic tail are released and transported into the nucleus in association with β-catenin, but not plakoglobin, *i.e.* γ-catenin (Wen, 2003). MUC1 oncoprotein also associates with p53 regulatory domain to function as a transcription coactivator at the p21 tumor suppressor gene and to promote p53-dependent cell fate, selection of growth arrest, and survival responses to genotoxic anticancer agents such as cisplatin (Wei, 2005). Mutations in p53 and APC synergize to promote mammary carcinogenesis in mice, suggesting that normal p53 is involved in the loss of APC-deficient epithelial cells (Méniel, 2005). This provides a potential mechanism for p53 deficiency-accelerated neoplasia in the absence of APC in this model since p53 down-regulates β-catenin (Sadot, 2001). Inherited mutations in the P53 gene occur in individuals with the Li-Fraumeni syndrome characterized by the development of multiple tumors including breast cancers (Malkin, 1990).

Endometrial and Ovarian Cancers

In the mouse uterus, estradiol rapidly up-regulates the expression of Wnt-4, Wnt-5a and frizzled-2. Progesterone alone or in combination with estradiol induces Wnt-4 activation in primary mammary epithelial cells (Hou 2004; Brisken 2000). This regulation of Wnt expression may explain, in part, the protective role of progesterone against endometrial neoplasia. Ias a matter of fact, it has been shown that fetal exposure to estrogens deregulates the expression of Wnt-7a, with the disruption of uterine architecture as a result (Miller, 1998). In contrast, progestogens up-regulate Wnt-7a expression in the endometrium, counteracting the disruption of its expression by estrogenic compounds (Oehler, 2002). This effect of progestogens may be responsible for the restoration of normal function of the endometrium and the reduction of the risk of uterine neoplasia. Alterations in Wnt pathways signalling, mainly due to β-catenin mutations, deregulation, and nuclear localization occur frequently in endometrial hyperplasia and cancer (Nei, 1999; Fukuchi, 1998). Mutations leading to deregulation of β-catenin were also found in nearly half of ovarian carcinomas (Zhai, 2002). Expression and localization of β-catenin were shown to be frequently altered in cervix cancer (Pereira-Suarez, 2002). The β-catenin target gene *bHLH* encoding the transcription factor ITF-2 is overexpressed in human ovarian cancer and induces neoplastic transformation (Kolligs, 2002).

Prostate Cancer

β-catenin signalling pathway involvement in prostate cancer was recently suggested. Analysis of genetic pathways activated by the androgen receptor (AR) in the normal ventral prostate of rats treated with 5α-dihydrotestosterone (DHT) demonstrated a modulation of members of the Wnt and Notch signalling pathways, together with other growth factors and peptide hormone signalling systems (Nantermet, 2004). In human prostate cancer, β-catenin is activated by mutation in 5% of tumors. Significantly, 25 -38% of metastatic androgen-independent prostate tumors were found to show anomalies in cytoplasmic and /or nuclear β-catenin. Recent analysis of Wnt gene expression in prostate cancer has shown that Wnt-11 is elevated in androgen-independent prostate cancer cell lines and in high grade prostate tumors (Zhu, 2004). Mutual inhibition of Wnt-11 and AR which occurs in LNCaP cells was lost in hormone-independent prostate cancer cells. These observations are consistent with a model in which androgen depletion activates Wnt-11 -dependent signals that inhibit androgen dependent but not androgen-independent cell growth. Anomalous intracellular distribution of β-catenin between the membrane and cytoplasmic /nuclear fractions has been recently demonstrated in primary prostate tumors, that correlated with the hormone-refractory (HRPC) status and Gleason score of the tumors (de la Taille, 2003). Accumulation of β-catenin has been detected in advanced, androgen-independent prostate tumors (de la Taille, 2003). In a subset of tumors, this is due to stabilizing β-catenin mutations (Cheshire and Isaacs, 2003). Wnt expression is widespread in prostate cancer cells, and Wnt-3a has been found to cooperate with the androgen receptor as inducer of gene expression and cell proliferation (Verras, 2004).

7.d. Other Human Solid Tumors

Lung cancer

The activation of β-catenin by Dvl may contribute to lung tumor formation, as Dvl-3 is frequently overexpressed in NSCLC (Uematsu, 2003). Altered Hox, Wnt, FGF regulatory interactions were shown to occur frequently in human lung cancer. Thus, Wnt-7a was lost concomitantly with HOXA1 in immortalized bronchial epithelial cells. A correlation between deletion of β-catenin, reduced Wnt-7a, and low expression of several Hox genes (HoxA1, A7, A9, A10) was also found in mesothelioma cells NCI-H28 (Calvo, 2000). As stated above, Wnt-7a expression was frequently reduced in human lung tumors, and several Hox genes were upregulated, suggesting direct Wnt-Hox relationships (Calvo, 2000). The same researchers have shown that Wnt-7a can induce E-caderin expression in lung cancer cells, while loss of Wnt-7a expression in lung cancer, described by them, could explain the loss of E-cadherin in such tumors and cell lines, contributing to progression and invasiveness (Ohira, 2003). In non-small-cell lung carcinoma (NSCLC) nuclear β-catenin accumulation in a subset of patients was correlated with increased cell proliferation and overexpression of E2F1 and MDM2, aterrant P53 expression, and diminished p27 KIP expression and APC levels. However, there were no mutations in β-catenin (exon 3) and APC (Kostinas, 2002). In contrast, le molecular pathways underlying the progression of lung tumors induced by the carcinogen N-nitrosobis (2-hydroxypropyl) amine in rats are quite different since 23-40% animals are characterized by adenocarcinomas with APC and β-catenin mutations (Tsujiuchi, 2000). Several reports have documented deletions at 5q of the APC gene, without evidence of APC point mutations (Cooper 1996; Fong, 1995; Wada, 1997; Wieland, 1994). Since aberrations of chromosome arm 1p are common in lung cancer (and other cancers), a genomic array analysis with complete tiling coverage of 1p was performed, to profile preinvasive and invasive NSCLC samples (Garnis, C, 2005). This study suggested the involvement of Wnt- 4 and other genes of Wnt and of Notch pathways in lung tumorigenesis, quite similar to their role in fetal lung development. Down-regulation of WIF-1 expression occurred in a large proportion of human lung cancers (83%), through promoter methylation (Mazieres, 2004).

Skin Cancer

Pilomatricoma, a common skin cancer, is caused by activating mutations in β-catenin (Chan, 1999), and enhanced β-catenin signalling together with loss of presenilin, was demonstrated in skin tumorigenesis (Xia, 2001). Frequent alterations in β-catenin intracellular localization were found in melanoma (Rimm, 1999). Also, aberrations of Wnt signalling pathway genes CTNNB1, APC, ICAT (inhibitor of β-catenin Tcf-4) were shown in malignant melanoma, the most frequent being a markedly reduced ICAT transcript levels, found in 28/36 (78%) tumors. Mutations in CTNNB1, APC, or ICAT were rare (Reifenberger, 2002). Wnt signalling pathways are also altered in other cancers such as medulloblastoma (Marino, 2005), and renal and bladder cancers; myeloma and B-cell lymphomas, and in other human diseases, including sporadic and FAP-associated aggressive fibromatosis, *i.e.* desmoid tumors (Tejpar, 1999 ; Moon, 2004; Logan, CY, 2004).

8. Therapeutic Interventions at Stem Cell Manipulations and Wnt Signalling

Apart from the purely academic interest, there are several domains where knowledge of Wnt signalling pathways and their implications in physiological and pathological processes can be of great importance: tissue repair (amplification and differentiation of stem cells *in vitro*), and cancer (diagnosis, reorientation of malignant cells to the differentiation programme, definition of new targets for therapy).

In the field of stem cell biology and clinical applications, important advances have been made concerning the effect of β-catenin/ LEF-1 /TCF in the maintenance and expansion of the pool of human stem cells in long term culture (Reya, 2003). In addition, the recent successful isolation of the active lipid -modified active Wnt molecules with the capability of inducing self-renewal of hematopoietic stem cells (HSC), demonstrated in lethally-irradiated mice reconstituted with Wnt-3a -stimulated human stem cells in culture (Willert, 2003), opens new opportunities for the manipulation of stem cells, expanding HSCs from a patient or a donor *in vitro*, for transplantation. Increased understanding of the commitment of the erythoid lineages by stem cell factors, cytokines, and stromal cells mimicking the marrow microenvironment allowed the large-scale generation of red blood cells in culture keeping determinants of mature erythrocytes in vivo (Giarratana, 2005). Recent findings related to stem cell plasticity, the ability of transplanted bone-marrow cells to transdifferentiate into intestinal subepithelial myofibroblasts, suggest that bone marrow cells may be used to deliver therapeutic genes to damaged tissues, for instance in mesenchymal diseases of the gastrointestinal tract such as fibrosis, mesenchymal tumors, and Crohn's disease. This plasticity can be a possible tool for using Wnt pathway genes for clinical purposes (Okamoto, 2002; Wagers, 2004). For cellular therapy, it will be of crucial importance to check the genetic and chromosomal integrity and functional differentiation programmes of human ES cells in long-term culture for safe transplantation in patients with tissue disorders and malignancies.

The great amount of information gathered over recent years concerning the different components of the Wnt signalling pathways, including alterations found in subsets of patients with various cancers, opens new clinical promises. More research is needed to clarify the participation of the Wnt pathways in the mechanisms regulating the growth of epithelial progenitors and exit from the proliferative compartment, migration and differentiation along colorectal crypts and gastric glands. Immortal progenitor stem cells reflecting intense self-renewal at the base of human colorectal crypts are considered as the crucial compartment responsible of the emergence and production of intestinal cancer cells. Transformed cells escape internal and local controls of the intestinal progenitor cell niche, regarding crypt cell proliferation, apoptosis and migration /differentiation. At a relatively early stage of colorectal cancer, APC mutations can be detected in DNA purified from feces (Traverso, 2002). Analysis of other alterations in Wnt pathway might be developed for early diagnosis of colon cancer and possibly other cancers. Feasibility is linked to the technology and cost of such procedures. Therapeutic interventions targeting the proliferation /differentiation balance at Wnt signalling and crossroads have the potential to induce a selective terminal differentiation and /or apoptosis in transformed cells. There are already examples of manipulations capable to reorient cancer cells to the appropriate terminal differentiation pathways. The use of retinoic acid (RA) is one such "success story" in hematopoietic malignancies, for instance promyelocytic leukemia (Degos, 2001). In this context, it is interesting to note that the activity of β-catenin-LEF /TCF signalling and RA -responsive promoters are closely linked (Easwaran, 1999). This was also accomplished in hormone-dependent breast cancer cells (Fritah, 2005) and monocytic differentiation of human leukemia cells by bile acids (Zimber, 2000), suggesting involvement of functional interactions between nuclear bile acid farnesoid receptors FXR and retinoic acid receptors RAR /RXR or vitamin D3 receptors. Both bile acids and Vit D3 play critical roles in gut pathophysiology and carcinogenesis (Palmer, 2003). Another aspect of therapeutic applications linked to the differentiation of malignant cells is related to the expression of unique determinants as targets for therapy. Retinoic acid treatment of mice bearing tumors driven by activation of the β-catenin /Wnt-1 pathway resulted in an increased expression of *Stra 6* gene encoding cell surface antigens in tumors, but not in normal tissues (Tice, 2002). Another approach tested was the activation of A3 adenosine receptors (A3AR) by the synthetic agonist IB-MECA that specifically affected various tumor cells, especially melanoma. Deregulation of Wnt signalling was shown to be involved (Fishman, 2002).

Regarding the pertinence of therapeutic strategies against the transforming activity of the paracrine and autocrine Wnt pathways (Bafico, 2004) and other interventions targeting GSK-3, two recent reports cast some doubts concerning their use against cancer progression. Indeed, GSK-3β inhibition by lithium, SB216763, as well as by dominant-negative GSK-3β mutants respectively conferred the invasive phenotype in human colorectal cancer cells and resistance to chemotherapy -induced apoptosis in HepG2 cells exposed to etoposide or camptothecin (Beurel, 2004; Le Floch, 2005). Conversely, Wnt-2 antibodies or Wnt-2 siRNA were shown to induce apoptosis in melanoma cells that overexpress Wnt-2, to inhibit the invasive potential of cancer cells (Le Floch, 2005) and partially reverse chemotherapy-induced drug resistance (You, 2004). Cancer cell invasion induced by Wnt-2 was also abolished by the soluble Wnt-1/ Wnt-2 inhibitors sFRP-3 and sFRP-1 (Bafico, 1999; Le Floch, 2004). Recently, promising small molecule Wnt inhibitors which disrupt β-catenin/ TCF transcription complex were shown to antagonize cell proliferation, duplication of the *Xenopus* embryonic dorsal axis, and expression of the Wnt targets c-myc or cyclin D1 (Lepourcelet, 2004). Restitution and enhancement of wt-APC upon exposure of colon cancer cells to DNA-alkylating/ damaging agents (N-methyl-N'-nitro-N-nitrosoguanidine, methylmethane sulfonate) may enhance the efficacy of new chemotherapeutic drugs and blockade of DNA repair through the long patch BER system (Narayan, 2005).

9. Conclusion

The extreme complexity, redundancy and versality of the Wnt signalling pathways are in line with their critical role in embryonic stem cell proliferation and diversification, cell fate and positioning during embryogenesis, development, regeneration, tissue repair, and production of lineage-restricted stem cells and progenitor cells. This situation is well illustrated and amplified by the diversity and the intricate interactions between the Wnt ligand -dependent and -independent Wnt signals and other major pathways involved in early development and carcinogenesis. Wnt pathways are critically involved in the homoeostasis of self-renewing progenitor cells in adult intestine and haematopoietic stem cells. Wnt, homeotic genes, and epithelium- mesenchyme interactions are required in endoderm formation, gut morphogenesis, and development of gut lymphatics and the systemic vasculature. However, the sequential orchestration of the Wnt pathways components during normal development is lost during the cacophony of the neoplastic transformation.

There is only little information on the role of gene imprinting mechanisms targeting canonical and noncanonical Wnt pathways, and their possible implication on maternal and paternal specification of developmental programmes or putative roles in cancer progression. Similarly, intermediary signalling elements and "black box" linking Wnt -activated Fz receptors, heterotrimeric G-proteins and downstream elements such as Dvl, GSK-3β, and nuclear functions are still missing and remain to be identified as potential therapeutic targets. Combinatory chemotherapy using DNA damaging agents, neutralizing antibodies or siRNA strategies targeting cell proliferation and essential survival mechanisms controlled by Wnt signalling elements and crosstalks may lead to promising advances in the treatment of cancer patients with diverse malignancies. The identification of the specific genetic, epigenetic, and phenotypic changes linked to chromosomal instability, oncogenic activations, and alterations in the Wnt pathways in primary and metastatic cancers could provide new informations on the complex web of the Wnt signals for the discovery of drugs targeting the Wnt connection in cancer patients and other disorders.

ACKNOWLEGEMENTS

Aided by INSERM, Paris VI University, AstraZeneca, IPSEN, AstraZeneca, and fellowship grants to NLF from La Ligue Contre le Cancer and Association pour la Recherche Contre le Cancer (ARC). We also thank Mrs. Jacqueline Binet, Rina Byer, and Michèle Groussard for retrieving information for this review. The authors regret that due to space and time limitations they were unable to refer to several pertinent papers on Wnt signaling. In Memory of Camille Claudel.

NOTE ADDED IN PROOF

More than 500 articles concerning Wnt/β-catenin pathway appeared since this review was sent for publication, adding more information on the diverse biological processes controlled by this major signalling pathway. These include:

1) Developmental aspects: epithelial-mesenchymal transition, planar cell polarity, and embryonic "morphogenesis";

2) Stem cells' biology and maintenance, mainly hematopoietic and intestinal epithelial stem cells, and also epithelial hair-follicle and corneal epithelial stem cells; cardiomyocytes, osteogenesis and bone repair processes;

3) The role of Wnt pathway orthologs in invertebrates (Drosophila, nematodes), and the early evolutionary role of this signaling pathway in Cnidaria were re-emphasized. In addition, new studies in non-mammalian vertebrates (Xenopus, Zebrafish) and in avian development were described;

4) The role of Wnt signaling pathway in diverse types of human cancer (prevention, progression, and invasiveness) and in several mouse models was further investigated and recapitulated. Down-regulation of HoxB13 was shown to be important in colorectal cancer, and to be mediated by the regulation of Tcf 4 stability (Jung, 2005). Upregulation of the *WNT2* gene, and inactivation of *APC* and *SFRP2* genes by promoter methylation were observed along the progression of Barrett's esophagus from low-grade dysplasia to adenocarcinomas (Clément, 2006). The Wnt targets EphB receptor genes *EphB2* and *B3* encode receptor tyrosine kinases implicated in cell compartimentalization and ordered cellular migrations during embryonic development and sorting of Paneth cells to the crypt base in the small intestine (Clevers and Battle, 2006).

Two forms of GSK3 localized in the cytoplasm and plasma membrane respectively inhibit and stimulate Wnt signaling and Xenopus axis duplication (Zeng, 2005). Interestingly, Wnt stimulation is mediated by sequential phosphorylation of LRP6 by GSK3 and casein kinase 1, allowing the engagement of LRP6 with Axin (Davidson, 2005). Inhibition of GSK and concomitant activation of Wnt signaling may provide a clinical means for the hematopoietic stem cell repopulation in transplant recipients showing poor engraftment (Trowbridge, 2006). These GSK3 inhibitors are currently in clinical use and evaluation in the treatment of diabetes and Alzelmer disease (Cohen, 2004). Also, activation of c-Src was shown to enhance the accumulation of nuclear β-catenin and its transcriptional activity through cyclin D1 (Karni, 2005). β-catenin/Lef/Tcf pathway is activated by plasminogen activators (tpA) and involved in cell cycle progression and proliferation in ECV 304 carcinoma cells (Schwalm-Maupas, 2005). A new role of β-catenin was suggested by using different Lef-1 mutants, showing that β-catenin may function as a nuclear import receptor for Lef-1, in addition to the Lef-1: importin alpha/beta-mediated transport (Asally, 2005). In agreement with our previous data, a positive interaction was found between the / LEF-1/ TCF4 and AP-1 transcription factor in b-catenin –independent and -dependent manner at the matrilysin and c-jun promoters (Rivat, 2003, Le Floch 2005 and Epub 2004, Nateris, 2005). In these reports, the interaction of c-Jun with LEF-1 or TCF4 was confirmed biochemically .

The potential role of Wnt pathway in cancer prevention strategies/treatments is envisaged in the cellular, animal, and human responses to the NSAIDs sulindac, indomethacin, and nitric-oxide-donating aspirin NO-ASA formulations (Koornstra, 2005; Kapitanovic, 2005; Kashfi, 2005).

Global gene expression techniques and proteomic analysis of specialized normal renewing tissue compartments and of human cancers have recently revealed new functions and target genes for Wnt/Frizzled components, and novel interactions with other major signaling pathways. Thus, proteomic analysis of conditioned media from human neonatal foreskin fibroblasts used in the maintenance of human embryonic stem cells suggested a role of several Wnt signaling proteins (Prowse, 2005). As for intestinal stem cells, proteomic analysis has suggested that BMP signaling inhibits ISC self-renewal by suppressing Wnt/β-catenin signaling, and that this cross-talk is mediated by PTEN/Akt pathway and by 14-3-3 Zeta (Tian, 2005). Functional genomic analysis using whole genome interference screens revealed a potential role of the small GTPase rab5 in Drosophila Wingless pathway, in vitro and in vivo, as rab5 inhibited Wg-dependent activation of target genes (i.e., senseless gene). Rab5 action on the Wnt receptors and coreceptors Fzd2 and Arrow (Lrp6) was postulated (DasGupta, 2005).

In situ bybridization study of adult mouse colon demonstrated an expression of TCF-3 in inverse gradient to that of TCF-4. Considering TCF-3 primary function as a transcriptional repressor, a possible functional interaction between TCF-3 and TCF-4 was suggested. Also, new mesenchymal-epithelial interactions of Wnt signaling components in the adult mouse small intestine and colon were implied (Gregorieff, 2005). A recent microarray analysis of transgenic-mouse livers carrying liver-specific non-mutated β-catenin overexpression, showed differential expression of approximately 400 genes and suggested that EGF-R is a direct target of Wnt/β-catenin pathway, and that its activation might contribute to the known mitogenic effects of increased β-catenin levels in the liver (Tan, 2005). FGF-20 and DKK1 were shown recently by Varmus and colleagues (Chamorro, 2005) as transcriptional targets of β-catenin, using the microarray technique in human epithelial cell line engineered to produce mutant β-catenin, and in ovarian endometrioid adenocarcinomas.

Overall, the research on Wnt signaling pathway is growing rapidly in recent years, clearly demonstrating its position as one of the most important regulators of embryonic and adult stem cells' physiology, embryonic development, adult tissue renewal and repair, and carcinogenesis.

References

Abraham SA, Nobukawa B, Giardiello FM, Hamilton SR, Wu TT (2000). Fundic Gland Polyps in Familial Adenomatous Polyposis : Neoplasms with Frequent Somatic Adenomatous Polyposis Coli Gene Alterations. *Am J Pathol 157*: 747-754.

Abraham SA, Nobukawa B, Giardiello FM, Hamilton SR, Wu TT (2001). Sporadic Fundic Gland Polyps : Common Gastric Polyps Arising Through Activating Mutations in the ß-Catenin Gene *Am J Pathol 158*: 1005-1010.

Adam L, Vadlamudi RK, McCrea P, Kumar R (2001). Tiam1 overexpression potentiates heregulin-induced lymphoid enhancer factor-1/beta -catenin nuclear signaling in breast cancer cells by modulating the intercellular stability. *J Biol Chem 276*: 28443-28450.

Aerne B, Ish-Horowicz D (2004). Receptor tyrosine phosphatase psi is required for Delta/Notch signalling and cyclic gene expression in the presomitic mesoderm. *Development. 131*:3391-3399.

Ahumada A, Slusarski DC, Liu X, Moon RT, Malbon CC, Wang HY (2002). Signaling of rat Frizzled-2 through phosphodiesterase and cyclic GMP. *Science 298*: 2006-2010.

Alonso L, Fuchs E (2003). Stem cells in the skin: waste not, Wnt not. *Genes Dev. 17*:1189-1200.

Al-Tassan N, Chmiel NH, Maynard J, Fleming N, Livingston AL, Williams GT, Hodges AK, Davies DR, David SS, Sampson JR, Cheadle JP (2002). Inherited variants of MYH associated with somatic G:C-->T:A mutations in colorectal tumors. *Nat Genet. 2*:227-32.

Amoyel M, Cheng YC, Jiang YJ, Wilkinson DG (2005). Wnt1 regulates neurogenesis and mediates lateral inhibition of boundary cell specification in the zebrafish hindbrain. *Development. 132*:775-785.

Asally M, Yoneda Y. (2005). *Exp Cell Res* 308:357-363.

Attoub S, Rivat C, Bedin M, Bruyneel E, Louvet C, André T, Mareel M, Mester J, and Gespach C (2002). The c-kit tyrosine kinase inhibitor STI571 for colorectal cancer therapy. *Cancer Res. 62* : 4879-4883.

Aulehla A, Wehrle C, Brand-Saberi B, Kemler R, Gossler A, Kanzler B, Herrmann BG (2003). Wnt3a Plays a Major Role in the Segmentation Clock Controlling Somitogenesis. *Dev Cell 4*:395-406.

Aulehla, A, Herrmann BG (2004). Segementation in vertebrates : clock and gradient finally joined. *Genes Dev 18*:2060-2067.

Austin TW, Solar GP, Ziegler FC, Liem L, Matthews W (1997). A role for the Wnt gene family in hematopoiesis: expansion of multilineage progenitor cells. *Blood 89*:3624-3635.

Axelrod JD, Miller JR, Shulman JM, Moon RT, Perrimon N (1998). Differential recruitment of Dishevelled provides signaling specificity in the planar cell polarity and Wingless signaling pathways. *Genes Dev 12*: 2610-2622.

Bafico A, Gazit A, Pramila T, Finch PW, Yaniv A, Aaronson SA (1999). Interaction of frizzled related protein (FRP) with Wnt ligands and the frizzled receptor suggests alternative mechanisms for FRP inhibition of Wnt signaling. *J Biol Chem 274*: 16180-16187.

Bafico A, Liu G, Yaniv A, Gazit A, Aaronson SA (2001). Novel mechanism of Wnt signalling inhibition mediated by Dickkopf-1 interaction with LRP6/Arrow. *Nat Cell Biol. 3*: 683-686.

Bafico A, Liu G, Goldin L, Harris V, Aaronson SA (2004). An autocrine mechanism for constitutive Wnt pathway activation in human cancer cells. *Cancer Cell 5*:497-506.

Bai YQ, Yamamoto H, Akiyama Y, Tanaka H, Takizawa T, Koike M, Kenji Yagi O, Saitoh K, Takeshita K, Iwai T, Yuasa Y (2002). Ectopic expression of homeodomain protein CDX2 in intestinal metaplasia and carcinomas of the stomach. *Cancer Lett. 176* :47-55.

Bara J, Forgue-Lafitte ME, Maurin N, Flejou JF, Zimber A (2003).Abnormal expression of gastric mucin in human and rat aberrant crypt foci during colon carcinogenesis. *Tumour Biol. 24*:109-15.

Barker N, Hurlstone A, Musisi H, Miles A, Bienz M, Clevers H (2001). The chromatin remodelling factor Brg-1 interacts with beta-catenin to promote target gene activation. *Embo J. 20*: 4935-4943.

Batlle E, Henderson JT, Beghtel H, van den Born MM, Sancho E, Huls G, Meeldijk J, Robertson J, van de Wetering M, Pawson T, Clevers H (2002). Beta-catenin and TCF mediate cell positioning in the intestinal epithelium by controlling the expression of EphB/ephrinB. *Cell 111*: 251-263.

Bauer A, Chauvet S, Huber O, Usseglio F, Rothbacher U, Aragnol D, Kemler R, Pradel J (2000). Pontin52 and reptin52 function as antagonistic regulators of beta-catenin signalling activity. *Embo J 19*: 6121-6130.

Beachy PA, Karhadkar SS, Berman DM (2004).Tissue repair and stem cell renewal in carcinogenesis. *Nature. 432*:324-31.

Behrens J, Jerchow BA, Wurtele M, Grimm J, Asbrand C, Wirtz R, Kuhl M, Wedlich D, Birchmeier W (1998). Functional interaction of an axin homolog, conductin, with beta-catenin, APC, and GSK3beta. *Science 280*: 596-599.

Bejsovec A (2000). Wnt signaling: an embarrassment of receptors. *Curr Biol 10*: R919-922.

Beurel E, Kornprobst M, Blivet-Van Eggelpoel MJ, Ruiz-Ruiz C, Cadoret A, Capeau J, Desbois-Mouthon C (2004). GSK-3beta inhibition by lithium confers resistance to chemotherapy-induced apoptosis through the repression of CD95 (Fas/APO-1) expression. *Exp Cell Res. 300*:354-64.

Bhanot P, Fish M, Jemison JA, Nusse R, Nathans J, Cadigan KM (1999). Frizzled and Dfrizzled-2 function as redundant receptors for Wingless during Drosophila embryonic development. *Development 126*: 4175-4186.

Bienz M (2002). The subcellular destinations of APC proteins. *Nat Rev Mol Cell Biol 3*: 328-338.

Bijur GN, Jope RS (2001). Proapoptotic stimuli induce nuclear accumulation of glycogen synthase kinase-3 beta. *J Biol Chem 276*: 37436-37442.

Bijur GN, Jope RS. (2003). Glycogen synthase kinase-3 beta is highly activated in nuclei and mitochondria. *Neuroreport 14*: 2415-2419.

Birchmeier W, Behrens J (1994). Cadherin expression in carcinomas: role in the formation of cell junctions and the prevention of invasiveness. *Biochim Biophys Acta 1198*: 11-26.

Blache P, van de Wetering M, Duluc I, Domon C, Berta P, Freund JN, Clevers H, Jay P (2004). SOX9 is an intestine crypt transcription factor, is regulated by the Wnt pathway, and represses the CDX2 and MUC2 genes. *J Cell Biol 166*: 37-47.

Boivin GP, Washington K, Yang K, Ward JM, Pretlow TP, Russell R, Besselsen DG, Godfrey VL, Doetschman T, Dove WF, Pitot HC, Halberg RB, Itzkowitz SH, Groden J, Coffey RJ (2003). Pathology of mouse models of intestinal cancer: consensus report and recommendations. *Gastroenterology 124*:762-77.

Bonde J, Hess DA, Nolta JA (2004). Recent advances in hematopoietic stem cell biology. *Curr Opin Hematol. 11*:392-398.

Boon EM, van der Neut R, van de Wetering M, Clevers H, Pals ST (2002). Wnt signaling regulates expression of the receptor tyrosine kinase met in colorectal cancer. *Cancer Res 62*: 5126-5128.

Boutros M, Paricio N, Strutt DI, Mlodzik M (1998). Dishevelled activates JNK and discriminates between JNK pathways in planar polarity and wingless signaling. *Cell 94*: 109-118.

Boutros M, Mihaly J, Bouwmeester T, Mlodzik M (2000). Signaling specificity by Frizzled receptors in Drosophila. *Science 288*: 1825-1828.

Brabletz T, Jung A, Dag S, Hlubek F, Kirchner T (1999). beta-catenin regulates the expression of the matrix metalloproteinase-7 in human colorectal cancer. *Am J Pathol 155*: 1033-1038.

Brandenberger R, Wei H, Zhang S, Lei S, Murage J, Fisk GJ, Li Y, Xu C, Fang R, Guegler K, Rao MS, Mandalam R, Lebkowski J, Stanton LW (2004). Transcriptome characterization elucidates signaling networks that control human ES cell growth and differentiation. *Nat Biotechnol. 22* :707-16.

Brannon M, Brown JD, Bates R, Kimelman D, Moon RT (1999). XCtBP is a XTcf-3 co-repressor with roles throughout Xenopus development. *Development 126*: 3159-3170.

Brembeck FH, Schwarz-Romond T, Bakkers J, Wilhelm S, Hammerschmidt M, Birchmeier W. (2004). Essential role of BCL9-2 in the switch between beta-catenin's adhesive and transcriptional functions. *Genes Dev 18*: 2225-2230.

Brisken, C., Heineman, A., Chavarria, T., Elenbaas, B., Tan, J., Dey, SK., McMahon, JA, McMahon, A P., and Weinberg, R A (2000). Essential function of Wnt-4 in mammary gland development downstream of progesterone signaling. *Genes Dev. 14*, 650-654.

Brocardo M, Nathke IS, Henderson BR. Redefining the subcellular location and transport of APC: new insights using a panel of antibodies (2005). *EMBO Rep. 6*:184-90.

Brott BK, Sokol SY (2002). Regulation of Wnt/LRP signaling by distinct domains of Dickkopf proteins. *Mol Cell Biol 22*: 6100-6110.

Brunner E, Peter O, Schweizer L, Basler K (1997). Pangolin encodes a Lef-1 homologue that acts downstream of Armadillo to transduce the Wingless signal in Drosophila. *Nature 385*: 829-833.

Bui, T. D., Zhang, L., Rees, M. C., Bicknell, R., and Harris, AL (1997). Expression and hormone regulation of Wnt2, 3, 4, 5a, 7a, 7b and 10b in normal human endometrium and endometrial carcinoma. *Br.J.Cancer 75*, 1131-1136.

Cadigan KM, Fish MP, Rulifson EJ, Nusse R (1998). Wingless repression of Drosophila frizzled 2 expression shapes the Wingless morphogen gradient in the wing. *Cell 93*: 767-777.

Cadoret A, Ovejero C, Saadi-Kheddouci S, Souil E, Fabre M, Romagnolo B, Kahn A, Perret C (2001). Hepatomegaly in Transgenic Mice Expressing an Oncogenic Form of ß-Catenin. *Cancer Res. 61*: 3245-3249.

Calvisi DF, Factor VM, Loi R, Thorgeirsson SS (2001). Activation of ß-Catenin during Hepatocarcinogenesis in Transgenic Mouse Models: Relationship to Phenotype and Tumor Grade. *Cancer Res. 61*: 2085-2091.

Calvo R, West J, Franklin W, Erickson P, Bemis L, Li E, Helfrich B, Bunn P, Roche J, Brambilla E, Rosell R, Gemmill RM, Drabkin HA (2000). Altered HOX and WNT7A expression in human lung cancer. *Proc Natl Acad Sci U S A. 97* : 12776-12781.

Capdevila J, Izpisua Belmonte JC (2001). Patterning mechanisms controlling vertebrate limb development. *Annu Rev Cell Dev Biol.* 17:87-132.

Capelluto DG, Kutateladze TG, Habas R, Finkielstein CV, He X, Overduin M (2002). The DIX domain targets dishevelled to actin stress fibres and vesicular membranes. *Nature 419*: 726-729.

Carron C, Pascal A, Djiane A, Boucaut JC, Shi DL, Umbhauer M (2003). Frizzled receptor dimerization is sufficient to activate the Wnt/beta-catenin pathway. *J Cell Sci 116*: 2541-2550.

Cavallo RA, Cox RT, Moline MM, Roose J, Polevoy GA, Clevers H, Peifer M, Bejsovec A. (1998). Drosophila Tcf and Groucho interact to repress Wingless signalling activity. *Nature 395*: 604-608.

Chamorro MN, Schwartz DR, et al (2005). *EMBO J* 24:73-84.

Chan EF, Gat U, McNiff JM and Fuchs E. A common human skin tumor is caused by activating mutations in β-catenins (1999). *Nat Genet 21*:410-3.

Chaussepied M, Ginsberg D (2005). E2F and Signal Transduction Pathways. *Cell Cycle 4(3)* in press.

Chen CM, Struhl G (1999). Wingless transduction by the Frizzled and Frizzled2 proteins of Drosophila. *Development 126*: 5441-5452.

Chen W, ten Berge D, Brown J, Ahn S, Hu LA, Miller WE, Caron MG, Barak LS, Nusse R, Lefkowitz RJ (2003). Dishevelled 2 recruits beta-arrestin 2 to mediate Wnt5A-stimulated endocytosis of Frizzled 4. *Science 301*: 1391-1394.

Chen AE, Ginty DD, Fan CM(2005) Protein kinase A signalling via CREB controls myogenesis induced by Wnt proteins. *Nature. 433*:317-22.

Chepko G, Dickson RB (2003). Ultrastructure of the putative stem cell niche in rat mammary epithelium. *Tissue Cell. 35*:83-93.

Choi J, Park SY, Costantini F, Jho EH, Joo CK. (2004). APC is down-regulated by the ubiquitin-proteasome pathway in a process facilitated by Axin. *J Biol Chem. 279*: 49188-49198

Choi SC, Han JK (2002). Xenopus Cdc42 regulates convergent extension movements during gastrulation through Wnt/Ca2+ signaling pathway. *Dev Biol 244*: 342-357.

Chesire DR, Isaacs WB (2003). β-catenin signaling in prostate cancer : an early perspective. *Endocrine Rel Cancer 10* : 537-60

Christian JL, Moon RT (1993). Interactions between Xwnt-8 and Spemann organizer signaling pathways generate dorsoventral pattern in the embryonic mesoderm of Xenopus. *Genes Dev 7*: 13-28.

Church V, Nohno T, Linker C, Marcelle C, Francis-West P (2002). Wnt regulation of chondrocyte differentiation. *J Cell Sci. 115* :4809-18.

Civenni G, Holbro T, Hynes NE (2003). Wnt1 and Wnt5a induce cyclin D1 expression through ErbB1 transactivation in HC11 mammary epithelial cells. *EMBO Rep 4*: 166-171.

Clément G, et al (2006). *Oncogene* Jan 16, Epub ahead of print.

Clements WM, Wang J, Sarnaik A, Kim OJ, MacDonald J, Fenoglio-Preiser C, Groden J, Lowy AM (2002). ß-Catenin Mutation Is a Frequent Cause of Wnt Pathway Activation in Gastric Cancer. *Cancer Res. 62*: 3503-3506.

Clevers H and Battle E (2006). *Cancer Res*. 66: 2-5.

Cliffe A, Hamada F, Bienz M (2003). A role of Dishevelled in relocating Axin to the plasma membrane during wingless signaling. *Curr Biol 13*: 960-966.

Coghlan MP, Culbert AA, Cross DA, Corcoran SL, Yates JW, Pearce NJ, Rausch OL, Murphy GJ, Carter PS, Roxbee Cox L, Mills D, Brown MJ, Haigh D, Ward RW, Smith DG, Murray KJ, Reith AD, Holder JC (2000). Selective small molecule inhibitors of glycogen synthase kinase-3 modulate glycogen metabolism and gene transcription. *Chem Biol 7*: 793-803.

Cohen P, Goedert M (2004). GSK3 inhibitors: development and therapeutic potential. *Nat Rev Drug Discov. 3*:479-87.

Cole A, Frame S, Cohen P (2004). Further evidence that the tyrosine phosphorylation of glycogen synthase kinase-3 (GSK3) in mammalian cells is an autophosphorylation event. *Biochem J 377*: 249-255.

Conacci-Sorrell ME, Ben-Yedidia T, Shtutman M, Feinstein E, Einat P, Ben-Ze'ev A (2002). Nr-CAM is a target gene of the beta-catenin/LEF-1 pathway in melanoma and colon cancer and its expression enhances motility and confers tumorigenesis. *Genes Dev 16*: 2058-2072.

Cong F, Schweizer L, Varmus H (2004a). Casein kinase Iepsilon modulates the signaling specificities of dishevelled. *Mol Cell Biol 24:* 2000-2011.

Cong F, Varmus H (2004b). Nuclear-cytoplasmic shuttling of Axin regulates subcellular localization of beta-catenin. *Proc Natl Acad Sci U S A 101*: 2882-2887.

Cooper CA, Bubb VJ, Smithson N, Carter RL, Gledhill S, Lamb D, Wyllie AH, Carey FA (1996). Loss of heterozygosity at 5q21 in non-small cell lung cancer: a frequent event but without evidence of apc mutation. *J Pathol. 180*:33-7.

Cossu G, Tajbakhsh S, Buckingham M. How is myogenesis initiated in the embryo? (1996). *Trends Genet. 12*:218-23.

Couso JP, Martinez Arias A (1994). Notch is required for wingless signaling in the epidermis of Drosophila. *Cell 79*: 259-272.

Crawford HC, Fingleton BM, Rudolph-Owen LA, Goss KJ, Rubinfeld B, Polakis P, Matrisian LM (1999). The metalloproteinase matrilysin is a target of beta-catenin transactivation in intestinal tumors. *Oncogene 18*: 2883-2891.

Crawford HC, Fingleton B, Gustavson MD, Kurpios N, Wagenaar RA, Hassell JA, Matrisian LM (2001). The PEA3 subfamily of Ets transcription factors synergizes with beta-catenin-LEF-1 to activate matrilysin transcription in intestinal tumors. *Mol Cell Biol 21*: 1370-1383.

Cui H, Onyango P, Brandenburg S, Wu Y, Hsieh CL, Feinberg AP (2002). Loss of imprinting in colorectal cancer linked to hypomethylation of H19 and IGF2. *Cancer Res. 62*:6442-6446.

Cui H, Cruz-Correa M, Giardiello FM, Hutcheon DF, Kafonek DR, Brandenburg S, Wu Y, He X, Powe NR, Feinberg AP (2003). Loss of IGF2 imprinting: a potential marker of colorectal cancer risk. *Science 299*:1753-1755.

Cui Y, Brown JD, Moon RT, Christian JL (1995). Xwnt-8b: a maternally expressed Xenopus Wnt gene with a potential role in establishing the dorsoventral axis. *Development 121*: 2177-2186.

da Costa LT, He TC, Yu J, Sparks AB, Morin PJ, Polyak K, Laken S, Vogelstein B, Kinzler KW (1999). CDX2 is mutated in a colorectal cancer with normal APC/beta-catenin signaling. *Oncogene. 18*:5010-5014.

Dajani R, Fraser E, Roe SM, Young N, Good V, Dale TC, Pearl LH (2001). Crystal structure of glycogen synthase kinase 3 beta: structural basis for phosphate-primed substrate specificity and autoinhibition. *Cell 105*: 721-732.

Daniels DL, Eklof Spink K, Weis WI (2001). beta-catenin: molecular plasticity and drug design. *Trends Biochem Sci 26*: 672-678.

Daniels DL, Weis WI (2002). ICAT inhibits beta-catenin binding to Tcf/Lef-family transcription factors and the general coactivator p300 using independent structural modules. *Mol Cell 10*: 573-584.

Darmoul D, Gratio V, Devaud H, Peiretti F, Laburthe M (2004). Activation of proteinase-activated receptor 1 promotes human colon cancer cell proliferation through epidermal growth factor receptor transactivation. *Mol Cancer Res. 2*:514-22.

DasGupta R, et al (2005). *Science* 308:826-833.

Davidson G, Wu W et al (2005). *Nature* 438: 867-872.

Davies JA, Fisher CE (2002). Genes and proteins in renal development. *Exp Nephrol. 10*:102-13.

Degos L, Wang ZY (2001). All trans retinoic acid in acute promyelocytic leukemia. *Oncogene. 20*:7140-5.

de Groot RP, Auwerx J, Bourouis M, Sassone-Corsi P (1993). Negative regulation of Jun/AP-1: conserved function of glycogen synthase kinase 3 and the Drosophila kinase shaggy. *Oncogene 8*: 841-847.

de La Coste A, Romagnolo B, Billuart P, Renard CA, Buendia MA, Soubrane O, Fabre M, Chelly J, Beldjord C, Kahn A, Perret C (1998). Somatic mutations of the beta-catenin gene are frequent in mouse and human hepatocellular carcinomas. *Proc Natl Acad Sci U S A 95*: 8847-8851.

de la Taille A, Rubin MA, Chen MW, Vacherot F, de Medina SG, Burchardt M, Buttyan R, Chopin D. (2003). Beta-catenin-related anomalies in apoptosis-resistant and hormone-refractory prostate cancer cells. *Clin Cancer Res. 9*:1801-7.

Deng G, Song GA, Pong E, Sleisenger M, Kim YS (2004). Promoter methylation inhibits APC gene expression by causing changes in chromatin conformation and interfering with the binding of transcription factor CCAAT-binding factor. *Cancer Res. 64*:2692-8.

de Santa Barbara P, van den Brink GR, Roberts DJ (2003). Development and differentiation of the intestinal epithelium. *Cell Mol Life Sci. 60*:1322-32.

De Wever O, Westbroek W, Verloes A, Bloemen N, Bracke M, Gespach C, Bruyneel E, Mareel M (2004a). Critical role of N-cadherin in myofibroblast invasion and migration in vitro stimulated by colon-cancer-cell-derived TGF-beta or wounding. *J Cell Sci. 117* : 4691-4703.

De Wever O, Nguyen QD, Van Hoorde L, Bracke M, Bruyneel E, Gespach C, Mareel M (2004b). Tenascin-C and SF/HGF produced by myofibroblasts in vitro provide convergent pro-invasive signals to human colon cancer cells through RhoA and Rac. *FASEB J. 18*:1016-1028.

Diehl JA, Cheng M, Roussel MF, Sherr CJ (1998). Glycogen synthase kinase-3beta regulates cyclin D1 proteolysis and subcellular localization. *Genes Dev 12*: 3499-3511.

Dikovskaya D, Zumbrunn J, Penman GA, Nathke IS (2001). The adenomatous polyposis coli protein: in the limelight out at the edge. *Trends Cell Biol 11*: 378-384.

Dimitriadis A, Vincan E, Mohammed IM, Roczo N, Phillips WA, Baindur-Hudson S (2001). Expression of Wnt genes in human colon cancers. *Cancer Lett 166*: 185-191.

Ding VW, Chen RH, McCormick F (2000). Differential regulation of glycogen synthase kinase 3beta by insulin and Wnt signaling. *J Biol Chem 275*: 32475-32481.

Djiane A, Riou J, Umbhauer M, Boucaut J, Shi D (2000). Role of frizzled 7 in the regulation of convergent extension movements during gastrulation in Xenopus laevis. *Development 127*: 3091-3100.

Dominguez I, Itoh K, Sokol SY (1995). Role of glycogen synthase kinase 3 beta as a negative regulator of dorsoventral axis formation in Xenopus embryos. *Proc Natl Acad Sci U S A 92*: 8498-8502.

Domon-Dell C, Freund JN (2002).Stimulation of Cdx1 by oncogenic beta-catenin/Tcf4 in colon cancer cells; opposite effect of the CDX2 homeoprotein. *FEBS Lett. 518* :83-7.

Duncan AW, Rattis FM, DiMascio LN, Congdon KL, Pazianos G, Zhao C, Yoon K, Cook JM, Willert K, Gaiano N, Reya T (2005). Integration of Notch and Wnt signaling in hematopoietic stem cell maintenance. *Immunol. 6*:314-22.

Du SJ, Purcell SM, Christian JL, McGrew LL, Moon RT (1995). Identification of distinct classes and functional domains of Wnts through expression of wild-type and chimeric proteins in Xenopus embryos. *Mol Cell Biol 15:* 2625-2634.

Dulaimi, E., Hillinck, J., Ibanez, d. C., I, Al Saleem, T., and Cairns, P (2004). Tumor suppressor gene promoter hypermethylation in serum of breast cancer patients. *Clin.Cancer Res. 10*, 6189-6193.

Duval A, Gayet J, Zhou XP, Iacopetta B, Thomas G, Hamelin R (1999). Frequent frameshift mutations of the TCF-4 gene in colorectal cancers with microsatellite instability. *Cancer Res 59*: 4213-4215.

Duval A, Iacopetta B, Ranzani GN, Lothe RA, Thomas G, Hamelin R (1999). Variable mutation frequencies in coding repeats of TCF-4 and other target genes in colon, gastric and endometrial carcinoma showing microsatellite instability. *Oncogene 18*:6806-9.

Easwaran V, Pishvaian M, Salimuddin, Byers S (1999). Cross-regulation of beta-catenin-LEF/TCF and retinoid signaling pathways. *Curr Biol. 9*:1415-8.

Eda A, Osawa H, Satoh K, Yanaka I, Kihira K, Ishino Y, Mutoh H, Sugano K (2003). Aberrant expression of CDX2 in Barrett's epithelium and inflammatory esophageal mucosa. *J Gastroenterol. 38*:14-22.

Efstathiou JA, Noda M, Rowan A, Dixon C, Chinery R, Jawhari A, Hattori T, Wright NA, Bodmer WF, Pignatelli M (1998). Intestinal trefoil factor controls the expression of the adenomatous polyposis coli-catenin and the E-cadherin-catenin complexes in human colon carcinoma cells. *Proc Natl Acad Sci U S A 95*: 3122-3127.

Eldar-Finkelman H, Seger R, Vandenheede JR, Krebs EG (1995). Inactivation of glycogen synthase kinase-3 by epidermal growth factor is mediated by mitogen-activated protein kinase/p90 ribosomal protein S6 kinase signaling pathway in NIH/3T3 cells. *J Biol Chem 270*: 987-990.

Emami S, Rodrigues S, Rodrigue CM, Floch N, Rivat C, Attoub, Bruyneel E and Gespach C (2004). Trefoil factor family (TFF) peptides and cancer progression. *Peptides 25*: 885-898.

Endo Y, Wolf V, Muraiso K, Kamijo K, Soon L, Uren A, Barshishat-Kupper M, Rubin JS (2005). Wnt-3a-dependent cell motility involves RhoA activation and is specifically regulated by dishevelled-2. *J Biol Chem 280*:777-86.

Faassen AE, Schrager JA, Klein DJ, Oegema TR, Couchman JR, McCarthy JB (1992). A cell surface chondroitin sulfate proteoglycan, immunologically related to CD44, is involved in type I collagen-mediated melanoma cell motility and invasion. *J Cell Biol 116*: 521-531.

Fagotto F, Gluck U, Gumbiner BM (1998). Nuclear localization signal-independent and importin/karyopherin-independent nuclear import of beta-catenin. *Curr Biol 8*: 181-190.

Fang X, Yu SX, Lu Y, Bast RC, Jr., Woodgett JR, Mills GB (2000). Phosphorylation and inactivation of glycogen synthase kinase 3 by protein kinase A. *Proc Natl Acad Sci U S A 97*: 11960-11965.

Farr GH, 3rd, Ferkey DM, Yost C, Pierce SB, Weaver C, Kimelman D (2000). Interaction among GSK-3, GBP, axin, and APC in Xenopus axis specification. *J Cell Biol 148*: 691-702.

Faux MC, Ross JL, Meeker C, Johns T, Ji H, Simpson RJ, Layton MJ, Burgess AW (2004). Restoration of full-length adenomatous polyposis coli (APC) protein in a colon cancer cell line enhances cell adhesion. *J Cell Sci. 117*:427-39.

Fekany-Lee K, Gonzalez E, Miller-Bertoglio V, Solnica-Krezel L (2000). The homeobox gene bozozok promotes anterior neuroectoderm formation in zebrafish through negative regulation of BMP2/4 and Wnt pathways. *Development. 127*:2333-45.

Ferkey DM, Kimelman D (2002). Glycogen synthase kinase-3 beta mutagenesis identifies a common binding domain for GBP and Axin. *J Biol Chem 277*: 16147-16152.

Finch PW, He X, Kelley MJ, Uren A, Schaudies RP, Popescu NC, Rudikoff S, Aaronson SA, Varmus HE, Rubin JS (1997). Purification and molecular cloning of a secreted, Frizzled-related antagonist of Wnt action. *Proc Natl Acad Sci U S A 94*: 6770-6775.

Fischer OM, Giordano S, Comoglio PM, Ullrich A (2004). Reactive oxygen species mediate Met receptor transactivation by G protein-coupled receptors and the epidermal growth factor receptor in human carcinoma cells. *J Biol Chem 279*: 28970-28978.

Fishman P, Madi L, Bar-Yehuda S, Barer F, Del Valle L, Khalili K (2002). Evidence for involvement of Wnt signaling pathway in IB-MECA mediated suppression of melanoma cells. *Oncogene. 21*:4060-4.

Fodde R, Edelmann W, Yang K, van Leeuwen C, Carlson C, Renault B, Breukel C, Alt E, Lipkin M, Khan PM, Kutcherlapati R (1994). A targeted chain-termination mutation in the mouse Apc gene results in multiple intestinal tumors. *Proc Natl Acad Sci U S A. 91*:8969-73.

Fodde R, Kuipers J, Rosenberg C, Smits R, Kielman M, Gaspar C, van Es JH, Breukel C, Wiegant J, Giles RH, Clevers H (2001). Mutations in the APC tumour suppressor gene cause chromosomal instability. *Nat Cell Biol. 3*:433-8.

Fong KM, Zimmerman PV, Smith PJ (1995). Tumor progression and loss of heterozygosity at 5q and 18q in non-small cell lung cancer. *Cancer Res. 55*:220-3.

Frame S, Cohen P, Biondi RM (2001). A common phosphate binding site explains the unique substrate specificity of GSK3 and its inactivation by phosphorylation. *Mol Cell 7*: 1321-1327.

Fraser E, Young N, Dajani R, Franca-Koh J, Ryves J, Williams RS, Yeo M, Webster MT, Richardson C, Smalley MJ, Pearl LH, Harwood A, Dale TC (2002). Identification of the Axin and Frat binding region of glycogen synthase kinase-3. *J Biol Chem 277*: 2176-2185.

Fritah A, Saucier C, Mester J, Redeuilh G, Sabbah M (2005). p21WAF1/CIP1 Selectively Controls the Transcriptional Activity of Estrogen Receptor {alpha}. *Mol Cell Biol. 25*:2419-30.

Fu VX, Schwarze SR, Kenowski ML, Leblanc S, Svaren J, Jarrard DF (2004). A loss of insulin-like growth factor-2 imprinting is modulated by CCCTC-binding factor down-regulation at senescence in human epithelial cells. *J Biol Chem. 279*:52218-52226.

Fuchs E, Merrill BJ, Jamora C, DasGupta R (2001). At the roots of a never-ending cycle. Dev *Cell. 1*:13-25.

Fukuchi T, Sakamoto M, Tsuda H, Maruyama K, Nozawa, Hirohashi S (1998). Beta-catenin mutation in carcinoma of the uterine endometrium. *Cancer Res. 58*: 3526-3528.

Fukumoto S, Hsieh CM, Maemura K, Layne MD, Yet SF, Lee KH, Matsui T, Rosenzweig A, Taylor WG, Rubin JS, Perrella MA, Lee ME (2001). Akt participation in the Wnt signaling pathway through Dishevelled. *J Biol Chem 276*: 17479-17483.

Galceran J, Sustmann C, Hsu SC, Folberth S, Grosschedl R. LEF1-mediated regulation of Delta-like1 links Wnt and Notch signaling in somitogenesis (2004). *Genes Dev. 18*:2718-23.

Garnis C, Campbell J, Davies JJ, Macaulay C, Lam S, Lam WL (2005). Involvement of multiple developmental genes on chromosome 1p in lung tumorigenesis. *Hum Mol Genet. 14*:475-82.

Gat U, DasGupta R, Degenstein L, Fuchs E (1998). De Novo hair follicle morphogenesis and hair tumors in mice expressing a truncated beta-catenin in skin. *Cell. 95*:605-14.

Gazit A, Yaniv A, Bafico A, Pramila T, Igarashi M, Kitajewski J, Aaronson SA (1999). Human frizzled 1 interacts with transforming Wnts to transduce a TCF dependent transcriptional response. *Oncogene 18*: 5959-5966.

Giarratana MC, Kobari L, Lapillonne H, Chalmers D, Kiger L, Cynober T, Marden MC, Wajcman H, Douay L.(2005). Ex vivo generation of fully mature human red blood cells from hematopoietic stem cells.*Nat Biotechnol. 23*:69-74.

Gieseler K, Graba Y, Mariol MC, Wilder EL, Martinez-Arias A, Lemaire P, Pradel J (1999). Antagonist activity of DWnt-4 and wingless in the Drosophila embryonic ventral ectoderm and in heterologous Xenopus assays. *Mech Dev 85*: 123-131.

Glinka A, Wu W, Delius H, Monaghan AP, Blumenstock C, Niehrs C (1998). Dickkopf-1 is a member of a new family of secreted proteins and functions in head induction. *Nature 391*: 357-362.

Golan T, Yaniv A, Bafico A, Liu G, Gazit A (2004). human Frizzled 6 (HFz6) acts as a negative regulator of the canonical Wnt. beta-catenin signaling cascade. *J Biol Chem 15*:14879-88.

Gonzalez-Sancho JM, Brennan KR, Castelo-Soccio LA, Brown AM. (2004). Wnt proteins inducc dishevelled phosphorylation via an LRP5/6- independent mechanism, irrespective of their ability to stabilize beta-catenin. *Mol Cell Biol 24*: 4757-4768.

Gottardi CJ, Gumbiner BM (2001). Adhesion signaling: how beta-catenin interacts with its partners. *Curr Biol 11*: R792-794.

Greco TL, Takada S, Newhouse MM, McMahon JA, McMahon AP, Camper SA (1996). Analysis of the vestigial tail mutation demonstrates that Wnt-3a gene dosage regulates mouse axial development. *Genes Dev 10*: 313-324.

Giudicelli F, Lewis J (2004). The vertebrate segmentation clock. *Curr Opin Genet Dev 14*:407-14.

Gregorieff A, Pinto D, et al (2005). Gastroenterology 129:626-638.

Gregory CA, Singh H, Perry AS, Prockop DJ (2003). The Wnt signaling inhibitor dickkopf-1 is required for reentry into the cell cycle of human adult stem cells from bone marrow. *J Biol Chem. 278*:28067-78.

Guger KA, Gumbiner BM (1995). beta-Catenin has Wnt-like activity and mimics the Nieuwkoop signaling center in Xenopus dorsal-ventral patterning. *Dev Biol 172*: 115-125.

Gunhaga L, Marklund M, Sjodal M, Hsieh JC, Jessell TM, Edlund T (2003). Specification of dorsal telencephalic character by sequential Wnt and FGF signaling. *Nat Neurosci. 6*:701-7.

Gunthert U, Hofmann M, Rudy W, Reber S, Zoller M, Haussmann I, Matzku S, Wenzel A, Ponta H, Herrlich P (1991). A new variant of glycoprotein CD44 confers metastatic potential to rat carcinoma cells. *Cell 65*: 13-24.

Gupta RA, Wang D, Katkuri S, Wang H, Dey SK, DuBois RN (2004). Activation of nuclear hormone receptor peroxisome proliferator-activated receptor-delta accelerates intestinal adenoma growth. *Nat Med. 10*:245-7.

Habas R, Kato Y, He X (2001). Wnt/Frizzled activation of Rho regulates vertebrate gastrulation and requires a novel Formin homology protein Daam1. *Cell 107*: 843-854.

Habas R, Dawid IB, He X (2003). Coactivation of Rac and Rho by Wnt/Frizzled signaling is required for vertebrate gastrulation. *Genes Dev 17*: 295-309.

Haegele L, Ingold B, Naumann H, Tabatabai G, Ledermann B, Brandner S (2003). Wnt signalling inhibits neural differentiation of embryonic stem cells by controlling bone morphogenetic protein expression. *Mol Cell Neurosci. 24*:696-708.

Hagen T, Di Daniel E, Culbert AA, Reith AD (2002). Expression and characterization of GSK-3 mutants and their effect on beta-catenin phosphorylation in intact cells. *J Biol Chem 277*: 23330-23335.

Harada N, Tamai Y, Ishikawa T, Sauer B, Takaku K, Oshima M, Taketo MM (1999). Intestinal polyposis in mice with a dominant stable mutation of the beta-catenin gene. *Embo J 18*: 5931-5942.

Hart MJ, de los Santos R, Albert IN, Rubinfeld B, Polakis P (1998). Downregulation of beta-catenin by human Axin and its association with the APC tumor suppressor, beta-catenin and GSK3 beta. *Curr Biol 8*: 573-581.

He TC, Sparks AB, Rago C, Hermeking H, Zawel L, da Costa LT, Morin PJ, Vogelstein B, Kinzler KW (1998). Identification of c-MYC as a target of the APC pathway. *Science 281*: 1509-1512.

He TC, Chan TA, Vogelstein B, Kinzler KW. (1999) PPARdelta is an APC-regulated target of nonsteroidal anti-inflammatory drugs. *Cell 99*: 335-345.

He X, Saint-Jeannet JP, Woodgett JR, Varmus HE, Dawid IB (1995). Glycogen synthase kinase-3 and dorsoventral patterning in Xenopus embryos. *Nature 374*: 617-622.

He X, Saint-Jeannet JP, Wang Y, Nathans J, Dawid I, Varmus H (1997). A member of the Frizzled protein family mediating axis induction by Wnt-5A. *Science 275*: 1652-1654.

He X, Semenov M, Tamai K, Zeng X (2004). LDL receptor-related proteins 5 and 6 in Wnt/beta-catenin signaling: arrows point the way. *Development 131*: 1663-1677.

Heasman J, Crawford A, Goldstone K, Garner-Hamrick P, Gumbiner B, McCrea P, Kintner C, Noro CY, Wylie C (1994). Overexpression of cadherins and underexpression of beta-catenin inhibit dorsal mesoderm induction in early Xenopus embryos. *Cell 79*: 791-803.

Hecht A, Litterst CM, Huber O, Kemler R (1999). Functional characterization of multiple transactivating elements in beta-catenin, some of which interact with the TATA-binding protein in vitro. *J Biol Chem 274*: 18017-18025.

Heisenberg CP, Tada M, Rauch GJ, Saude L, Concha ML, Geisler R, Stemple DL, Smith JC, Wilson SW (2000). Silberblick/Wnt11 mediates convergent extension movements during zebrafish gastrulation. *Nature 405*: 76-81.

Henderson BR, Fagotto F (2002). The ins and outs of APC and beta-catenin nuclear transport. *EMBO Rep 3*: 834-839.

Hino S, Michiue T, Asashima M, Kikuchi A (2003). Casein kinase I epsilon enhances the binding of Dvl-1 to Frat-1 and is essential for Wnt-3a-induced accumulation of beta-catenin. *J Biol Chem 278*: 14066-14073.

Hiscox S, Jiang WG (1999). Hepatocyte growth factor/scatter factor disrupts epithelial tumour cell-cell adhesion: involvement of beta-catenin. *Anticancer Res 19*: 509-517.

Hoeflich KP, Luo J, Rubie EA, Tsao MS, Jin O, Woodgett JR (2000). Requirement for glycogen synthase kinase-3beta in cell survival and NF-kappaB activation. *Nature 406*: 86-90.

Hofmann K (2000). A superfamily of membrane-bound O-acyltransferases with implications for wnt signaling. *Trends Biochem Sci. 25*:111-2.

Hofmann M, Schuster-Gossler K, Watabe-Rudolph M, Aulehla A, Herrmann BG, Gossler A (2004). WNT signaling, in synergy with T/TBX6, controls Notch signaling by regulating Dll1 expression in the presomitic mesoderm of mouse embryos. *Genes Dev. 18*:2712-7.

Holcombe RF, Marsh JL, Waterman ML, Lin F, Milovanovic T, Truong T (2002). Expression of Wnt ligands and Frizzled receptors in colonic mucosa and in colon carcinoma. *Mol Pathol 55*: 220-226.

Hoppler S, Brown JD, Moon RT (1996). Expression of a dominant-negative Wnt blocks induction of MyoD in Xenopus embryos. *Genes Dev 10*: 2805-2817.

Hoschuetzky H, Aberle H, Kemler R. (1994) Beta-catenin mediates the interaction of the cadherin-catenin complex with epidermal growth factor receptor. *J Cell Biol 127*: 1375-1380.

Hou, X., Tan, Y., Li, M., Dey, S. K., and Das, S. K (2004). Canonical Wnt signaling is critical to estrogen-mediated uterine growth. *Mol.Endocrinol. 18*, 3035-3049.

Hovanes K, Li TW, Munguia JE, Truong T, Milovanovic T, Lawrence Marsh J, Holcombe RF, Waterman ML (2001). Beta-catenin-sensitive isoforms of lymphoid enhancer factor-1 are selectively expressed in colon cancer. *Nat Genet 28:* 53-57.

Hsu W, Zeng L, Costantini F (1999). Identification of a domain of Axin that binds to the serine/threonine protein phosphatase 2A and a self-binding domain. *J Biol Chem 274*: 3439-3445.

Huang H, Fujii H, Sankila A, Mahler-Araujo BM, Matsuda M, Cathomas G, Ohgaki H (1999). ß-Catenin Mutations Are Frequent in Human Hepatocellular Carcinomas Associated with Hepatitis C Virus Infection. *Am J Pathol 155*: 1795-1801.

Huguet EL, McMahon JA, McMahon AP, Bicknell R, Harris AL (1994). Differential expression of human Wnt genes 2, 3, 4, and 7B in human breast cell lines and normal and disease states of human breast tissue. *Cancer Res 54*: 2615-2621.

Ikeda S, Kishida S, Yamamoto H, Murai H, Koyama S, Kikuchi A (1998). Axin, a negative regulator of the Wnt signaling pathway, forms a complex with GSK-3beta and beta-catenin and promotes GSK-3beta-dependent phosphorylation of beta-catenin. *Embo J 17*: 1371-1384.

Ikeya M, Lee SM, Johnson JE, McMahon AP, Takada S (1997). Wnt signalling required for expansion of neural crest and CNS progenitors. *Nature 389*: 966-970.

Inagawa S, Itabashi M, Adachi S, Kawamoto T, Hori M, Shimazaki J, Yoshimi F, Fukao K (2002). Expression and Prognostic Roles of ß-Catenin in Hepatocellular Carcinoma: Correlation with Tumor Progression and Postoperative Survival. *Clin. Cancer Res. 8*: 450-456.

Ishikawa A, Kitajima S, Takahashi Y, Kokubo H, Kanno J, Inoue T, Saga Y (2004). Mouse Nkd1, a Wnt antagonist, exhibits oscillatory gene expression in the PSM under the control of Notch signaling. *Mech Dev. 12*:1443-53.

Ishitani T, Ninomiya-Tsuji J, Nagai S, Nishita M, Meneghini M, Barker N, Waterman M, Bowerman B, Clevers H, Shibuya H, Matsumoto K (1999). The TAK1-NLK-MAPK-related pathway antagonizes signalling between beta-catenin and transcription factor TCF. *Nature 399*: 798-802.

Ishitani T, Kishida S, Hyodo-Miura J, Ueno N, Yasuda J, Waterman M, Shibuya H, Moon RT, Ninomiya-Tsuji J, Matsumoto K (2003). The TAK1-NLK mitogen-activated protein kinase cascade functions in the Wnt-5a/Ca(2+) pathway to antagonize Wnt/beta-catenin signaling. *Mol Cell Biol 23*: 131-139.

Jagla K, Jagla T, Heitzler P, Dretzen G, Bellard F, Bellard M (1997). Ladybird, a tandem of homeobox genes that maintain late wingless expression in terminal and dorsal epidermis of the Drosophila embryo. *Development. 124*:91-100.

Jho EH, Zhang T, Domon C, Joo CK, Freund JN, Costantini F (2002). Wnt/beta-catenin/Tcf signaling induces the transcription of Axin2, a negative regulator of the signaling pathway. *Mol Cell Biol 22*: 1172-1183.

Jin, Z., Tamura, G., Tsuchiya, T., Sakata, K., Kashiwaba, M., Osakabe, M., and Motoyama, T (2001). Adenomatous polyposis coli (APC) gene promoter hypermethylation in primary breast cancers. *Br.J Cancer 85*, 69-73.

Jope RS, Johnson GVW (2004). The glamour and gloom of glycogen synthase kinase-3. *TRENDS in Biochemical Sciences. 29*: 95-102.

Jung C, Kim RS, et al (2005). *Br J Cancer* 92:2233-2239.

Kadowaki T, Wilder E, Klingensmith J, Zachary K, Perrimon N (1996). The segment polarity gene porcupine encodes a putative multitransmembrane protein involved in Wingless processing. *Genes Dev 10*: 3116-3128.

Kanai Y, Ochiai A, Shibata T, Oyama T, Ushijima S, Akimoto S, Hirohashi S (1995). c-erbB-2 gene product directly associates with beta-catenin and plakoglobin. *Biochem Biophys Res Commun 208*: 1067-1072.

Kapitanovic S, Cacer T, et al (2005). *Exp Mol Pathol.* – Epub ahead of print.

Karasawa T, Yokokura H, Kitajewski J, Lombroso PJ (2002). Frizzled-9 is activated by Wnt-2 and functions in Wnt/beta -catenin signaling. *J Biol Chem 277*: 37479-37486.

Karni R, Gus Y, et al (2005). *Mol Cell Biol* 25:5031-5039.

Kashfi K, Rigas B, et al (2005). *Biochem Soc. Trans* 33:701-704.

Katoh, M (2003). Expression and regulation of WNT1 in human cancer: up-regulation of WNT1 by beta-estradiol in MCF-7 cells. *Int.J.Oncol. 22*, 209-212.

Kawano Y, Kypta R (2003). Secreted antagonists of the Wnt signalling pathway. *J Cell Sci 116*: 2627-2634.

Kaykas A, Yang-Snyder J, Heroux M, Shah KV, Bouvier M, Moon RT (2004). Mutant Frizzled 4 associated with vitreoretinopathy traps wild-type Frizzled in the endoplasmic reticulum by oligomerization. *Nat Cell Biol 6*: 52-58.

Kaytor MD, Orr HT (2002). The GSK3 beta signaling cascade and neurodegenerative disease. *Curr Opin Neurobiol 12*: 275-278.

Kilian B, Mansukoski H, Barbosa FC, Ulrich F, Tada M, Heisenberg CP (2003). The role of Ppt/Wnt5 in regulating cell shape and movement during zebrafish gastrulation. *Mech Dev 120*: 467-476.

Kim D, Kim S, Koh H, Yoon SO, Chung AS, Cho KS, Chung J (2001). Akt/PKB promotes cancer cell invasion via increased motility and metalloproteinase production. *Faseb J 15*: 1953-1962.

Kim JS, Crooks H, Dracheva T, Nishanian TG, Singh B, Jen J, Waldman T (2002). Oncogenic beta-catenin is required for bone morphogenetic protein 4 expression in human cancer cells. *Cancer Res 62*: 2744-2748.

Kim S, Lee J, Park J, Chung J (2003). BP75, bromodomain-containing M(r) 75,000 protein, binds dishevelled-1 and enhances Wnt signaling by inactivating glycogen synthase kinase-3 beta. *Cancer Res 63*: 4792-4795.

Kirikoshi H, Sagara N, Saitoh T, Tanaka K, Sekihara H, Shiokawa K, Katoh M (1999). Molecular cloning and characterization of human Frizzled-4 on chromosome 11q14-q21. *Biochem Biophys Res Commun 264*: 955-961.

Kirikoshi, H. and Katoh, M (2002). Expression and regulation of WNT10B in human cancer: up-regulation of WNT10B in MCF-7 cells by beta-estradiol and down-regulation of WNT10B in NT2 cells by retinoic acid. *Int.J Mol.Med. 10*, 507-511.

Kishida M, Koyama S, Kishida S, Matsubara K, Nakashima S, Higano K, Takada R, Takada S, Kikuchi A (1999a). Axin prevents Wnt-3a-induced accumulation of beta-catenin. *Oncogene 18*: 979-985.

Kishida S, Yamamoto H, Ikeda S, Kishida M, Sakamoto I, Koyama S, Kikuchi A (1998). Axin, a negative regulator of the wnt signaling pathway, directly interacts with adenomatous polyposis coli and regulates the stabilization of beta-catenin. *J Biol Chem 273*: 10823-10826.

Kishida S, Yamamoto H, Hino S, Ikeda S, Kishida M, Kikuchi A (1999b). DIX domains of Dvl and axin are necessary for protein interactions and their ability to regulate beta-catenin stability. *Mol Cell Biol 19*: 4414-4422.

Kobayashi S, Kohda T, Ichikawa H, Ogura A, Ohki M, Kaneko-Ishino T, Ishino F (2002). Paternal expression of a novel imprinted gene, Peg12/Frat3, in the mouse 7C region homologous to the Prader-Willi syndrome region. *Biochem Biophys Res Commun. 290*:403-408.

Koh TJ, Bulitta CJ, Fleming JV, Dockray GJ, Varro A, Wang TC (2000). Gastrin is a target of the beta-catenin/TCF-4 growth-signaling pathway in a model of intestinal polyposis. *J Clin Invest 106*: 533-539.

Koike J, Takagi A, Miwa T, Hirai M, Terada M, Katoh M (1999). Molecular cloning of Frizzled-10, a novel member of the Frizzled gene family. *Biochem Biophys Res Commun 262*: 39-43.

Kokubu C, Heinzmann U, Kokubu T, Sakai N, Kubota T, Kawai M, Wahl MB, Galceran J, Grosschedl R, Ozono K, Imai K (2004). Skeletal defects in ringelschwanz mutant mice reveal that Lrp6 is required for proper somitogenesis and osteogenesis. *Development. 131*:5469-80.

Kolligs FT, Nieman MT, Winer I, Hu G, Van Mater D, Feng Y, Smith IM, Wu R, Zhai Y, Cho KR, Fearon ER (2002). ITF-2, a downstream target of the Wnt/TCF pathway, is activated in human cancers with beta-catenin defects and promotes neoplastic transformation. *Cancer Cell 1*: 145-155.

Koornstra JJ, Rijcken FEM, et al (2005). *Cancer Epidemiol Biomarkers Prev* 14:1608-1612.

Korinek V, Barker N, Moerer P, van Donselaar E, Huls G, Peters PJ, Clevers H (1998). Depletion of epithelial stem-cell compartments in the small intestine of mice lacking Tcf-4. *Nat Genet 19*: 379-383.

Kotelevets L, Noe V, Bruyneel E, Myssiakine E, Chastre E, Mareel M, Gespach C (1998). Inhibition by platelet-activating factor of Src- and hepatocyte growth factor-dependent invasiveness of intestinal and kidney epithelial cells. Phosphatidylinositol 3'-kinase is a critical mediator of tumor invasion. *J Biol Chem 273*: 14138-14145.

Kotsinas A, Evangelou K, Zacharatos P, Kittas C, Gorgoulis VG (2002). Proliferation, but Not Apoptosis, Is Associated with Distinct ß-Catenin Expression Patterns in Non-Small-Cell Lung Carcinomas : Relationship with Adenomatous Polyposis Coli and G1-to S-Phase Cell-Cycle Regulators . *Am J Pathol 161*: 1619-1634.

Kouzmenko AP, Takeyama K, Ito S, Furutani T, Sawatsubashi S, Maki A, Suzuki E,Kawasaki Y, Akiyama T, Tabata T, Kato S (2004). Wnt /beta-catenin and estrogen signaling converge in vivo. *J Biol Chem. 279*:40255-40268.

Kratochwil K, Galceran j, Tontsch S, Roth W, Grosschedl R (2002). FGF4, a direct target of LEF1 and Wnt signaling, can rescue the arrest of tooth organogenesis in Lef1-/- mice. *Genes Dev 16*:3173-85.

Krupnik VE, Sharp JD, Jiang C, Robison K, Chickering TW, Amaravadi L, Brown DE, Guyot D, Mays G, Leiby K, Chang B, Duong T, Goodearl AD, Gearing DP, Sokol SY, McCarthy SA (1999). Functional and structural diversity of the human Dickkopf gene family. *Gene 238*: 301-313.

Krylova O, Messenger MJ, Salinas PC (2000). Dishevelled-1 regulates microtubule stability: a new function mediated by glycogen synthase kinase-3beta. J Cell Biol 151: 83-94.

Kuhl M, Sheldahl LC, Malbon CC, Moon RT (2000). Ca(2+)/calmodulin-dependent protein kinase II is stimulated by Wnt and Frizzled homologs and promotes ventral cell fates in Xenopus. *J Biol Chem 275*: 12701-12711.

Kuhl M, Geis K, Sheldahl LC, Pukrop T, Moon RT, Wedlich D (2001). Antagonistic regulation of convergent extension movements in Xenopus by Wnt/beta-catenin and Wnt/Ca2+ signaling. *Mech Dev 106*: 61-76.

Kusserow A, Pang K, Sturm C, Hrouda M, Lentfer J, Schmidt HA, Technau U, von Haeseler A, Hobmayer B, Martindale MQ, Holstein TW (2005). Unexpected complexity of the Wnt gene family in a sea anemone. *Nature 433*:156-60.

Landesman Y, Sokol SY (1997). Xwnt-2b is a novel axis-inducing Xenopus Wnt, which is expressed in embryonic brain. *Mech Dev 63*: 199-209.

Lawrence PA, Casal J, Struhl G (2002). Towards a model of the organisation of planar polarity and pattern in the Drosophila abdomen. *Development 129*: 2749-2760.

Lee E, Salic A, Kruger R, Heinrich R, Kirschner MW (2003). The Roles of APC and Axin Derived from Experimental and Theoretical Analysis of the Wnt Pathway. PLoS Biol 1: E10.

Lee JS, Ishimoto A, Yanagawa S (1999). Characterization of mouse dishevelled (Dvl) proteins in Wnt/Wingless signaling pathway. *J Biol Chem 274*: 21464-21470.

Lee SM, Tole S, Grove E, McMahon AP (2000). A local Wnt-3a signal is required for development of the mammalian hippocampus. *Development 127*: 457-467.

Le Floch N, Rivat C, De Wever O, Bruyneel E, Mareel M, Dale T, and Gespach C (2005). The proinvasive activity of Wnt-2 is mediated through a noncanonical Wnt pathway coupled to GSK-3□ and c-jun/ AP-1 signaling. *FASEB J. 19*:144 - 146.

Lejeune S, Huguet EL, Hamby A, Poulsom R, Harris AL (1995). Wnt5a cloning, expression, and up-regulation in human primary breast cancers. *Clin Cancer Res 1*: 215-222.

Leung JY, Kolligs FT, Wu R, Zhai Y, Kuick R, Hanash S, Cho KR, Fearon ER (2002). Activation of AXIN2 expression by beta-catenin-T cell factor. A feedback repressor pathway regulating Wnt signaling. *J Biol Chem. 277*:21657-65.

Lepourcelet M, Chen YN, France DS, Wang H, Crews P, Petersen F, Bruseo C, Wood AW, Shivdasani RA (2004). Small-molecule antagonists of the oncogenic Tcf/beta-catenin protein complex. *Cancer Cell 5*: 91-102.

Li L, Yuan H, Weaver CD, Mao J, Farr GH 3rd, Sussman DJ, Jonkers J, Kimelman D, Wu D (1999). Axin and Frat1 interact with dvl and GSK, bridging Dvl to GSK in Wnt-mediated regulation of LEF-1. *Embo J 18*: 4233-4240.

Li L, Akashi K (2003). Unraveling the molecular components and genetic blueprints of stem cells. *Biotechniques. 35*:1233-1239.

Li Y, Bharti A, Chen D, Gong J, Kufe D (1998). Interaction of glycogen synthase kinase 3beta with the DF3/MUC1 carcinoma-associated antigen and beta-catenin. *Mol Cell Biol 18*: 7216-7224.

Li Y, Kuwahara H, Ren J, Wen G, Kufe D (2001a). The c-Src tyrosine kinase regulates signaling of the human DF3/MUC1 carcinoma-associated antigen with GSK3 beta and beta-catenin. *J Biol Chem 276*: 6061-6064.

Li Y, Ren J, Yu W, Li Q, Kuwahara H, Yin L, Carraway KL, Kufe D (2001b). The epidermal growth factor receptor regulates interaction of the human DF3/MUC1 carcinoma antigen with c-Src and beta-catenin. *J Biol Chem 276*: 35239-35242.

Li Y, Yu WH, Ren J, Chen W, Huang L, Kharbanda S, Loda M, Kufe D (2003). Heregulin targets gamma-catenin to the nucleolus by a mechanism dependent on the DF3/MUC1 oncoprotein. *Mol Cancer Res 1*: 765-775.

Lickert H, Domon C, Huls G, Wehrle C, Duluc I, Clevers H, Meyer BI, Freund JN, Kemler R. Wnt/(beta)-catenin signaling regulates the expression of the homeobox gene Cdx1 in embryonic intestine (2000). *Development. 127*:3805-13.

Lin K, Wang S, Julius MA, Kitajewski J, Moos M, Jr., Luyten FP (1997). The cysteine-rich frizzled domain of Frzb-1 is required and sufficient for modulation of Wnt signaling. *Proc Natl Acad Sci U S A 94*: 11196-11200.

Lin, SY, Xia, W., Wang, JC., Kwong, K. Y., Spohn, B., Wen, Y., Pestell, R. G., and Hung, M C (2000). Beta-catenin, a novel prognostic marker for breast cancer: its roles in cyclin D1 expression and cancer progression. *Proc.Natl.Acad.Sci.U.S.A 97*, 4262-4266.

Lipton L, Halford SE, Johnson V, Novelli MR, Jones A, Cummings C, Barclay E, Sieber O, Sadat A, Bisgaard ML, Hodgson SV, Aaltonen LA, Thomas HJ, Tomlinson IP (2003). Carcinogenesis in MYH-associated polyposis follows a distinct genetic pathway. *Cancer Res. 63*:7595-9.

Liu C, Kato Y, Zhang Z, Do VM, Yankner BA, He X (1999a). beta-Trcp couples beta-catenin phosphorylation-degradation and regulates Xenopus axis formation. *Proc Natl Acad Sci U S A 96*: 6273-6278.

Liu C, Li Y, Semenov M, Han C, Baeg GH, Tan Y, Zhang Z, Lin X, He X (2002). Control of beta-catenin phosphorylation/degradation by a dual-kinase mechanism. *Cell 108*: 837-847.

Liu D, el-Hariry I, Karayiannakis AJ, Wilding J, Chinery R, Kmiot W, McCrea PD, Gullick WJ, Pignatelli M (1997). Phosphorylation of beta-catenin and epidermal growth factor receptor by intestinal trefoil factor. *Lab Invest 77*: 557-563.

Liu SJ, Zhang JY, Li HL, Fang ZY, Wang Q, Deng HM, Gong CX, Grundke-Iqbal I, Iqbal K, Wang JZ (2004). Tau becomes a more favorable substrate for GSK-3 when it is prephosphorylated by PKA in rat brain. *J Biol Chem. 279* : 50078-88.

Liu T, Liu X, Wang H, Moon RT, Malbon CC (1999b). Activation of rat frizzled-1 promotes Wnt signaling and differentiation of mouse F9 teratocarcinoma cells via pathways that require Galpha(q) and Galpha(o) function. *J Biol Chem 274*: 33539-33544.

Liu T, DeCostanzo AJ, Liu X, Wang Hy, Hallagan S, Moon RT, Malbon CC (2001). G protein signaling from activated rat frizzled-1 to the beta-catenin-Lef-Tcf pathway. *Science 292*: 1718-1722.

Liu W, Dong X, Mai M, Seelan RS, Taniguchi K, Krishnadath KK, Halling KC, Cunningham JM, Boardman LA, Qian C, Christensen E, Schmidt SS, Roche PC, Smith DI, Thibodeau SN (2000). Mutations in AXIN2 cause colorectal cancer with defective mismatch repair by activating beta-catenin/TCF signalling. *Nat Genet 26*: 146-147.

Liu X, Liu T, Slusarski DC, Yang-Snyder J, Malbon CC, Moon RT, Wang H (1999c). Activation of a frizzled-2/beta-adrenergic receptor chimera promotes Wnt signaling and differentiation of mouse F9 teratocarcinoma cells via Galphao and Galphat. *Proc Natl Acad Sci U S A 96*: 14383-14388.

Logan CY, Nusse R (2004). The Wnt signaling pathway in development and disease. *Annu Rev Cell Dev Biol. 20*:781-810.

Lohnes D (2003).The Cdx1 homeodomain protein: an integrator of posterior signaling in the mouse. *Bioessays. 25*:971-80.

Lu Z, Ghosh S, Wang Z, Hunter T. (2003) Downregulation of caveolin-1 function by EGF leads to the loss of E-cadherin, increased transcriptional activity of beta-catenin, and enhanced tumor cell invasion. *Cancer Cell 4*: 499-515.

Lustig B, Jerchow B, Sachs M, Weiler S, Pietsch T, Karsten U, van de Wetering M, Clevers H, Schlag PM, Birchmeier W, Behrens J (2002). Negative feedback loop of Wnt signaling through upregulation of conductin/axin2 in colorectal and liver tumors. *Mol Cell Biol 22*: 1184-1193.

Lyons JP, Mueller UW, Ji H, Everett C, Fang X, Hsieh JC, Barth AM, McCrea PD (2004). Wnt-4 activates the canonical beta-catenin-mediated Wnt pathway and binds Frizzled-6 CRD: functional implications of Wnt/beta-catenin activity in kidney epithelial cells. *Exp Cell Res 298*: 369-387.

Malkin D, Li FP, Strong LC, Fraumeni JF Jr, Nelson CE, Kim DH, Kassel J, Gryka MA, Bischoff FZ, Tainsky MA, et al (1990). Germ line p53 mutations in a familial syndrome of breast cancer, sarcomas, and other neoplasms. *Science. 250*:1233-1238.

Maloof JN, Whangbo J, Harris JM, Jongeward GD, Kenyon C (1999). A Wnt signaling pathway controls hox gene expression and neuroblast migration in C. elegans. *Development 126*: 37-49.

Mann B, Gelos M, Siedow A, Hanski ML, Gratchev A, Ilyas M, Bodmer WF, Moyer MP, Riecken EO, Buhr HJ, Hanski C (1999). Target genes of beta-catenin-T cell-factor/lymphoid-enhancer-factor signaling in human colorectal carcinomas. *Proc Natl Acad Sci U S A 96*: 1603-1608.

Mao B, Wu W, Li Y, Hoppe D, Stannek P, Glinka A, Niehrs C (2001a). LDL-receptor-related protein 6 is a receptor for Dickkopf proteins. *Nature 411*: 321-325.

Mao B, Wu W, Davidson G, Marhold J, Li M, Mechler BM, Delius H, Hoppe D, Stannek P, Walter C, Glinka A, Niehrs C (2002). Kremen proteins are Dickkopf receptors that regulate Wnt/ □-catenin signalling. *Nature 417*: 664-667.

Mao B, Niehrs C (2003). Kremen2 modulates Dickkopf 2 activity during Wnt/LRP6 signaling. *Gene 302*: 179-183.

Mao J, Wang J, Liu B, Pan W, Farr GH 3rd, Flynn C, Yuan H, Takada S, Kimelman D, Li L, Wu D (2001b). Low-density lipoprotein receptor-related protein-5 binds to Axin and regulates the canonical Wnt signaling pathway. *Mol Cell 7*: 801-809.

Marchenko GN, Marchenko ND, Leng J, Strongin AY (2002). Promoter characterization of the novel human matrix metalloproteinase-26 gene: regulation by the T-cell factor-4 implies specific expression of the gene in cancer cells of epithelial origin. *Biochem J 363*: 253-262.

Marino S (2005). Medulloblastoma: developmental mechanisms out of control. *Trends Mol Med. 11*:17-22.

Marlow F, Topczewski J, Sepich D, Solnica-Krezel L (2002). Zebrafish Rho kinase 2 acts downstream of Wnt11 to mediate cell polarity and effective convergence and extension movements. *Curr Biol 12*: 876-884.

Matsubayashi H, Sese S, Lee JS, Shirakawa T, Iwatsubo T, Tomita T, Yanagawa S (2004). Biochemical characterization of the Drosophila wingless signaling pathway based on RNA interference. *Mol Cell Biol 24*: 2012-2024.

Mayer K, Hieronymus T, Castrop J, Clevers H, Ballhausen WG (1997). Ectopic activation of lymphoid high mobility group-box transcription factor TCF-1 and overexpression in colorectal cancer cells. *Int J Cancer 72*: 625-630.

Mazelin L, Bonod-Bidaud C, Pays L, Bernet A, Gespach C, Bredesen DE, Scoazec JY, and Mehlen P (2004). Netrin-1 controls colorectal tumorigenesis by regulating apoptosis. *Nature 431* :80-84.

Mazieres J, He B, You L, Xu Z, Lee AY, Mikami I, Reguart N, Rosell R, McCormick F, Jablons DM (2004). Wnt inhibitory factor-1 is silenced by promoter hypermethylation in human lung cancer. *Cancer Res. 64*:4717-20.

McMahon AP, Moon RT (1989). Ectopic expression of the proto-oncogene int-1 in Xenopus embryos leads to duplication of the embryonic axis. *Cell 58*: 1075-1084.

McMahon AP, Bradley A (1990). The Wnt-1 (int-1) proto-oncogene is required for development of a large region of the mouse brain. *Cell 62*: 1073-1085.

McMahon AP, Joyner AL, Bradley A, McMahon JA (1992). The midbrain-hindbrain phenotype of Wnt-1-/Wnt-1- mice results from stepwise deletion of engrailed-expressing cells by 9.5 days postcoitum. *Cell 69*: 581-595.

McManus DT, Olaru A, Meltzer SJ (2004). Biomarkers of esophageal adenocarcinoma and Barrett's esophagus. *Cancer Res. 64*:1561-9.

Meniel V, Hay T, Douglas-Jones A, Sansom OJ, Clarke AR (2005). Mutations in Apc and p53 synergize to promote mammary neoplasia. *Cancer Res. 65*:410-6.

Meijer L, Flajolet M, Greengard P (2004). Pharmacological inhibitors of glycogen synthase kinase 3. *Trends Pharmacol Sci. 9*:471-80.

Miller C, Degenhardt K, Sassoon DA (1998). Fetal exposure to DES results in de-regulation of Wnt7a during uterine morphogenesis. *Nat Genet. 20*:228-30.

Milovanovic T, Planutis K, Nguyen A, Marsh JL, Lin F, Hope C, Holcombe RF (2004). Expression of Wnt genes and frizzled 1 and 2 receptors in normal breast epithelium and infiltrating breast carcinoma. Int J Oncol. 25 :1337-42.

Miwa N, Furuse M, Tsukita S, Niikawa N, Nakamura Y, Furukawa Y (2000). Involvement of claudin-1 in the beta-catenin/Tcf signaling pathway and its frequent upregulation in human colorectal cancers. *Oncol Res 12*: 469-476.

Mogensen MM, Tucker JB, Mackie JB, Prescott AR, Nathke IS (2002). The adenomatous polyposis coli protein unambiguously localizes to microtubule plus ends and is involved in establishing parallel arrays of microtubule bundles in highly polarized epithelial cells. *J Cell Biol. 157*:1041-8.

Molenaar M, van de Wetering M, Oosterwegel M, Peterson-Maduro J, Godsave S, Korinek V, Roose J, Destree O, Clevers H (1996). XTcf-3 transcription factor mediates beta-catenin-induced axis formation in Xenopus embryos. *Cell 86*: 391-399.

Monga SP, Mars WM, Pediaditakis P, Bell A, Mule K, Bowen WC, Wang X, Zarnegar R, Michalopoulos GK (2002). Hepatocyte growth factor induces Wnt-independent nuclear translocation of beta-catenin after Met-beta-catenin dissociation in hepatocytes. *Cancer Res 62*: 2064-2071.

Monkley SJ, Delaney SJ, Pennisi DJ, Christiansen JH, Wainwright BJ (1996). Targeted disruption of the Wnt2 gene results in placentation defects. *Development 122*: 3343-3353.

Moon RT, Campbell RM, Christian JL, McGrew LL, Shih J, Fraser S (1993). Xwnt-5A: a maternal Wnt that affects morphogenetic movements after overexpression in embryos of Xenopus laevis. *Development 119*: 97-111.

Moon RT, Bowerman B, Boutros M, Perrimon N (2002). The promise and perils of Wnt signaling through β-catenin. *Science 296* : 16446-6.

Moon RT, Kohn AD, De Ferrari GV, Kaykas A (2004). WNT and beta-catenin signalling: diseases and therapies. *Nat Rev Genet. 5*:691-701.

Moore KA. Recent advances in defining the hematopoietic stem cell niche (2004). Curr Opin Hematol. 11:107-11.

Morfini G, Szebenyi G, Elluru R, Ratner N, Brady ST (2002). Glycogen synthase kinase 3 phosphorylates kinesin light chains and negatively regulates kinesin-based motility. *Embo J 21*: 281-293.

Mori H, Tomari T, Koshikawa N, Kajita M, Itoh Y, Sato H, Tojo H, Yana I, Seiki M (2002). CD44 directs membrane-type 1 matrix metalloproteinase to lamellipodia by associating with its hemopexin-like domain. *Embo J 21*: 3949-3959.

Morin PJ, Sparks AB, Korinek V, Barker N, Clevers H, Vogelstein B, Kinzler KW (1997). Activation of beta-catenin-Tcf signaling in colon cancer by mutations in beta-catenin or APC. *Science 275*: 1787-1790.

Moucadel V, Totaro MS, Dell CD, Soubeyran P, Dagorn JC, Freund JN, Iovanna JL (2002). The homeobox gene Cdx1 belongs to the p53-p21(WAF)-Bcl-2 network in intestinal epithelial cells. *Biochem Biophys Res Commun. 297*:607-15.

Mukai F, Ishiguro K, Sano Y, Fujita SC (2002). Alternative splicing isoform of tau protein kinase I/glycogen synthase kinase 3beta. *J Neurochem 81*: 1073-1083.

Muller H, Samanta R, Wieschaus E (1999a). Wingless signaling in the Drosophila embryo: zygotic requirements and the role of the frizzled genes. *Development 126*: 577-586.

Muller T, Choidas A, Reichmann E, Ullrich A (1999b). Phosphorylation and free pool of beta-catenin are regulated by tyrosine kinases and tyrosine phosphatases during epithelial cell migration. *J Biol Chem 274*: 10173-10183.

Muller WJ, Neville MC (2001). Introduction: Signaling in mammary development and tumorigenesis. *J Mammary Gland Biol Neoplasia. 6* :1-5.

Nagahata T, Shimada T, Harada A, Nagai H, Onda M, Yokoyama S, Shiba T, Jin E, Kawanami O, Emi M (2003). Amplification, up-regulation and over-expression of DVL-1, the human counterpart of the Drosophila disheveled gene, in primary breast cancers. *Cancer Sci 94*: 515-518.

Nantermet PV, Xu J, Yu Y, Hodor P, Holder D, Adamski S, Gentile MA, Kimmel DB, Harada S, Gerhold D, Freedman LP, Ray WJ (2004). Identification of genetic pathways activated by the androgen receptor during the induction of proliferation in the ventral prostate gland. *J Biol Chem. 279*:1310-22.

Narayan S, Roy D (2003). Role of APC and DNA mismatch repair genes in the development of colorectal cancers. *Mol Cancer. 2*:41.

Narayan S, Jaiswal AS, Balusu R. Tumor Suppressor APC Blocks DNA Polymerase {beta}-dependent Strand Displacement Synthesis during Long Patch but Not Short Patch Base Excision Repair and Increases Sensitivity to Methylmethane Sulfonate (2005). *J Biol Chem. 280*:6942-9.

Nateris AS (2005). *Nature* 437: 281-285.

Näthke IS (2004). The adenomatous polyposis coli protein : The Achilles heel of the gut epithelium. *Annu Rev Cell Dev Biol. 20*:337-366.

Nei H, Saito T, Yamasaki H, Mizumoto H, Ito E, Kudo R (1999). Nuclear localization of beta-catenin in normal and carcinogenic endometrium. *Mol Carcinog. 25*:207-18.

Nguyen QD, Faivre S, Bruyneel E, Rivat C, Seto M, Endo T, Mareel M, Emami S, and Gespach C (2002).. RhoA- and RhoD- dependent regulatory switch of G□-subunits signaling by PAR-1 receptors in cellular invasion. *The FASEB Journal 16* : 565-576 (2002).

Niemann C, Watt FM (2002). Designer skin : lineage commitment in postnatal epidermis. *Trends Cell Biol 12*:185-92.

Ninomiya I, Endo Y, Fushida S, Sasagawa T, Miyashita T, Fujimura T, Nishimura G, Tani T, Hashimoto T, Yagi M, Shimizu K, Ohta T, Yonemura Y, Inoue M, Sasaki T, Miwa K (2000). Alteration of beta-catenin expression in esophageal squamous-cell carcinoma. *Int J Cancer 85*:757-61.

Noordermeer J, Klingensmith J, Perrimon N, Nusse R.(1994). dishevelled and armadillo act in the wingless signalling pathway in Drosophila. *Nature 367*: 80-83.

Nusse R & Varmus HE (1982). Many tumors induced by the mouse mammary tumor virus contain a provirus integrated in the same region of the host genome. *Cell 31* : 99-109.

Nusse R, van Ooyen A, Cox D, Fung YK, Varmus H (1984). Mode of proviral activation of a putative mammary oncogene (int-1) on mouse chromosome 15. *Nature 307*: 131-136.

Nusse R, Brown A, Papkoff J, Scambler P, Shackleford , McMahon A, Moon R (1991). A new nomenclature for int-1 and related genes : the Wnt gene family. *Cell 64* : 231.

Nusse R. (2003) Wnts and Hedgehogs: lipid-modified proteins and similarities in signaling mechanisms at the cell surface. *Development 130*: 5297-5305.

Oda T, Kanai Y, Oyama T, Yoshiura K, Shimoyama Y, Birchmeier W, Sugimura T, Hirohashi S (1994). E-cadherin gene mutations in human gastric carcinoma cell lines. *Proc Natl Acad Sci U S A 91*: 1858-1862.

Oehler, M K, MacKenzie, I Z, Wallwiener, D, Bicknell, R, and Rees, M C (2002). Wnt-7a is upregulated by norethisterone in human endometrial epithelial cells: a possible mechanism by which progestogens reduce the risk of estrogen-induced endometrial neoplasia. *Cancer Lett. 186*, 75-81.

Okamoto R, Yajima T, Yamazaki M, Kanai T, Mukai M, Okamoto S, Ikeda Y, Hibi T, Inazawa J, Watanabe M (2002). Damaged epithelia regenerated by bone marrow-derived cells in the human gastrointestinal tract. *Nat Med. 8*:1011-7.

Olayioye MA, Neve RM, Lane HA, Hynes NE (2000). The ErbB signaling network: receptor heterodimerization in development and cancer. *Embo J 19*: 3159-3167.

Olson DJ, Gibo DM (1998). Antisense wnt-5a mimics wnt-1-mediated C57MG mammary epithelial cell transformation. *Exp Cell Res 241*: 134-141.

Ongusaha PP, Kwak JC, Zwible AJ, Macip S, Higashiyama S, Taniguchi N, Fang L, Lee SW (2004). HB-EGF is a potent inducer of tumor growth and angiogenesis. *Cancer Res. 64*:5283-90.

Ouko, L., Ziegler, T. R., Gu, L. H., Eisenberg, L. M., and Yang, V. W. (2004). Wnt11 signaling promotes proliferation, transformation, and migration of IEC6 intestinal epithelial cells. *J Biol.Chem. 279*, 26707-26715.

Palmer HG, Sanchez-Carbayo M, Ordonez-Moran P, Larriba MJ, Cordon-Cardo C, Munoz A (2003). Genetic signatures of differentiation induced by 1alpha,25-dihydroxyvitamin D3 in human colon cancer cells. *Cancer Res. 63*:7799-806.

Park BK, Zeng X, Glazer RI (2001). Akt1 induces extracellular matrix invasion and matrix metalloproteinase-2 activity in mouse mammary epithelial cells. *Cancer Res 61*: 7647-7653.

Peek RM Jr, Blaser MJ (2002). Helicobacter pylori and gastrointestinal tract adenocarcinomas. *Nat Rev Cancer. 2*:28-37.

Peifer M, Sweeton D, Casey M, Wieschaus E (1994). wingless signal and Zeste-white 3 kinase trigger opposing changes in the intracellular distribution of Armadillo. *Development 120*: 369-380.

Penton A, Wodarz A, Nusse R (2002). A mutational analysis of dishevelled in Drosophila defines novel domains in the dishevelled protein as well as novel suppressing alleles of axin. *Genetics 161*: 747-762.

Penzo-Mendez A, Umbhauer M, Djiane A, Boucaut JC, Riou JF (2003). Activation of Gbetagamma signaling downstream of Wnt-11/Xfz7 regulates Cdc42 activity during Xenopus gastrulation. *Dev Biol 257*: 302-314.

Pereira-Suarez AL, Meraz MA, Lizano M, Estrada-Chavez C, Hernandez F, Olivera P, Perez E, Padilla P, Yaniv M, Thierry F, Garcia-Carranca A (2002). Frequent alterations of the beta-catenin protein in cancer of the uterine cervix. *Tumour Biol. 23*:45-53.

Peters JM, McKay RM, McKay JP, Graff JM (1999). Casein kinase I transduces Wnt signals. *Nature 401*: 345-350.

Philipp-Staheli J, Kim KH, Payne SR, Gurley KE, Liggitt D, Longton G, Kemp CJ (2002). Pathway-specific tumor suppression. Reduction of p27 accelerates gastrointestinal tumorigenesis in Apc mutant mice, but not in Smad3 mutant mice. *Cancer Cell 1*:355-68.

Pinson KI, Brennan J, Monkley S, Avery BJ, Skarnes WC (2000). An LDL-receptor-related protein mediates Wnt signalling in mice. *Nature 407*: 535-538.

Pinto D, Gregorieff A, Begthel H, Clevers H (2003). Canonical Wnt signals are essential for homeostasis of the intestinal epithelium. *Genes Dev 17*: 1709-1713.

Pires-daSilva A, Sommer RJ (2003). The evolution of signalling pathways in animal development. *Nat Rev Genet. 4*:39-49.

Polakis P (1997). The adenomatous polyposis coli (APC) tumor suppressor. *Biochim Biophys Acta. 1332*:F127-147.

Polakis P (2000). Wnt signaling and cancer. *Genes Dev 14*: 1837-1851.

Prawitt D, Enklaar T, Gartner-Rupprecht B, Spangenberg C, Oswald M, Lausch E, Schmidtke P, Reutzel D, Fees S, Lucito R, Korzon M, Brozek I, Limon J, Housman DE, Pelletier J, Zabel B (2005). Microdeletion of target sites for insulator protein CTCF in a chromosome 11p15 imprinting center in Beckwith-Wiedemann syndrome and Wilms' tumor. *Proc Natl Acad Sci U S A. 102*:4085-4090.

Prowse AB, McQuade LR, et al. (2005). *Proteomics* 5:978-989.

Prunier C, Mazars A, Noë V, Bruyneel E, Mareel M, Gespach C, and Atfi A (1999). Evidence that SMAD 2 is a tumor suppressor implicated in the control of cellular invasion. *J Biol Chem. 274*: 22919-22922.

Prunier C, Hocevan BA, Howe P (2004). Wnt Signaling : Physiology and Pathology. Growth Factors, 22 : 141-150.

Reichling T, Heppner Goss K, Carson DJ, Holdcraft RW, Ley-Ebert C, Witte D, Aronow BJ, and Groden J (2005). Transcriptional profiles of intestinal tumors in APC Min mice are unique from those of embryonic intestine and identify novel gene targets dysregulated in human colorectal tumors. *Cancer Res. 65*: 166-176.

Reifenberger J, Knobbe CB, Wolter M, Blaschke B, Schulte KW, Pietsch T, Ruzicka T, Reifenberger G (2002). Molecular genetic analysis of malignant melanomas for aberrations of the WNT signaling pathway genes CTNNB1, APC, ICAT and BTRC. *Int J Cancer. 100*:549-56.

Ren J, Li Y, Kufe D (2002). Protein kinase C delta regulates function of the DF3/MUC1 carcinoma antigen in beta-catenin signaling. *J Biol Chem 277*: 17616-17622.

Reya T, Duncan AW, Ailles L, Domen J, Scherer DC, Willert K, Hintz L, Nusse R, Weissman IL (2003). A role for Wnt signalling in self-renewal of haematopoietic stem cells. *Nature 423*: 409-14.

Riley BB, Chiang MY, Storch EM, Heck R, Buckles GR, Lekven AC (2004). Rhombomere boundaries are Wnt signaling centers that regulate metameric patterning in the zebrafish hindbrain. *Dev Dyn. 231*:278-91.

Rimm DL, Caca K, Hu G, Harrison FB, Fearon ER (1999). Frequent Nuclear/Cytoplasmic Localization of ß-Catenin without Exon 3 Mutations in Malignant Melanoma. *Am J Pathol 154*: 325-329.

Rivat C, Le Floch N, Sabbah M, Teyrol I, Redeuilh G, Bruyneel E, Mareel M, Matrisian LM, Crawford HC, Gespach C, Attoub S (2003). Synergistic cooperation between the AP-1 and LEF-1 transcription factors in activation of the matrilysin promoter by the src oncogene: implications in cellular invasion. *FASEB J. 17*:1721-3.

Rivat C, Rodrigues S, Bruyneel E, Piétu G, Robert A, Redeuilh G, Brake M, Gespach C, and Attoub S (2005). Implication of STAT3 signaling in human colonic cancer cells during Intestinal trefoil factor (TFF3) and VEGF- mediated cellular invasion and tumor growth. *Cancer Research 65* : 195-202.

Roberts MS, Woods AJ, Dale TC, Van Der Sluijs P, Norman JC (2004). Protein kinase B/Akt acts via glycogen synthase kinase 3 to regulate recycling of alpha v beta 3 and alpha 5 beta 1 integrins. *Mol Cell Biol 24*: 1505-1515.

Robinson GW, Hennighausen L, Johnson PF (2000). Side-branching in the mammary gland: the progesterone-Wnt connection. *Genes Dev. 14*:889-94.

Rocheleau CE, Downs WD, Lin R, Wittmann C, Bei Y, Cha YH, Ali M, Priess JR, Mello CC (1997). Wnt signaling and an APC-related gene specify endoderm in early C. elegans embryos. *Cell 90*: 707-716.

Rockman SP, Currie SA, Ciavarella M, Vincan E, Dow C, Thomas RJ, Phillips WA (2001). Id2 is a target of the beta-catenin/T cell factor pathway in colon carcinoma. *J Biol Chem 276*: 45113-45119.

Rodrigues S, Nguyen QD, Faivre S, Bruyneel E, Thim L, Westley B, May F, Flatau G, Mareel M, Gespach C, and Emami S (2001). Activation of cellular invasion by trefoil peptides and src is mediated by cyclooxygenase- and thromboxane A2 receptor-dependent signaling pathways. *The FASEB Journal, 15* : 1517-1528.

Rodrigues S, Attoub S, Nguyen QD, Bruyneel E, Rodrigue CM, Westley BR, May FE, Thim L, Mareel M, Emami S, Gespach C (2003). Selective abrogation of the proinvasive activity of the trefoil peptides pS2 and spasmolytic polypeptide by disruption of the EGF receptor signaling pathways in kidney and colonic cancer cells. *Oncogene. 22*:4488-97.

Roose J, Molenaar M, Peterson J, Hurenkamp J, Brantjes H, Moerer P, van de Wetering M, Destree O, Clevers H (1998). The Xenopus Wnt effector XTcf-3 interacts with Groucho-related transcriptional repressors. *Nature 395*: 608-612.

Roose J, Huls G, van Beest M, Moerer P, van der Horn K, Goldschmeding R, Logtenberg T, Clevers H (1999). Synergy between tumor suppressor APC and the beta-catenin-Tcf4 target Tcf1. *Science 285*: 1923-1926.

Rosin-Arbesfeld R, Cliffe A, Brabletz T, Bienz M (2003). Nuclear export of the APC tumour suppressor controls beta-catenin function in transcription. *EMBO J. 22*:1101-13.

Rousset R, Mack JA, Wharton KA Jr, Axelrod JD, Cadigan KM, Fish MP, Nusse R, Scott MP (2001). Naked cuticle targets dishevelled to antagonize Wnt signal transduction. *Genes Dev 15*: 658-671.

Rubinfeld B, Albert I, Porfiri E, Fiol C, Munemitsu S, Polakis P (1996). Binding of GSK3beta to the APC-beta-catenin complex and regulation of complex assembly. *Science 272*: 1023-1026.

Sachdev S, Bruhn L, Sieber H, Pichler A, Melchior F, Grosschedl R (2001). PIASy, a nuclear matrix-associated SUMO E3 ligase, represses LEF1 activity by sequestration into nuclear bodies. *Genes Dev. 15*:3088-103.

Sadot E, Geiger B, Oren M, Ben-Ze'ev A (2001). Down-regulation of beta-catenin by activated p53. *Mol Cell Biol. 21*:6768-81.

Sagara N, Toda G, Hirai M, Terada M, Katoh M (1998). Molecular cloning, differential expression, and chromosomal localization of human frizzled-1, frizzled-2, and frizzled-7. *Biochem Biophys Res Commun 252*: 117-122.

Saitoh T, Hirai M, Katoh M (2001a). Molecular cloning and characterization of human Frizzled-8 gene on chromosome 10p11.2. *Int J Oncol 18*: 991-996.

Saitoh T, Hirai M, Katoh M (2001b). Molecular cloning and characterization of human Frizzled-5 gene on chromosome 2q33.3-q34 region. *Int J Oncol 19*: 105-110.

Saitoh T, Mine T, Katoh M (2002). Frequent up-regulation of WNT5A mRNA in primary gastric cancer. *Int J Mol Med. 9* :515-9.

Saitoh, T, Katoh, M (2002). Expression and regulation of WNT5A and WNT5B in human cancer: up-regulation of WNT5A by TNFalpha in MKN45 cells and up-regulation of WNT5B by beta-estradiol in MCF-7 cells. *Int.J Mol.Med. 10*, 345-349.

Sakamoto I, Kishida S, Fukui A, Kishida M, Yamamoto H, Hino S, Michiue T, Takada S, Asashima M, Kikuchi A (2000). A novel beta-catenin-binding protein inhibits beta-catenin-dependent Tcf activation and axis formation. *J Biol Chem 275*: 32871-32878.

Sakatani T, Kaneda A, Iacobuzio-Donahue CA, Carter MG, Witzel SD, Okano H, Ko MS, Ohlsson R, Longo DL, Feinberg AP (2005). Loss of Imprinting of Igf2 Alters Intestinal Maturation and Tumorigenesis in Mice. *Science. 307* : 1976-1978.

Sala CF, Formenti E, Terstappen GC, Caricasole A (2000). Identification, gene structure, and expression of human frizzled-3 (FZD3). *Biochem Biophys Res Commun 273*: 27-34.

Sancho E, Batlle E, Clevers H (2004). Signaling pathways in intestinal development and cancer. *Annu Rev Cell Dev Biol. 20*:695-723.

Saneyoshi T, Kume S, Amasaki Y, Mikoshiba K (2002). The Wnt/calcium pathway activates NF-AT and promotes ventral cell fate in Xenopus embryos. *Nature 417*: 295-299.

Sastry SK, Burridge K (2000). Focal adhesions: a nexus for intracellular signaling and cytoskeletal dynamics. *Exp Cell Res 261*: 25-36.

Sato A, Kojima T, Ui-Tei K, Miyata Y, Saigo K (1999). Dfrizzled-3, a new Drosophila Wnt receptor, acting as an attenuator of Wingless signaling in wingless hypomorphic mutants. *Development 126*: 4421-4430.

Sato H, Takino T, Okada Y, Cao J, Shinagawa A, Yamamoto E, Seiki M (1994). A matrix metalloproteinase expressed on the surface of invasive tumour cells. *Nature 370*: 61-65.

Sato N, Meijer L, Skaltsounis L, Greengard P, Brivanlou AH. Maintenance of pluripotency in human and mouse embryonic stem cells through activation of Wnt signaling by a pharmacological GSK-3-specific inhibitor (2004). *Nat Med. 10*:55-63.

Satoh S, Daigo Y, Furukawa Y, Kato T, Miwa N, Nishiwaki T, Kawasoe T, Ishiguro H, Fujita M, Tokino T, Sasaki Y, Imaoka S, Murata M, Shimano T, Yamaoka Y, Nakamura Y (2000). AXIN1 mutations in hepatocellular carcinomas, and growth suppression in cancer cells by virus-mediated transfer of AXIN1. *Nat Genet 24*: 245-250.

Schlesinger A, Shelton CA, Maloof JN, Meneghini M, Bowerman B.(1999). Wnt pathway components orient a mitotic spindle in the early Caenorhabditis elegans embryo without requiring gene transcription in the responding cell. *Genes Dev. 13*:2028-38.

Schroeder JA, Adriance MC, Thompson MC, Camenisch TD, Gendler SJ (2003). MUC1 alters beta-catenin-dependent tumor formation and promotes cellular invasion. *Oncogene 22*: 1324-1332.

Schwalm-Maupas F, Robinet C, et al (2005). *Cancer Res* 65:526-532.

Schwarz-Romond T, Asbrand C, Bakkers J, Kuhl M, Schaeffer HJ, Huelsken J, Behrens J, Hammerschmidt M, Birchmeier W (2002). The ankyrin repeat protein Diversin recruits Casein kinase Iepsilon to the beta-catenin degradation complex and acts in both canonical Wnt and Wnt/JNK signaling. *Genes Dev 16*: 2073-2084.

Semenov MV, Tamai K, Brott BK, Kuhl M, Sokol S, He (2001). Head inducer Dickkopf-1 is a ligand for Wnt coreceptor LRP6. *Curr Biol 11*: 951-961.

Shannon JM, Hyatt BA (2004). Epithelial-mesenchymal interactions in the developing lung. *Annu Rev Physiol. 66*:625-45.

Shea CM, Edgar CM, Einhorn TA, Gerstenfeld LC (2003). BMP treatment of C3H10T1/2 mesenchymal stem cells induces both chondrogenesis and osteogenesis. *J Cell Biochem. 90*:1112-27.

Sheldahl LC, Park M, Malbon CC, Moon RT (1999). Protein kinase C is differentially stimulated by Wnt and Frizzled homologs in a G-protein-dependent manner. *Curr Biol 9*: 695-698.

Sheldahl LC, Slusarski DC, Pandur P, Miller JR, Kuhl M, Moon RT (2003). Dishevelled activates Ca2+ flux, PKC, and CamKII in vertebrate embryos. *J Cell Biol 161*: 769-777.

Sherr CJ (1996). Cancer cell cycles. *Science 274*: 1672-1677.

Shibamoto S, Hayakawa M, Takeuchi K, Hori T, Oku N, Miyazawa K, Kitamura N, Takeichi M, Ito F (1994). Tyrosine phosphorylation of beta-catenin and plakoglobin enhanced by hepatocyte growth factor and epidermal growth factor in human carcinoma cells. *Cell Adhes Commun 1*: 295-305.

Shimokawa T, Furukawa Y, Sakai M, Li M, Miwa N, Lin YM, Nakamura Y (2003). Involvement of the FGF18 gene in colorectal carcinogenesis, as a novel downstream target of the beta-catenin/T-cell factor complex. *Cancer Res 63*: 6116-6120.

Shtutman M, Zhurinsky J, Simcha I, Albanese C, D'Amico M, Pestell R, Ben-Ze'ev A (1999). The cyclin D1 gene is a target of the beta-catenin/LEF-1 pathway. *Proc Natl Acad Sci U S A 96*: 5522-5527.

Shulman JM, Perrimon N, Axelrod JD (1998). Frizzled signaling and the developmental control of cell polarity. *Trends Genet 14*: 452-458.

Sieber OM, Howarth KM, Thirlwell C, Rowan A, Mandir N, Goodlad RA, Gilkar A, Spencer-Dene B, Stamp G, Johnson V, Silver A, Yang H, Miller JH, Ilyas M, Tomlinson IP (2004). Myh deficiency enhances intestinal tumorigenesis in multiple intestinal neoplasia (ApcMin/+) mice. *Cancer Res. 64*:8876.

Siegfried E, Wilder EL, Perrimon N (1994). Components of wingless signalling in Drosophila. *Nature 367*: 76-80.

Silver SJ, Rebay I (2005). Signaling circuitries in development: insights from the retinal determination gene network. *Development. 132*:3-13.

Slusarski DC, Corces VG, Moon RT (1997a). Interaction of Wnt and a Frizzled homologue triggers G-protein-linked phosphatidylinositol signalling. *Nature 390*: 410-413.

Slusarski DC, Yang-Snyder J, Busa WB, Moon RT (1997b). Modulation of embryonic intracellular Ca2+ signaling by Wnt-5A. *Dev Biol 182*: 114-120.

Smalley MJ, Sara E, Paterson H, Kishida M, Sakamoto I, Koyama S, Kikuchi A (1999). Interaction of axin and Dvl-2 proteins regulates Dvl-2-stimulated TCF-dependent transcription. *Embo J 18*: 2823-2835.

Smalley MJ, Dale TC (2001). Wnt signaling and mammary tumorigenesis. *J Mammary Gland Biol Neoplasia 6*:37-52.

Sokol SY, Klingensmith J, Perrimon N, Itoh K (1995). Dorsalizing and neuralizing properties of Xdsh, a maternally expressed Xenopus homolog of dishevelled. *Development 121*: 1637-1647.

Sokol SY (1996). Analysis of Dishevelled signalling pathways during Xenopus development. *Curr Biol 6*: 1456-1467.

Sonoshita M, Takaku K, Sasaki N, Sugimoto Y, Ushikubi F, Narumiya S, Oshima M, Taketo MM (2001). Acceleration of intestinal polyposis through prostaglandin receptor EP2 in Apc(Delta 716) knockout mice. *Nat Med. 7*:1048-1051.

Spiegelman VS, Slaga TJ, Pagano M, Minamoto T, Ronai Z, Fuchs SY (2000). Wnt/beta-catenin signaling induces the expression and activity of betaTrCP ubiquitin ligase receptor. *Mol Cell 5*: 877-882.

Staal FJ, Clevers HC (2005). WNT signalling and haematopoiesis: a WNT-WNT situation. *Nat Rev Immunol. 5*:21-30.

Strovel ET, Wu D, Sussman DJ (2000). Protein phosphatase 2Calpha dephosphorylates axin and activates LEF-1-dependent transcription. *J Biol Chem 275*: 2399-2403.

Stuart RO, Bush KT, Nigam SK (2003). Changes in gene expression patterns in the ureteric bud and metanephric mesenchyme in models of kidney development. *Kidney Int. 64*:1997-2008.

Strutt DI, Weber U, Mlodzik M (1997). The role of RhoA in tissue polarity and Frizzled signalling. *Nature 387*: 292-295.

Suzuki T, Yano H, Nakashima Y, Nakashima O, Kojiro M (2002). Beta-catenin expression in hepatocellular carcinoma: a possible participation of beta-catenin in the dedifferentiation process. *J Gastroenterol Hepatol. 17*:994-1000.

Suzuki H, Watkins DN, Jair KW, Schuebel KE, Markowitz SD, Dong Chen W, Pretlow TP, Yang B, Akiyama Y, Van Engeland M, Toyota M, Tokino T, Hinoda Y, Imai K, Herman JG, Baylin SB (2004). Epigenetic inactivation of SFRP genes allows constitutive WNT signaling in colorectal cancer. *Nat Genet. 36*:417-22.

Tada M, Smith JC (2000). Xwnt11 is a target of Xenopus Brachyury: regulation of gastrulation movements via Dishevelled, but not through the canonical Wnt pathway. *Development 127*: 2227-2238.

Takada S, Stark KL, Shea MJ, Vassileva G, McMahon JA, McMahon AP (1994). Wnt-3a regulates somite and tailbud formation in the mouse embryo. *Genes Dev 8*: 174-189.

Takahashi M, Tsunoda T, Seiki M, Nakamura Y, Furukawa Y (2002). Identification of membrane-type matrix metalloproteinase-1 as a target of the beta-catenin/Tcf4 complex in human colorectal cancers. *Oncogene 21*: 5861-5867.

Takayasu H, Horie H, Hiyama E, Matsunaga T, Hayashi Y, Watanabe Y, Suita S, Kaneko M, Sasaki F, Hashizume K, Ozaki T, Furuuchi K, Tada M, Ohnuma N, Nakagawara A (2001). Frequent Deletions and Mutations of the ß-Catenin Gene Are Associated with Overexpression of Cyclin D1 and Fibronectin and Poorly Differentiated Histology in Childhood Hepatoblastoma. *Clin. Cancer Res. 7*: 901-908.

Takemaru K, Yamaguchi S, Lee YS, Zhang Y, Carthew RW, Moon RT (2003). Chibby, a nuclear beta-catenin-associated antagonist of the Wnt/Wingless pathway. *Nature 422*: 905-909.

Takemaru KI, Moon RT (2000). The transcriptional coactivator CBP interacts with beta-catenin to activate gene expression. *J Cell Biol 149*: 249-254.

Tamai K, Semenov M, Kato Y, Spokony R, Liu C, Katsuyama Y, Hess F, Saint-Jeannet JP, He X. (2000). LDL-receptor-related proteins in Wnt signal transduction. *Nature 407*: 530-535.

Tamai K, Zeng X, Liu C, Zhang X, Harada Y, Chang Z, He X (2004). A mechanism for Wnt coreceptor activation. *Mol Cell 13*: 149-156.

Tan X, Apte U, et al (2005). *Gastroenterology* 129:285-302.

Tanaka K, Kitagawa Y, Kadowaki T (2002). Drosophila segment polarity gene product porcupine stimulates the posttranslational N-glycosylation of wingless in the endoplasmic reticulum. *J Biol Chem 277*: 12816-12823.

Taniguchi K, Roberts LR, Aderca IN, Dong X, Qian C, Murphy LM, Nagorney DM, Burgart LJ, Roche PC, Smith DI, Ross JA, Liu W (2002). Mutational spectrum of beta-catenin, AXIN1, and AXIN2 in hepatocellular carcinomas and hepatoblastomas. *Oncogene. 21*:4863-71.

Tejpar S, Nollet F, Li C, Wunder JS, Michils G, dal Cin P, Van Cutsem E, Bapat B, van Roy F, Cassiman JJ, Alman BA (1999). Predominance of beta-catenin mutations and beta-catenin dysregulation in sporadic aggressive fibromatosis (desmoid tumor). *Oncogene 47*:6615-6620.

ter Haar E, Coll JT, Austen DA, Hsiao HM, Swenson L, Jain J (2001). Structure of GSK3beta reveals a primed phosphorylation mechanism. *Nat Struct Biol 8*: 593-596.

Terris B, Pineau P, Bregeaud L, Valla D, Belghiti J, Tiollais P, Degott C, Dejean A (1999). Close correlation between beta-catenin gene alterations and nuclear accumulation of the protein in human hepatocellular carcinomas. *Oncogene. 18*:6583-8.

Tetsu O, McCormick F (1999). Beta-catenin regulates expression of cyclin D1 in colon carcinoma cells. *Nature 398*: 422-426.

Thesleff I (2003). Epithelial-mesenchymal signalling regulating tooth morphogenesis. *J. Cell Sci. 116*: 1647-1648.

Thomas GM, Frame S, Goedert M, Nathke I, Polakis P, Cohen P (1999). A GSK3-binding peptide from FRAT1 selectively inhibits the GSK3-catalysed phosphorylation of axin and beta-catenin. *FEBS Lett 458*: 247-251.

Thomas KR, Capecchi MR (1990). Targeted disruption of the murine int-1 proto-oncogene resulting in severe abnormalities in midbrain and cerebellar development. *Nature 346*: 847-850.

Thorpe CJ, Schlesinger A, Bowerman B.(2000). Wnt signalling in Caenorhabditis elegans: regulating repressors and polarizing the cytoskeleton. *Trends Cell Biol.10*:10-7.

Tian Q, He XC et al (2005). *Cell Cycle* 4:215-216.

Tice DA, Szeto W, Soloviev I, Rubinfeld B, Fong SE, Dugger DL, Winer J, Williams PM, Wieand D, Smith V, Schwall RH, Pennica D, Polakis P (2002). Synergistic induction of tumor antigens by Wnt-1 signaling and retinoic acid revealed by gene expression profiling. *J Biol Chem. 277*:14329-35.

Tokuhara M, Hirai M, Atomi Y, Terada M, Katoh M (1998). Molecular cloning of human Frizzled-6. *Biochem Biophys Res Commun 243*: 622-627.

Tolwinski NS, Wehrli M, Rives A, Erdeniz N, DiNardo S, Wieschaus E (2003). Wg/Wnt signal can be transmitted through arrow/LRP5,6 and Axin independently of Zw3/Gsk3beta activity. *Dev Cell 4*: 407-418.

Torres MA, Yang-Snyder JA, Purcell SM, DeMarais AA, McGrew LL, Moon RT (1996). Activities of the Wnt-1 class of secreted signaling factors are antagonized by the Wnt-5A class and by a dominant negative cadherin in early Xenopus development. *J Cell Biol 133*: 1123-1137.

Townsley FM, Cliffe A, Bienz M (2004). Pygopus and Legless target Armadillo/beta-catenin to the nucleus to enable its transcriptional co-activator function. *Nat Cell Biol 6*: 626-633.

Traverso G, Shuber A, Levin B, Johnson C, Olsson L, Schoetz DJ Jr, Hamilton SR, Boynton K, Kinzler KW, Vogelstein B (2002). Detection of APC mutations in fecal DNA from patients with colorectal tumors. *N Engl J Med. 346*:311-20.

Trowbridge J, et al (2006). *Nature Medicine* 12: 89-98.

Tsai RY, Kittappa R, McKay RD (2002). Plasticity, niches, and the use of stem cells. *Dev Cell. 2* :707-12.

Tsujiuchi T, Tsutsumi M, Sasaki Y, Murata N, Konishi Y (2000). Mutations of adenomatous polyposis coli and beta-catenin genes during progression of lung tumors induced by N-nitrosobis(2-hydroxypropyl)amine in rats. *Cancer Res. 60* :6611-6.

Tumbar T, Guasch G, Greco V, Blanpain C, Lowry WE, Rendl M, Fuchs E (2004). Defining the epithelial stem cell niche in skin. *Science. 303*:359-63.

Uematsu K, He B, You L, Xu Z, McCormick F, Jablons DM (2003). Activation of the Wnt pathway in non small cell lung cancer: evidence of dishevelled overexpression. *Oncogene 22*: 7218-7221.

Umbhauer M, Djiane A, Goisset C, Penzo-Mendez A, Riou JF, Boucaut JC, Shi DL (2000). The C-terminal cytoplasmic Lys-thr-X-X-X-Trp motif in frizzled receptors mediates Wnt/beta-catenin signalling. *Embo J 19*: 4944-4954.

Umeki K, Shiota G, Kawasaki H (1999). Clinical significance of c-met oncogene alterations in human colorectal cancer. *Oncology 56*: 314-321.

Ungar AR, Kelly GM, Moon RT (1995). Wnt4 affects morphogenesis when misexpressed in the zebrafish embryo. *Mech Dev 52*: 153-164.

Uren A, Reichsman F, Anest V, Taylor WG, Muraiso K, Bottaro DP, Cumberledge S, Rubin JS (2000). Secreted frizzled-related protein-1 binds directly to Wingless and is a biphasic modulator of Wnt signaling. *J Biol Chem 275*: 4374-4382.

Uyttaendael H, Soriano JV, Montesano R, Kitajewski J (1998). Notch4 and Wnt-1 Proteins Function to Regulate Branching Morphogenesis of Mammary Epithelial Cells in an Opposing Fashion.*Dev Biol 196*:204-17.

Van Den Berg DJ, Sharma AK, Bruno E, Hoffman R (1998). Role of members of the Wnt gene family in human hematopoiesis. *Blood 92* :3189-202.

van den Brink GR, Bleuming SA, Hardwick JC, Schepman BL, Offerhaus GJ, Keller JJ, Nielsen C, Gaffield W, van Deventer SJ, Roberts DJ, Peppelenbosch MP (2004). Indian Hedgehog is an antagonist of Wnt signaling in colonic epithelial cell differentiation. *Nat Genet. 36* :277-82.

van de Wetering M, Cavallo R, Dooijes D, van Beest M, van Es J, Loureiro J, Ypma A, Hursh D, Jones T, Bejsovec A, Peifer M, Mortin M, Clevers H (1997). Armadillo coactivates transcription driven by the product of the Drosophila segment polarity gene dTCF. *Cell 88*: 789-799.

van de Wetering M, Sancho E, Verweij C, de Lau W, Oving I, Hurlstone A, van der Horn K, Batlle E, Coudreuse D, Haramis AP, Tjon-Pon-Fong M, Moerer P, van den Born M, Soete G, Pals S, Eilers M, Medema R, Clevers H (2002). The beta-catenin/TCF-4 complex imposes a crypt progenitor phenotype on colorectal cancer cells. *Cell 111*: 241-250.

van der Flier A, Sonnenberg A (2001). Function and interactions of integrins. *Cell Tissue Res 305*: 285-298.

van Es JH, Jay P, Gregorieff A, van Gijn ME, Jonkheer S, Hatzis P, Thiele A, van den Born M, Begthel H, Brabletz T, Taketo MM, Clevers H (2005). Wnt signalling induces maturation of Paneth cells in intestinal crypts. *Nat Cell Biol. Mar 20*; [Epub ahead of print]

Van Nhieu JT, Renard CA, Wei Y, Cherqui D, Zafrani ES, Buendia MA (1999). Nuclear Accumulation of Mutated ß-Catenin in Hepatocellular Carcinoma Is Associated with Increased Cell Proliferation. *Am J Pathol 155*: 703-710.

van Weeren PC, de Bruyn KM, de Vries-Smits AM, van Lint J, Burgering BM (1998). Essential role for protein kinase B (PKB) in insulin-induced glycogen synthase kinase 3 inactivation. Characterization of dominant-negative mutant of PKB. *J Biol Chem 273*: 13150-13156.

Veeman MT, Slusarski DC, Kaykas A, Louie SH, Moon RT (2003). Zebrafish prickle, a modulator of noncanonical Wnt/Fz signaling, regulates gastrulation movements. *Curr Biol 13*: 680-685.

Verras M, Brown J, Li X, Nusse R, Sun Z (2004). Wnt3a growth factor induces androgen receptor-mediated transcription and enhances cell growth in human prostate cancer cells. *Cancer Res. 64* :8860-6

Vider BZ, Zimber A, Chastre E, Prevot S, Gespach C, Estlein D, Wolloch Y, Tronick SR, Gazit A, Yaniv A (1996). Evidence for the involvement of the Wnt 2 gene in human colorectal cancer. *Oncogene 12*: 153-158.

von Kries JP, Winbeck G, Asbrand C, Schwarz-Romond T, Sochnikova N, Dell'Oro A, Behrens J, Birchmeier W (2000). Hot spots in beta-catenin for interactions with LEF-1, conductin and APC. *Nat Struct Biol. 7*:800- 807.

Wada M, Miller CW, Yokota J, Lee E, Mizoguchi H, Koeffler HP (1997). Molecular analysis of the adenomatous polyposis coli gene in sarcomas, hematological malignancies and noncolonic, neoplastic tissues. *J Mol Med. 75*:139-44.

Wagers AJ, Weissman IL (2004). Plasticity of adult stem cells. *Cell.* 2004 Mar 5;116(5):639-48.

Waltzer L, Bienz M. (1998). Drosophila CBP represses the transcription factor TCF to antagonize Wingless signalling. *Nature 395*: 521-525.

Wang D, Mann JR, Dubois RN (2004). WNT and Cyclooxygenase-2 cross-talk accelerates adenoma Growth. *Cell Cycle. 23*:1512-1515.

Wang HY, Malbon CC (2003). Wnt signaling, Ca2+, and cyclic GMP: visualizing Frizzled functions. *Science 300*: 1529-1530.

Wang S, Krinks M, Moos M (1997). Frzb-1, an antagonist of Wnt-1 and Wnt-8, does not block signaling by Wnts -3A, -5A, or -11. *Biochem Biophys Res Commun 236*: 502-504.

Weeraratna AT, Jiang Y, Hostetter G, Rosenblatt K, Duray P, Bittner M, Trent JM (2002). Wnt5a signaling directly affects cell motility and invasion of metastatic melanoma. *Cancer Cell 1*: 279-288.

Wehrli M, Dougan ST, Caldwell K, O'Keefe L, Schwartz S, Vaizel-Ohayon D, Schejter E, Tomlinson A, DiNardo S (2000). arrow encodes an LDL-receptor-related protein essential for Wingless signalling. *Nature 407*: 527-530.

Wharton KA, Jr., Zimmermann G, Rousset R, Scott MP (2001). Vertebrate proteins related to Drosophila Naked Cuticle bind Dishevelled and antagonize Wnt signaling. *Dev Biol 234*: 93-106.

Wharton KA (2003). Runnin' with the Dvl: proteins that associate with Dsh/Dvl and their significance to Wnt signal transduction. *Dev Biol 253*: 1-17.

Wei X, Xu H, Kufe D (2005). Human MUC1 oncoprotein regulates p53-responsive gene transcription in the genotoxic stress response. *Cancer Cell. 7*:167-78.

Wen Y, Caffrey TC, Wheelock MJ, Johnson KR, Hollingsworth MA (2003). Nuclear association of the cytoplasmic tail of MUC1 and beta-catenin. *J Biol Chem 278*: 38029-38039.

Wiechens N, Heinle K, Englmeier L, Schohl A, Fagotto F (2004). Nucleo-cytoplasmic shuttling of Axin, a negative regulator of the Wnt-beta-catenin Pathway. *J Biol Chem 279*: 5263-5267.

Wieland I, Bohm M (1994). Frequent allelic deletion at a novel locus on chromosome 5 in human lung cancer. *Cancer Res. 54*: 1772-1774.

Wielenga VJ, Smits R, Korinek V, Smit L, Kielman M, Fodde R, Clevers H, Pals ST (1999). Expression of CD44 in Apc and Tcf mutant mice implies regulation by the WNT pathway. *Am J Pathol 154*: 515-523.

Willert K, Brink M, Wodarz A, Varmus H, Nusse R (1997). Casein kinase 2 associates with and phosphorylates dishevelled. *Embo J 16*: 3089-3096.

Willert K, Shibamoto S, Nusse R (1999). Wnt-induced dephosphorylation of axin releases beta-catenin from the axin complex. *Genes Dev 13*: 1768-1773.

Willert K, Brown JD, Danenberg E, Duncan AW, Weissman IL, Reya T, Yates JR, Nusse R (2003). Wnt proteins are lipid-modified and can act as stem cell growth factors. *Nature 423*: 448-452.

Wilson SI, Rydstrom A, Trimborn T, Willert K, Nusse R, Jessell TM, Edlund T (2001). The status of Wnt signalling regulates neural and epidermal fates in the chick embryo. *Nature. 411*:325-30.

Winter CG, Wang B, Ballew A, Royou A, Karess R, Axelrod JD, Luo L (2001). Drosophila Rho-associated kinase (Drok) links Frizzled-mediated planar cell polarity signaling to the actin cytoskeleton. *Cell 105*: 81-91.

Wodarz A, Nusse R (1998). Mechanisms of Wnt signaling in development. *Annu Rev Cell Dev Biol 14*: 59-88.

Wolda SL, Moody CJ, Moon RT (1993). Overlapping expression of Xwnt-3A and Xwnt-1 in neural tissue of Xenopus laevis embryos. *Dev Biol 155*: 46-57.

Wong GT, Gavin BJ, McMahon AP (1994) Differential transformation of mammary epithelial cells by Wnt genes. *Mol Cell Biol 14*: 6278-6286.

Wong HC, Bourdelas A, Krauss A, Lee HJ, Shao Y, Wu D, Mlodzik M, Shi DL, Zheng J (2003). Direct binding of the PDZ domain of Dishevelled to a conserved internal sequence in the C-terminal region of Frizzled. *Mol Cell 12*: 1251-1260.

Wood RD, Mitchell M, Sgouros J, Lindahl T (2001). Human DNA repair genes. *Science 291*: 1284-1289.

Woodfield RJ, Hodgkin MN, Akhtar N, Morse MA, Fuller KJ, Saqib K, Thompson NT, Wakelam MJ (2001). The p85 subunit of phosphoinositide 3-kinase is associated with beta-catenin in the cadherin-based adhesion complex. *Biochem J 360*: 335-344.

Woodgett JR (1990). Molecular cloning and expression of glycogen synthase kinase-3/factor A. *Embo J 9*: 2431-2438.

Wu J, Klein TJ, Mlodzik M (2004). Subcellular localization of frizzled receptors, mediated by their cytoplasmic tails, regulates signaling pathway specificity. *PLoS Biol 2*: E158.

Wu W, Glinka A, Delius H, Niehrs C (2000). Mutual antagonism between dickkopf1 and dickkopf2 regulates Wnt/beta-catenin signalling. *Curr Biol 10*: 1611-1614.

Xia X, Qian S, Soriano S, Wu Y, Fletcher AM, Wang XJ, Koo EH, Wu X, Zheng H (2001). Loss of presenilin 1 is associated with enhanced beta-catenin signaling and skin tumorigenesis. *Proc Natl Acad Sci U S A. 98*:10863-8.

Xing Y, Clements WK, Kimelman D, Xu W (2003). Crystal structure of a beta-catenin/axin complex suggests a mechanism for the beta-catenin destruction complex. *Genes Dev 17*: 2753-2764.

Yamada T, Mori Y, Hayashi R, Takada M, Ino Y, Naishiro Y, Kondo T, Hirohashi S (2003). Suppression of intestinal polyposis in Mdr1-deficient ApcMin/+ mice. *Cancer Res 63*:895-901.

Yamamoto H, Ihara M, Matsuura Y, Kikuchi A (2003). Sumoylation is involved in □-catenin-dependent activaation of TCF-4. *EMBO J 22* : 2047-2059.

Yamanaka H, Moriguchi T, Masuyama N, Kusakabe M, Hanafusa H, Takada R, Takada S, Nishida E (2002). JNK functions in the non-canonical Wnt pathway to regulate convergent extension movements in vertebrates. *EMBO Rep 3*: 69-75.

Yan D, Wallingford JB, Sun TQ, Nelson AM, Sakanaka C, Reinhard C, Harland RM, Fantl WJ, Williams LT (2001a). Cell autonomous regulation of multiple Dishevelled-dependent pathways by mammalian Nkd. *Proc Natl Acad Sci U S A 98*: 3802-3807.

Yan D, Wiesmann M, Rohan M, Chan V, Jefferson AB, Guo L, Sakamoto D, Caothien RH, Fuller JH, Reinhard C, Garcia PD, Randazzo FM, Escobedo J, Fantl WJ, Williams LT (2001b). Elevated expression of axin2 and hnkd mRNA provides evidence that Wnt/beta -catenin signaling is activated in human colon tumors. *Proc Natl Acad Sci U S A 98*: 14973-14978.

Yau TO, Chan CY, Chan KL, Lee MF, Wong CM, Fan ST, Ng IO (2005). HDPR1, a novel inhibitor of the WNT/beta-catenin signaling, is frequently downregulated in hepatocellular carcinoma: involvement of methylation-mediated gene silencing. *Oncogene. 24*:1607-14.

Yook JI, Li XY, Ota I, Fearon ER, Weiss SJ (2005). Wnt-dependent regulation of the E-cadherin repressor snail. *J Biol Chem. Jan 11*; [Epub ahead of print].

Yoshikawa Y, Fujimori T, McMahon AP, Takada S (1997). Evidence that absence of Wnt-3a signaling promotes neuralization instead of paraxial mesoderm development in the mouse. *Dev Biol 183*: 234-242.

Yost C, Farr GH, 3rd, Pierce SB, Ferkey DM, Chen MM, Kimelman D (1998). GBP, an inhibitor of GSK-3, is implicated in Xenopus development and oncogenesis. *Cell 93*: 1031-1041.

You L, He B, Xu Z, Uematsu K, Mazieres J, Fujii N, Mikami I, Reguart N, McIntosh JK, Kashani-Sabet M, McCormick F, Jablons DM (2004). An anti-Wnt-2 monoclonal antibody induces apoptosis in malignant melanoma cells and inhibits tumor growth. *Cancer Res. 64*:5385-9.

Yu Q, Stamenkovic I (1999). Localization of matrix metalloproteinase 9 to the cell surface provides a mechanism for CD44-mediated tumor invasion. *Genes Dev 13*: 35-48.

Yu Q, Stamenkovic I (2000). Cell surface-localized matrix metalloproteinase-9 proteolytically activates TGF-beta and promotes tumor invasion and angiogenesis. *Genes Dev 14*: 163-176.

Yu WH, Woessner JF, Jr., McNeish JD, Stamenkovic I (2002). CD44 anchors the assembly of matrilysin/MMP-7 with heparin-binding epidermal growth factor precursor and ErbB4 and regulates female reproductive organ remodeling. *Genes Dev 16*: 307-323.

Yuan H, Mao J, Li L, Wu D (1999). Suppression of glycogen synthase kinase activity is not sufficient for leukemia enhancer factor-1 activation. *J Biol Chem. 274*:30419-23.

Yuasa Y (2003). Control of gut differentiation and intestinal-type gastric carcinogenesis. *Nat Rev Cancer. 8*:592-600.

Zamurovic N, Cappellen D, Rohner D, Susa M (2004). Coordinated activation of notch, Wnt, and transforming growth factor-beta signaling pathways in bone morphogenic protein 2-induced osteogenesis. Notch target gene Hey1 inhibits mineralization and Runx2 transcriptional activity. *J Biol Chem. 279*:37704-15.

Zeng W, Wharton KA Jr, Mack JA, Wang K, Gadbaw M, Suyama K, Klein PS, Scott MP (2000). naked cuticle encodes an inducible antagonist of Wnt signalling. *Nature 403*: 789-795.

Zeng X, Tamai K et al (2005). *Nature* 438: 873-877.

Zhai Y, Wu R, Schwartz DR, Darrah D, Reed H, Kolligs FT, Nieman MT, Fearon ER, Cho KR (2002). Role of ß-Catenin /T-Cell Factor-Regulated Genes in Ovarian Endometrioid Adenocarcinomas. *Am J Pathol 160*: 1229-1238.

Zhang F, Phiel CJ, Spece L, Gurvich N, Klein PS. (2003). Inhibitory phosphorylation of glycogen synthase kinase-3 (GSK-3) in response to lithium. Evidence for autoregulation of GSK-3. *J Biol Chem 278*: 33067-33077.

Zhang X, Gaspard JP, Chung DC (2001). Regulation of vascular endothelial growth factor by the Wnt and K-ras pathways in colonic neoplasia. *Cancer Res 61*: 6050-6054.

Zheng M, Messerschmidt D, Jungblut B, Sommer RJ (2005). Conservation and diversification of Wnt signaling function during the evolution of nematode vulva development. *Nat Genet. 37*:300-304.

Zhou BP, Deng J, Xia W, Xu J, Li YM, Gunduz M, Hung MC (2004). Dual regulation of Snail by GSK-3beta-mediated phosphorylation in control of epithelial-mesenchymal transition. *Nat Cell Biol. 6*:931-40.

Zhu H, Mazor M, Kawano Y, Walker MM, Leung HY, Armstrong K, Waxman J, Kypta RM (2004). Analysis of Wnt Gene Expression in Prostate Cancer: Mutual Inhibition by WNT11 and the Androgen Receptor. *Cancer Res. 64*: 7918-7926.

Zimber A, Chedeville A, Abita JP, Barbu V, Gespach C (2000). Functional interactions between bile acids, all-trans retinoic acid, and 1,25-dihydroxy-vitamin D3 on monocytic differentiation and myeloblastin gene down-regulation in HL60 and THP-1 human leukemia cells. *Cancer Res. 60*:672-8.

Zimber A, Nguyen QD, Gespach C (2004). Nuclear bodies and compartments: functional roles and cellular signalling in health and disease. *Cell Signal.16*:1085-104.

INDEX

A

B

C

I

K

L

M

N

O

P

R

S

T

U

V

W

X

Y

Z